Green Chemistry - New Perspectives

Edited by Brajesh Kumar and Alexis Debut

Published in London, United Kingdom

Green Chemistry - New Perspectives
http://dx.doi.org/10.5772/intechopen.98002
Edited by Brajesh Kumar and Alexis Debut

Contributors
Vichai Puripunyavanich, Lamai Maikaeo, Mayuree Limtiyayotin, Piyanuch Orpong, Brajesh Kumar, Katherine Guzmán, Marcelo Grijalva, Alexis Debut, Luis Cumbal, Endang Tri Wahyuni, Eko Sri Kunarti, Bhisma K. Patel, Suresh Rajamanickam, L. Raju Chowhan, Biplob Borah, Bilal Ahmad Mir, Himanshu Sharma, Surbhi Dhadda, Prakash Giri Goswami, Sarika Verma, Kamna Chaturvedi, Deeksha Malvi, Manish Dhangar, Harsh Bajpai, Ranjan K. Kumar Mohapatra, Avanish Kumar Srivastava, Alexandre H. Pinto, Dylan R. Cho, Anton O. Oliynyk, Julian R. Silverman, Nikhat Farhana, Ashwini Somayaji, Prajitha Biju, Tahreen Taj, Mohammed Asif Iqbal, Mohammed Gulzar Ahmed, Abdullah Khan, Roshan Sayeed, Natasha Naval Aggarwal, Nishmitha Gretta D'Souza, Sudhina Makuttan, Abdul Rahamanulla, Binitha N. Narayanan, Vijayasree Haridas, Zahira Yaakob, Juan E. Tacoronte, Joseph Cruel Sigüenza, Carla Bernal Villavicencio, María Elizabeth Canchingre, Christie Durán García, Ioana Stanciu

First published in London, United Kingdom, 2022 by IntechOpen
IntechOpen is the global imprint of INTECHOPEN LIMITED, registered in England and Wales, registration number: 11086078, 5 Princes Gate Court, London, SW7 2QJ, United Kingdom

British Library Cataloguing-in-Publication Data
A catalogue record for this book is available from the British Library

Additional hard and PDF copies can be obtained from orders@intechopen.com

Green Chemistry - New Perspectives
Edited by Brajesh Kumar and Alexis Debut
p. cm.
Print ISBN 978-1-80355-777-9
Online ISBN 978-1-80355-778-6
eBook (PDF) ISBN 978-1-80355-779-3

For EU product safety concerns:
IN TECH d.o.o., Prolaz Marije Krucifikse Kozulić 3, 51000 Rijeka, Croatia,
info@intechopen.com or visit our website at intechopen.com.

Meet the editors

Dr. Brajesh Kumar is currently an assistant professor and head of the Department of Chemistry, at TATA College, Kolhan University, Chaibasa, India. He received a Ph.D. in chemistry from the University of Delhi, India, in 2009. He is a pioneering researcher in the fields of green chemistry, nano-synthesis, environmental remediation, photocatalysis, nanomedicine, polymer and natural product extraction. Dr. Kumar has several national and international fellowships to his credit and has worked as a faculty member in various universities throughout India, Ecuador, and South Korea. He is a holder of two registered patents as well as a member of the American Chemical Society and the Indian Society of Chemists and Biologists (ISCB). He has published numerous (>75) SCI/SCIE/Scopus-indexed research articles (h-index =31, citations 3079) and is an active reviewer of more than seventy journals. He is also included in the World's Top 2% Scientists list (2021 & 2022) published by Stanford University, USA.

Dr. Alexis Debut completed his Ph.D. at Université des Sciences et Technologies de Lille, France in 2000. From 2000 to 2007 he worked as a researcher in the industry at Pirelli Group, and then at Prysmian Group as Telecom Business Head of Logistics for Italy. Since 2008 he has been working at the University of the Armed Forces ESPE, Ecuador. He is one of the founders of the country's Center for Nanoscience and Nanotechnology (CENCINAT), a national-level reference. He is head of the Nanomaterials Characterization Laboratory at CENCINAT. He has published more than 120 SCOPUS-indexed papers and conferences, reviewed more than 60 papers for international journals, and has more than 10 international patents.

Contents

Preface **IX**

Section 1
Green Chemistry - Fundaments **1**

Chapter 1 **3**
Green Methods of Chemical Analysis and Pollutant Removal
by Endang Tri Wahyuni and Eko Sri Kunarti

Chapter 2 **21**
Fundamental Principles to Address Green Chemistry and Green Engineering for Sustainable Future
by Nikhat Farhana, Mohammed Gulzar Ahmed, Mohammed Asif Iqbal, Natasha Naval Aggarwal, Prajitha Biju, Ashwini Somayaji, Abdul Rahamanulla, Nishmitha Gretta D'Souza, Sudhina Makuttan, Tahreen Taj, Abdullah Khan and Roshan Sayeed

Section 2
Green Organic Synthesis **37**

Chapter 3 **39**
Green Synthesis of Chalcone Derivatives Using Chalcones as Precursor
by Surbhi Dhadda, Prakash Giri Goswami and Himanshu Sharma

Chapter 4 **57**
Mechanochemistry in Organocatalysis: A Green and Sustainable Route toward the Synthesis of Bioactive Heterocycles
by Biplob Borah and L. Raju Chowhan

Chapter 5 **81**
Eco-Sustainable Catalytic System for Green Oxidation of Spirostanic Alcohols Using Hypervalent Iodine (III) Tempo-4-n-Acetoxyamine System
by Joseph Cruel Sigüenza, Carla Bernal Villavicencio, María Elizabeth Canchingre, Christie Durán García and Juan E. Tacoronte

Chapter 6 95
Potassium Persulfate as an Eco-Friendly Oxidant for Oxidative Transformations
by Bilal Ahmad Mir and Suresh Rajamanickam

Chapter 7 113
Thermally Activated Delayed Fluorescence (TADF) Compounds as Photocatalyst in Organic Synthesis: A Metal-Free Greener Approach
by Suresh Rajamanickam and Bhisma K. Patel

Chapter 8 139
Green Chemistry Applied to Transition Metal Chalcogenides through Synthesis, Design of Experiments, Life Cycle Assessment, and Machine Learning
by Alexandre H. Pinto, Dylan R. Cho, Anton O. Oliynyk and Julian R. Silverman

Section 3
Green Chemistry - Miscellenous 173

Chapter 9 175
Biopolymers
by Ioana Stanciu

Chapter 10 185
New Frontier of Plant Breeding Using Gamma Irradiation and Biotechnology
by Vichai Puripunyavanich, Lamai Maikaeo, Mayuree Limtiyayothin and Piyanuch Orpong

Section 4
Green Nanoparticle Synthesis 209

Chapter 11 211
Green Preparation of Fe_2O_3 Doped Gum Acacia Derived Porous Carbon/Graphene Ternary Nanocomposite as a Supercapacitor Electrode
by Vijayasree Haridas, Zahira Yaakob and Binitha N. Narayanan

Chapter 12 229
Recent Advances in the Green Synthesis of Lanthanide-Based Organic Compounds for Broad Application Spectrum in Different Sectors: A Review
by Kamna Chaturvedi, Deeksha Malvi, Manish Dhangar, Harsh Bajpai, Ranjan K. Mohapatra, Avanish Kumar Srivastava and Sarika Verma

Chapter 13 247
Ascorbic Acid-assisted Green Synthesis of Silver Nanoparticles: pH and Stability Study
by Katherine Guzmán, Brajesh Kumar, Marcelo Grijalva, Alexis Debut and Luis Cumbal

Preface

The pioneers of modern chemistry built the foundations of inventing, making and changing the chemical products that play an essential role in modern achievements. Green chemistry deals with chemical methodologies and approaches that reduce the use of raw materials or by-products resulting from a chemical reaction in manufacturing, including solvents or catalysts generating hazardous materials that are dangerous to human health or harmful to the environment. The use of sustainable, environmentally friendly processes and new, green chemistry technologies has become the choice for many processes involved in chemical synthesis, nano synthesis, extraction techniques, environmental remediation, and energy.

In this book, researchers from all over the world provide insights into the importance of green chemistry in different ways. Chapter 1 by Endang Tri Wahyuni and Eko Sri Kunarti highlights various chemical analysis and pollutant removal methods that follow some of the 12 principles of green chemistry. In Chapter 2, Nikhat Farhana focuses on eco-friendly protocols that are replacing traditional methods of synthesis followed in chemistry to synthesize life-saving drugs, and preventing the outflow of industrial waste. Surbhi Dhadda et al., in Chapter 3, discuss how recyclable ionic liquids (ILs) catalyze ring closure reactions of chalcones to obtain several heterocyclic rings, including pyrazoles, pyrans, and pyrimidines under ultrasonification. Among the very important features of these reactions are their short routine, high yields, environment-friendliness, high functional group tolerance, formation of a single product, high atom economy, and the absence of a need for column purification. Chapter 4 by Biplob Borah and L. Raju Chowhan provides a critical overview of the application of mechano-chemical techniques for the synthesis of five- and six-membered heterocycles, as well as complex-fused heterocycles and spiro-heterocycles under organocatalytic conditions. Sigüenza et al. in Chapter 5 report on the oxidation of spirostanic steroidal alcohols to their carbonyl analogues using the hypervalent iodine (III)/TEMPO-4-N-acetoxyamine system. In Chapter 6 by Bilal Ahmad Mir and Suresh Rajamanickam, oxidative reactions facilitated by potassium persulphate (K2S2O8) in the absence of a metal catalyst are reviewed, together with the quick and environmentally friendly synthesis of novel chemical species.

Recently, TADF materials have been used as an alternative to metal photocatalysts. In Chapter 7, Rajamanickam and Patel summarize the use of low-cost, less toxic and greener TADF materials, particularly 4CzIPN, as photocatalysts for various radical-based organic transformation reactions. Pinto et al. describe opportunities to make a solution-based green synthesis of transition metal chalcogenides and different experimental planning and analysis techniques, such as the design of experiments, life cycle assessment, and machine learning, in Chapter 8. Ioana Stanciu, in Chapter 9, details the classification, types, and important applications of biomaterials for different everyday applications. Radiation-induced mutation breeding, as reported by Puripunyavanich et al. in Chapter 10, is a remarkable method that can lead to genetic variations resulting in superior mutant cultivars with new and useful traits.

The fields of nanoscience and nanotechnology will continue to grow due to the numerous benefits for our daily lives, including in human health, food processing, environmental safety and device engineering. In Chapter 11, Haridas et al. describe the eco-friendly production of a high-performance supercapacitor electrode material using a combination of Fe_2O_3, gum acacia-derived porous carbon, and ball mill-synthesized graphene. Chapter 12 by Sarika Verma et al., highlights the various green methods of synthesis and discrete applications of inner transition compounds, including the fabrication of lanthanide-doped nanophosphors, rare-earth zirconates, metal oxide nanoparticles, nanocrystal-based photoluminescent materials doped with rare-earth ions, self-assembled nano-spherical dysprosium MOFs, and nucleotide-based lanthanide coordination polymers. In the final chapter, Guzmán et al. discuss the efficient eco-friendly in-situ synthesis of non-aggregated, quasi-spherical silver nanoparticles with an average size of 22.4± 13.2 nm using a mixture of ascorbic acid and citric acid at pH10.

Finally, we are grateful to Kumari Smita for her helpful comments on several chapters and her excellent support in the conceptualization of this book.

Dr. Brajesh Kumar
Department of Chemistry,
TATA College,
Kolhan University,
Chaibasa, Jharkhand, India

Alexis Debut
Professor,
Centro de Nanociencia y Nanotecnologia (CENCINAT),
Universidad de las Fuerzas Armadas ESPE,
Sangolqui, Ecuador

Section 1

Green Chemistry - Fundaments

Chapter 1

Green Methods of Chemical Analysis and Pollutant Removal

Endang Tri Wahyuni and Eko Sri Kunarti

Abstract

This chapter deals with chemical analysis and pollutant removal methods that follow some of the 12 principles of Green Chemistry. In this chapter, the 12 principles of the Green Chemistry along with the short description are highlighted. Several chemical analysis methods are presented, that are both used for chemical identification and concentration determination, whether conventionally or instrumentally. The conventional chemical analysis methods evaluated in this chapter include volumetric and gravimetric, while the instrumental ones presented are limited to atomic absorption spectrometry (AAS) and X-ray fluorescence (XRF) for determination of the analyte concentration, and Infrared spectrometry (IR) and X-ray diffraction (XRD) for chemical identification. Additionally, the pollutant removal methods involving conventional and advanced processes, are reviewed. The conventional chemical removal methods such as precipitation, coagulation, and adsorption are illustrated. The advanced methods in removing chemical pollutants discussed in this chapter are photocatalytic degradation, photo-oxidation/reduction, Fenton and Photo-Fenton, and ozonation. In the description of the chemical analysis and the chemical pollutant removal methods, the evaluation of the unsuitableness or suitableness toward some of the Green Chemistry principles are also accompanied. In addition, the ways to make the less green methods to be greener are also proposed.

Keywords: green chemistry principles, chemical analysis, pollutant removal, method greening

1. Introduction

The principles of Green Chemistry consist of 12 aspects, including [1–5]:

1. Pollution Prevention/Waste minimization,

2. Atom Economy,

3. Less Hazardous Chemical Synthesis.

4. Designing Safer Chemicals,

5. Safer Solvents and Auxiliaries,
6. Design for Energy Efficiency,
7. Use of Renewable Feedstocks,
8. Reduce Derivatives,
9. Catalysis,
10. Design for Degradation,
11. Real-time analysis for Pollution Prevention, and.
12. Inherently Safer Chemistry for Accident Prevention.

Principle no.1 refers that chemical pollution in the environment has to be prevented by minimizing waste from industrial production, chemical analysis laboratories, hospital activities, and many others. Some believe that it is better to prevent waste generation than to treat or clean up waste after it has been generated.

In principle no.2, it is presented that synthetic methods should be designed to maximize the incorporation of all materials used in the process into the final product. Hence no waste or minimum waste has resulted.

Principle no.3 suggests that whenever practicable, synthetic methodologies should be designed to use and generate substances that possess little or no toxicity to human health and the environment.

Based on principle no.4, it is illustrated that chemical products should be designed to preserve efficacy of the function while reducing toxicity. Green Chemists make sure that the things that we synthesize not only do what they are supposed to do, but they do it safely.

In Principle no.5, it is described that the use of auxiliary substances (solvents, separation agents, etc.) should be made unnecessary whenever possible and, when used, innocuous.

It is suggested by Principle no.6, that energy requirements should be recognized for their environmental and economic impacts and should be minimized. Synthetic methods should be conducted at ambient temperature and pressure.

Principle no.7 refers that raw material or feedstock should be renewable rather than depleting whenever technically and economically practical. Green chemists look for alternative sources for making materials. Renewable feedstocks (corn, potatoes, biomass) can be used to make many products: fuels (ethanol and bio-diesel), plastics, and more.

It is implied by principle no. 8, that unnecessary derivatization (blocking group, protection/deprotection, temporary modification of physical/chemical processes) should be avoided whenever possible.

Principle no.9 refers that in chemistry and biology, catalysis is the acceleration (increase in rate) of a chemical reaction by means of a substance, called a catalyst, which is itself not consumed by the overall reaction. Using catalysts can reduce energy, increases efficiency, and reduces by-product formation, which further generates energy efficiency and waste minimization.

Principle no.10 expresses that chemical products should be designed so that at the end of their function they do not persist in the environment and instead break down into innocuous degradation product. Design for degradation means that when green chemists design a new chemical (i.e., a pharmaceutical drug or medicine) or material (i.e., a new plastic) – they design it so that it breaks down at the end of its useful lifetime.

In principle no.11, it is messaged that real-time analysis for a chemist is the process of checking the progress of chemical reactions as it happens. Analytical methodologies need to be further developed to allow for real-time in-process monitoring and control prior to the formation of hazardous substances. Knowing when your product is "done" can save a lot of waste, time, and energy.

Principle no.12 infers that substance and the form of a substance used in a chemical process should be chosen so as to minimize the potential for chemical accidents, including releases, explosions, and fires.

The 12 principles of the Green Chemistry enable people to protect the planet from chemical threats and energy crisis, as well as to find creative ways to reduce chemical waste, conserve energy, and replace hazardous substances [2]. Hence, all human activities should be based on or considered to the all or some of the 12 principles of the Green Chemistry. Some of the activities involving chemicals are material production/ synthesis, chemical analysis, and chemical pollutant removal/treatment. In this Chapter, only chemical analysis and chemical pollutant removal or treatment methods are presented.

Many chemical analysis methods are recognized that are frequently used in a variety of fields including environment, health, food, mining, even archeology [6]. The analysis of chemical methods is usually conducted for identification of a certain chemical or some chemicals as well as for determination of the chemical concentration in the sample(s) [6–12]. The chemical analysis methods widely used involve simple as well as advanced technologies [6–12]. The conventional methods usually use more chemicals and auxiliaries [11], hence further resulting in the toxic chemical waste and wastewater, which create pollution [8, 9]. In contrast, the instrumental chemical analysis methods need less chemicals but may consume more energy [9, 10, 12]. The chemical waste and wastewater and inefficient energy are opposite to the principles of the Green Chemistry, which are waste minimization or pollution prevention, safer solvents and auxiliaries, and efficient energy [8, 9].

In order to reduce chemical waste, conserve energy, and replace hazardous substances, evaluation of some chemical analysis methods is required. It is important, therefore, to recognize the chemical analysis methods that are less suitable and suitable to the principles of the Green Chemistry. The ways to make the chemical analysis methods to be green are also essential to be explored and further to be used.

In addition, a lot of human activities involving chemical processes such as industry, mining, medical, and transportation, always result in chemical waste, that can be formed as gas/particulate, liquid and solid. The chemical waste or wastewater disposed of into the environment without any proper treatment lead to serious pollution [13–18].

The high air pollution can generate a variety of adverse health outcomes. It increases the risk of respiratory infections, heart disease, and lung cancer [13, 14]. The sources of air pollution vary from small units of cigarettes and natural sources such as volcanic activities to large volume of emissions from motor engines of automobiles and industrial activities. Both short and long-term exposure to air pollutants have been associated with health impacts. The most health-harmful pollutants – closely associated with excessive premature mortality – are fine PM2.5 particles that penetrate deep into lung passageways [14].

The serious water pollution due to the inadequately treated or treated industrial wastewater effluents may cause eutrophication in the receiving water bodies and also form a favorable condition for toxin-producing waterborne pathogens [15–18]. The chemicals in wastewater usually comprise of heavy metals and organic compounds [14, 17].

The release of heavy metals into wastewater through human and industrial activities has become a major problem both for humans and aquatic lives. Some negative impacts of heavy metals on aquatic ecosystems include the death of aquatic life, algal blooms, habitat destruction from sedimentation, debris, increased water flow, and other short- and long-term toxicity from chemical contaminants [17, 18]. Severe effects on human health may include reduced growth and development, cancer, organ damage, and nervous system damage [17, 18]. Among the heavy metals, hexavalent chromium is ranked among the top sixteen toxic pollutants that have harmful effects on human health. High chromium dosage has been reported to cause damage to human kidney and the liver, and at low concentration, it causes skin irritation and ulceration. Exposure to high chromium concentration also causes cancer in the digestive tract and lungs [18].

Persistent organic pollutants (POPs) are organic compounds of anthropogenic origin that resist degradation and accumulate in the food chain, and in extreme cases, death [13, 14]. Owing to their toxicity, they can pose a threat to humans and the environment. Some of the POPs polluting water are pentachlorophenol, DDT, hexachlorocyclohexanes, hexachlorobenzene, heptachlor, polychlorinated dibenzo-p-dioxins, polycyclic aromatic hydrocarbons, polychlorinated terphenyls, polybrominated diphenylethers, polybrominated dibenzo-p-dioxins, dibenzofurans, and short-chain chlorinated paraffins [14].

Therefore, removal of chemicals from wastewater before reaching ecosystem is urgent. Many methods for waste treatment are frequently reported, including conventional and advanced methods [19–21]. The conventional methods usually need more chemicals and so that dispose of chemical waste than the advanced methods [21]. The advanced methods use more energy such as light and high temperature than the conventional methods [21].

Using many chemicals and high energy is unexpected because these against the principles of the Green Chemistry [1–5]. It is still necessary to expose the chemical pollutant treatment methods that do not fully follow and follow the principles of the Green Chemistry. By knowing the greener pollutant removal methods, people can choose to use them, and further can prevent the environmental pollution and energy crisis.

Under the circumstances, in the following sections, some chemical analysis and pollutant removal methods that have unsuitableness or suitableness procedures toward some of the Green Chemistry principles are described, and the ways to substitute the less green with the greener methods are also presented. The chemical analysis and pollutant removal methods discussed are presented in the table below (**Table 1**).

2. Green chemical analysis methods

Green chemical analysis is an analysis procedure that avoids or reduces the undesirable environmental side effects of chemical analysis while preserving the classic analytical parameters of accuracy, sensitivity, selectivity, and precision [9, 10]. The

No	The methods	Function	Greenness
1	Volumetric	Quantitative chemical analysis	Less
2	Atomic absorption spectrophotometric	Quantitative chemical analysis	Less
3	X-Ray Fluorescence	Quantitative chemical analysis	Green
4	X-Ray diffraction	Identification	Green
5	Fourier Transform Infrared	Identification	Green

Table 1.
The chemical analysis and pollutant removal methods.

goal of green analytical chemistry is to use analytical procedures that generate less hazardous waste and that are safer to use and more benign to the environment. The main analytical result is related to an increase in analysis reliability, higher precision, and time-saving, which very positively combines with a substantial reduction of waste [9, 10]. The Green chemical analysis should apply at least four Green Chemistry principles, from the 12 principles [10], which are:

1. waste minimization or pollution prevention or prevention of waste generation (no. 1).

2. safer solvents and auxiliaries (no. 5),

3. design for energy efficiency (no. 6), and

4. safer chemistry to minimize the potential of chemical accidents (no. 12).

The chemical analysis methods can be categorized into conventional and instrumental methods, which are used whether for identification and concentration determination purposes [6, 7, 11, 12]. In this section, the conventional analysis method that is evaluated regarding greenless or greenness is volumetric, since it is widely used in environmental and food fields. Meanwhile, the instrumental methods discussed are limited to atomic absorption spectrophotometric the (AAS), X-ray fluorescence (XRF) Infrared spectrometry (IR), and X-ray diffraction (XRD), due to their intensively use in various fields.

2.1 Volumetric method

Volumetric is a chemical analysis method based on the reaction between analytes with the respective standard solution placed in a burette. This method is usually performed with large volume, and sometimes uses hazardous auxiliary. The solutions both standard and analyte, at the end of the process, become harmful wastewater. Although categorized into old or conventional method, volumetric is still frequently used as a standard method in environmental, food, and mineral analysis [8, 11].

In the environmental field, volumetric is placed as a standard method for chemical oxygen demand (COD) assay. COD level represents the quantity of organic and oxidizable inorganic chemicals polluting sample water. A commonly used oxidant in the COD assay is potassium dichromate ($K_2Cr_2O_7$) in combination with boiling sulfuric acid (H_2SO_4) [11]. It is clear that this procedure uses the toxic and carcinogenic

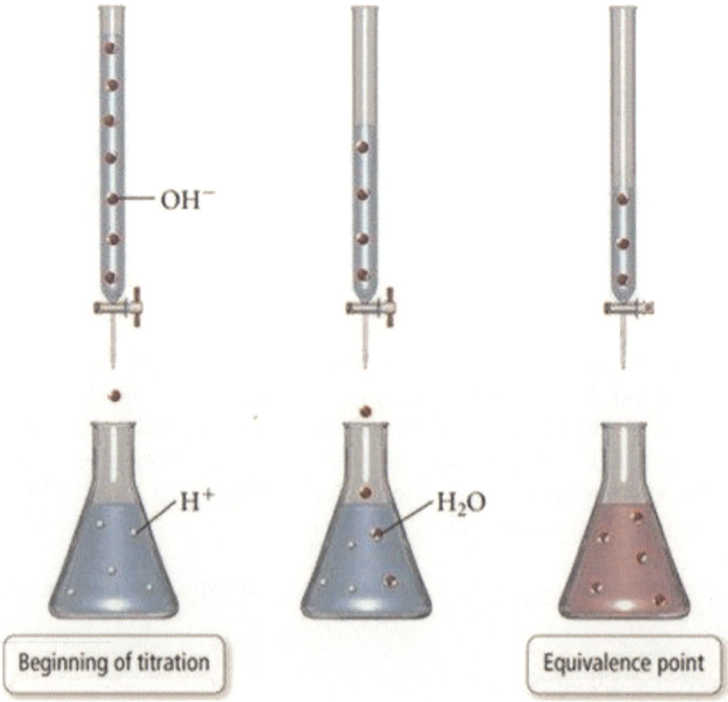

Figure 1.
Titration technique [8].

$K_2Cr_2O_7$ and corrosive sulfuric acid as the auxiliaries, which is against the Green Chemistry Principle number 5 [1–5]. Further, a large volume of $K_2Cr_2O_7$ and $H_2S_2O_4$ solutions are usually used, which results in a large volume of the corresponded chemical wastewater. This resultant of the wastewater will advance to create environmental pollution, which clearly is not suitable to the Green Chemistry Principle number 1 (**Figure 1**) [1–5].

Volumetric method is also used in food analysis, that is to determine saponification number. The saponification number represents an indication of the nature of the fatty acid's constituent of fat in coconut oil, olive oil, and sesame oil. In this procedure, KOH or NaOH, HCl, ethanol, and ether have to be used in large volumes. The use of the corrosive NaOH/KOH will leave the poison waste, which is unsuitable with principle no 1. In addition, since the procedure also uses the hazardous solvents, the procedure is contradiction with principle no 5.

The COD measurement is essential in monitoring the environmental quality and the determination of saponification number has an important role in food quality, so greenings the procedures are required. Greening analysis methods generally can be conducted in several ways [9, 10], such as:

1. modifying an old method to incorporate procedures that either use less hazardous chemicals or use lesser amounts of hazardous chemicals.

2. developing new analytical methodologies; instrumental methods in analysis is a decrease in sample volume needed for analysis.

3. use of direct techniques of analysis,

    - i.e., different laser-spectroscopic methods

    - or solventless processes of analysis

In the case of COD determination, the greening procedure can be conducted by reducing the quantity or volume of the reagent or substituting the toxic reagent with the saver or less toxic one. The strong oxidant but toxic $K_2Cr_2O_7$ can be substituted with $KMnO_4$ [4]. The other way is by applying a smaller volume of the reagents, which hence results in low volume of the wastewater or minimize the wastewater. Using instrumental method to determine saponification number, such as gas chromatography [6], which is greener, is also possible.

2.2 Spectrophotometric methods

2.2.1 Atomic absorption spectrophotometric (AAS)

Atomic absorption spectrophotometric (AAS) method provides concentration data of metals dissolved in the solution. Accordingly, solid samples such as soils, food, minerals, etc. have to be destructed to form a clear solution containing dissolved metal ions [12, 13].

In the AAS method, the dissolved metals have to exist in atomic form. The atomization of the metal ions requires high temperature, which can be from flame or from electric thermal, and flameless conditions [13]. In AAS method, chemicals are not required [12, 13] that can prevent waste generation, following principle no.1 [1–5]. It is clearly suggested that the method falls into a green chemical analysis method.

The atomization of most metal ions usually takes place at high temperature, about 2000–2500°C which can be provided by flame [13], as illustrated by **Figure 2**, and also can be from electric from graphite furnace. The high-temperature flame, in addition to consumes high energy also generates potential accidents, which are contrary to the principle number 6 and 12, respectively. The high electricity consumption is not in accordance to the principle no 6. It is concluded that based on the energy aspect, AAS is included as a less green method. Additionally, among the

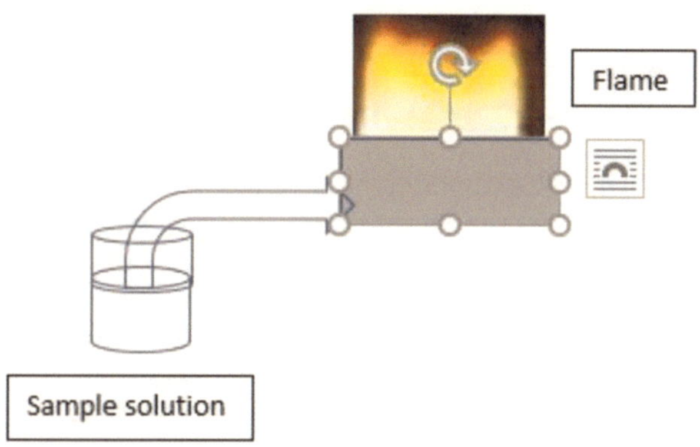

Figure 2.
Flame for atomization in AAS analysis.

metal ions, mercury is the liquid metal at room temperature allowing it to evaporate at high temperature. Accordingly, the atomization of mercury cannot be conducted at high temperature, but to be performed by reducing it to form an atomic phase at room temperature, then called flameless atomization [13]. Accordingly, flameless AAS seems to be greener than the flame one in terms of energy efficiency.

2.2.2 X-ray fluorescence (XRF) and X-ray diffraction (XRD)

XRF method is used for the determination of the elemental concentration, whether metals, metalloids, and non-metals. This method can be used for measuring solid, liquid, and aqueous solutions. The solid samples can be directly measured, and no needs to be prepared into the aqueous solutions, and hence no chemicals are used. Additionally, all metals, metalloids, and non-metals in the samples can be directly measured without any atomization to form elements [13]. It is clear hence that XRF is greener method than AAS to get the same information.

XRD can only be used for crystalline solid samples and that can be directly measured. In this method, chemicals are not required, avoiding it to result in waste. The samples have to be powdered with 100–250 mesh in size. The information given by this instrument is the type of crystal samples [13]. This method does not result in any chemical waste preventing environmental pollution.

However, some believe that the X-ray is a hazardous ray, but in the XRF and XRD instruments, the ray is strictly prevented to irradiate objects including people surrounding. Hence, these methods are in accordance with the principle of Green Chemistry no 12.

2.2.3 Fourier transform infrared (FTIR)

FTIR is a spectrophotometric method required to detect the characteristic bonds in molecules, which can further be used for the identification of the molecules. The samples analyzed can be liquid or solid. In the analysis proses, the liquid samples are placed in cuvettes, while the solid powdered is pelleted with KBr matrix [13]. This method does not need any chemicals and is operated with the low energy infrared. It is obvious that this spectrophotometric method meets the principles no. 1 and no. 6 of the green chemistry.

3. Green methods for pollutant removal

Human activities in hospitals, mining, variety of industries, and other fields almost always result in chemical waste and wastewater. The chemical waste can be toxic heavy metals, hazardous dyes, and persistent organic compounds. These unexpected chemicals adhere human health and ecosystem, which are essential to be treated or removed before entering the environment [14–18].

Several pollutant treatment/removal methods are recognized that are related to conventional and advanced technologies [20–23]. The conventional methods are represented by coagulation and adsorption, and the advanced methods discussed in this chapter consist of photocatalytic -degradation and photo-oxidation, categorized into advance oxidation processes (AOPs).

3.1 Coagulation

Coagulation is essentially a chemical process. It is the destabilization of colloids by the addition of chemicals to neutralize the negative charges of the colloids and to consolidate suspended contaminants for easy removal from water [23–26]. The chemicals are known as coagulants that fall into two categories that are inorganic and organic materials. Frequently used inorganic coagulants include aluminum sulfate, aluminum sulfate, aluminum chloride, and ferric sulfate [24]. Examples of common organic coagulants are polyamines, melamine-formaldehyde, and tannins [25]. Generally speaking, anionic coagulants are suitable to catch mineral particles, while cationic coagulants can capture organic colloids. Inorganic coagulants are usually cost-effective and can be used in a wider variety of applications [24–26]. However, the inorganic coagulants are usually health hazardous and transferred into hazardous sludge in large volume. This is used for removing particles, colloids, or oily materials in suspension. The process of coagulation is illustrated in **Figure 2**. From figure, it can be seen that at the end of the process, large amount of toxic sludge is produced, from the colloidal pollutant and the chemical coagulant. This sludge can be categorized as solid toxic waste. Hence coagulation is opposite to the principle of Green Chemistry no 1.

To make the method greener, the toxic solid waste has to be treated properly, such as by solidification method. In the solidification, the solid waste is mixed with limestone and cement to form a compact and stable solid. The compact and stable solid waste can be avoided from releasing into the environment (**Figure 3**).

3.2 Adsorption

Adsorption is a process that leads to transfer of a molecule or an ion from a fluid bulk to solid surface. This can occur because of physical forces or chemical bonds. In the simple term, adsorption is the attraction of ions or molecules onto the surface of a solid [27, 28]. Adsorption takes place when ions or molecules in a liquid bind themselves to the surface of a solid substance. The solids are called adsorbents, which have a very high internal surface area that permits adsorption. The adsorbent materials known are natural or synthetic zeolites, natural clay minerals, silica gel, activated aluminum, and silicic acid [28].

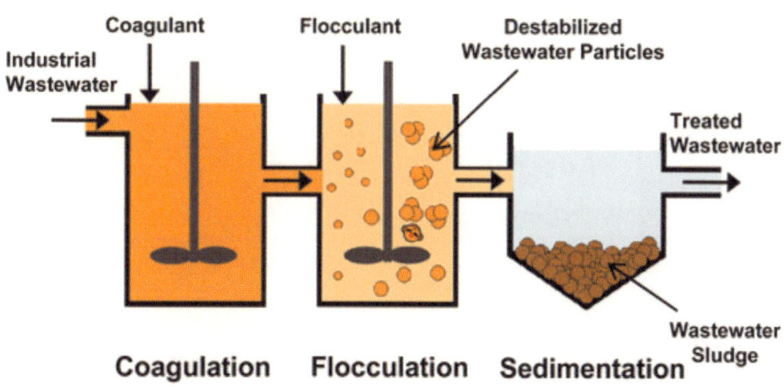

Figure 3.
Coagulation process [24].

Adsorption is believed as a simple and effective method to remove chemical or toxic pollutants. This method is most commonly implemented for the removal of low concentration of non-degradable organic compounds from groundwater, drinking water preparation, process water or tertiary cleansing after, for example, biological water purification [28, 29].

Furthermore, green adsorbents have also been developed, including bio sorbents prepared from Andean Sacha inchi (Plukenetia volubilis L.) shell biomass [30] and agricultural waste [31]. The adsorbents were prepared from the waste that are hazardous material free and low cost. It is obvious that such adsorbents well agree with the Green Chemistry principle no. 1, no. 3, and no. 5.

The adsorbents are usually non-toxic and low-cost materials. However, after a period of time (from minutes to hours) of the adsorption process, the adsorbent has been saturated with toxic pollutants, generating hazardous sludge or solid waste (**Figure 4**). It is clear as well that adsorption is less green method. The greening method can be conducted by converting the hazardous solid waste into a compact and stable solid material, preventing it to release into the environment.

3.3 Advance oxidation processes (AOPs)

Advanced oxidation processes are based on the generation of OH radicals that are very reactive, non-specific, and strong oxidant. The strength of the OH radical is indicated by the high standard reduction potential (E), as 2.80 V, which is higher than the standard reduction potential (E) of ozone (2.07 V), known as strong oxidizing agent [32]. AOPs are considered powerful methods for degradation of various organic pollutants due to their ability for removing almost any organic contaminant. A great number of methods are classified under the broad definition of AOPs based on the oxidizing agents applied [32–50].

Most of them use a combination of strong oxidizing agents (e.g., H_2O_2, O_3) with catalysts (e.g., transition metal ions) and irradiation (e.g., ultraviolet, visible) [32–47]. A combination of H_2O_2 and Fe(II) ion transition metal known as Fenton agent is used Fenton process. When the Fenton process is accompanied by ultraviolet or visible light, the process is named as photo-Fenton. The process involving TiO_2 photocatalyst and ultraviolet light irradiation is drawn as a photocatalysis process. Using O_3 as oxidant in the degradation process is called ozonation. Oxidizing agents from metals, metal oxides, and graphene can also be included in the AOPs.

Figure 4.
Simple illustration of adsorption process.

3.3.1 Fenton/photo-Fenton processes

Fenton oxidation process is a catalytic reaction of H_2O_2 with ferrous ions, that predominantly produces OH radicals as the central oxidizing species, and ferric ions as shown in Eq. (1). Then the ferric ions are reduced back by H_2O_2 into ferrous ions, as presented by Eq. (2) [32–36].

$$Fe^{2+} + H_2O_2 \rightarrow Fe^{3+} + {}^{-}OH + \bullet OH \quad (1)$$

$$Fe^{3+} + H_2O_2 \rightarrow Fe^{2+} + HO_2\bullet + H^{+} \quad (2)$$

The above reaction results in the continuous support of Fe^{2+} iron for the direct Fenton reaction, thus minimizing the required Fe^{2+} concentration, enhancing the catalytic oxidation cycle, and providing additional •OH [33].

Photo-Fenton process involves a combination of Fenton reagents (Fe^{2+} + H_2O_2) and UV–visible radiation (λ < 600 nm) that gives rise to extra OH radicals by two additional reactions. The reaction of OH radicals' formation due to the photodecomposition of H_2O_2 by UV light, as presented in Eq. (3) [33].

$$Fe(OH)^{2+} + light \rightarrow Fe^{2+} + \bullet OH\ (\lambda < 580\ nm) \quad (3)$$

$$H_2O_2 + light \rightarrow 2^{\bullet}OH\ (\lambda < 300\ nm) \quad (4)$$

Fenton, as well as Photo-Fenton type processes, are favored by acidic pH conditions, in the range of pH 2.8–3.0. However, the Fenton process produces a large amount of ferric hydroxide sludge at higher pH, which requires additional separation and disposal of solid waste (9). Accordingly, for the wastewater with higher pH in many cases, the acidification of the reaction medium is a necessity.

The application of UV-C and even UV-A (near UV) radiation during the Fenton (= photo-Fenton) process causes a dramatic increase in the •OH formation efficiency [33–36]. A large number of the •OH enables the use of lower ferrous catalyst concentrations, preventing the solid waste of the ferric hydroxide sludge. It seems that the photo-Fenton process is greener, in terms of waste minimization (principle no.1), and the effect of using UV light on the prevention of precipitation is found to be significant.

The low efficiency affecting photo-Fenton processes at neutral pH is mainly due to iron precipitation, and can be therefore prevented by properly adding iron complexing agents. As pointed out in reaction Eq. (5), such compounds (L) should be able to form stable complexes with Fe(III), which (i) significantly absorb UV–vis light and (ii) undergo photochemical reductions leading to Fe(II) ions [34–36]:

$$\left[Fe^{3+}L\right] + h\upsilon \rightarrow \left[Fe^{3+}L\right]* \rightarrow Fe^{2+} + L^{\bullet} \quad (5)$$

Iron complexing agents used for preventing precipitation of ferric hydroxide are Polycarboxylates and amino polycarboxylates that can form stable complexes with Fe (III), absorb light in the near-UV and the visible regions more efficiently than aquo-complexes [35, 36], and undergo photoreduction through a ligand-to-metal charge transfer (LMCT) generating Fe(II) ions [36]. The iron complexing compounds should be photo-degraded during photo-Fenton process, to avoid chemical waste formation. It is clearly seen that the addition of complexing compound can make Fenton and photo-Fenton greener.

3.3.2 Photocatalysis over TiO_2

The photocatalytic process using TiO_2 photocatalyst is very promising for application in water purification and wastewater treatment because many organic compounds can be decomposed and mineralized by the proceeding oxidation and reduction processes on TiO_2 surface [37–47]. The most commonly tested compounds for decomposition through photocatalysis are phenols, chlorophenols, pesticides, herbicides, benzenes, alcohols, dyes, pharmaceutics, humic acids, organic acids, and others [37]. Additionally, photocatalysis process over TiO_2 for reducing the toxic Cr (VI) into the harmless Cr(III) as well for oxidizing the hazardous Pb(II) into the safer PbO_2 are also assessed [37].

TiO_2 is the most commonly used photocatalyst, because it is non-toxic, chemically stable, cheap, and very efficient. In photocatalysis, light of energy greater than the bandgap of the semiconductor excites an electron from the valence band to the conduction band. In the case of anatase TiO_2, the bandgap is 3.2 eV, therefore UV light ($\lambda \leq 387$ nm) is required [37–47]. The absorption of a photon excites an electron to the conduction band (e CB) generating a positive hole in the valence band ($h^+{}_{VB}$) and an electron in the valence band ($e^-{}_{CB}$), as presented as Eq. (6). The hole can interact with a water molecule, as seen in Eq. (7).

$$TiO_2 + \text{UV light} \rightarrow TiO_2\left(h^+{}_{VB} + e^-{}_{CB}\right) \tag{6}$$

$$h^+{}_{VB} + H_2O \rightarrow \bullet OH + H^+ \tag{7}$$

However, it has some disadvantages: one of these is a relatively high value of the bandgap, around 3.2 eV, which limits its use under UV light. The other weaknesses are : high dispersion in the water which causes difficulties in sedimentation, and sensitive to the recombination of photoinduced electrons and holes, which decreases its photocatalytic activity [39–47]. The weakness of using UV light allows it to consume high energy (ignores principle no. 6) and the hazardous UV light is potential to cause an accident if exposes to a person for long time (less suitable to the principle no 12). Clearly, the method has not followed fully the principles of the Green Chemistry.

Therefore, an effort has been focused to overcome this deficiency of using UV light, by doping TiO_2 crystal structure with either metal elements [38–44], or non-metal elements [45–47]. Doping process is hoped to narrow the bandgap that falls into visible region. Metal elements that have been doped into TiO_2 include Ag [38–40], Au [41] Cu [42], and Fe [43, 44], while non-elemental dopants are N [45], S [46], and C [47]. The doping TiO_2 has been frequently reported to be able to decrease their bandgap from 3.2 to smaller than 3.0 eV. The narrowing gap is illustrated in **Figure 5**.

The gap decrease is able to enhance its photoactivity significantly under visible light irradiation. The use of visible light for replacing the UV light, enables photocatalysis process to be greener method. The photocatalysis process for complete degradation of organic pollutants will form smaller and saver molecules, which is in line with waste minimization. It is clear that this method obeys principle no. 2, and is in line with the green method.

3.3.3 By using nanomaterial oxidizing agents

Several metal [48] and metal oxide nanomaterials including iron oxide [49], graphene oxide [48], as well as graphene bounded with metals [50] have shown strong

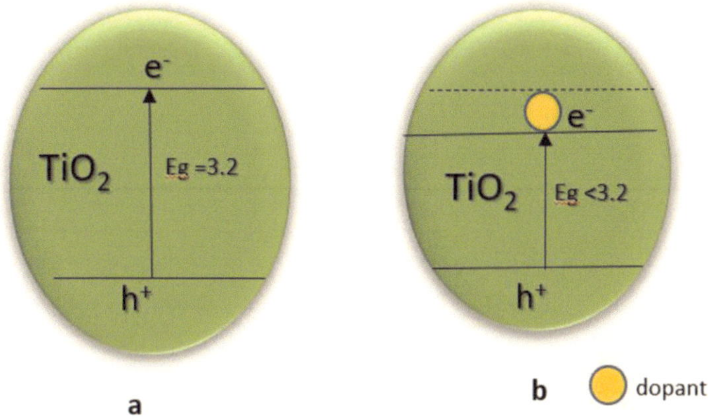

Figure 5.
The simple illustration of a) un-doped TiO_2, and b) doped TiO_2.

oxidizing power. Iron oxide nanoparticles have been prepared by using citrus extract for confinement of the particle growth. This method has produced in the nanoparticles providing a larger surface, an advance to result in effective degradation of some dyes [49]. The nanocomposite of graphene oxide bound with metal by using biomass as a template has also been reported [48]. This oxidizing agent has been proven to show effective degradation of the organic pollutants. Furthermore, the use of part of plants as reducing agents as well as a template for oxidizing agent nanomaterial has also been developed. One of the examples is graphene-supported silver nanocomposite [50]. The reducing agent from the biomass, replacing the toxic chemical can be categorized as the green reducing agent. The use of citrus and biomass waste as a template and reducing agent replacing the hazardous chemicals, allow the method as a greener one, due to the agreement with the principle no. 1, no. 5 and no. 12.

4. Conclusions

Several chemical analysis and pollutant removal methods are believed as very important and required by many fields. Some of the methods are recognized not obey some of the principles of the Green Chemistry. The greening methods of chemical analysis and chemical pollutant removal are essential, which can be conducted by reducing the quantity (mass and volume), substituting the toxic chemicals with the harmless or less toxic chemicals, modifying, and replacing them with the greener methods.

Conflict of interest

The authors declare no conflict of interest.

Author details

Endang Tri Wahyuni* and Eko Sri Kunarti
Faculty of Mathematic and Natural Sciences Gadjah, Chemistry Department, Mada University, Yogyakarta, Indonesia

*Address all correspondence to: endang_triw@ugm.ac.id

References

[1] Anastas PT, Warner JC. Green Chemistry: Theory and Practice. New York: Oxford University Press; 1998. p. 135

[2] Clark JH. Green chemistry: Challenges and opportunities. Green Chemistry. 1999;**1**:1-8. DOI: 10.1039/A807961G

[3] Anastas P, Eghbali N. Green chemistry: Principles and practice. Chemical Society Reviews. 2010;**39**: 301-312. DOI: 10.1039/B918763B

[4] Ahluwalia VK, Kidwai M. Basic principles of green chemistry. In: New Trends in Green Chemistry. Dordrecht: Springer; 2004. p. 263

[5] Chen T-S, Kim H, Pan S-Y, Tseng P-C, Lin Y-P. Chiang P-C Implementation of green chemistry principles in circular economy system towards sustainable development goals: Challenges and perspectives. Science Total Environment. 2020;**716**:136998

[6] Trim HH. Analytical Chemistry Methods and Applications. 1st ed. New Jersey, USA: Apple Academic Press; 2021. p. 380

[7] Fifield FW, Kealey D. Principles and Practice of Analytical Chemistry. 5th ed. New Jersey, USA: Wiley-Blackwel; 2000. p. 578

[8] Taleuzzaman M, Gilani SJ. First step analysis in quality control -volumetric analysis. Global Journal of Pharmaceutical Science. 2017;**1**(3): 9348-9360

[9] Armenta S, Garrigues S, de la Guardia M. Green analytical chemistry. Trends in Analytical Chemistry. 2008; **27**(6):497-451

[10] Anastas PT. Green chemistry and the role of analytical methodology development. Critical Reviews in Analytical Chemistry. 1999;**29**:167-175

[11] Amanatidou E, Trikoilidou E, Samiotis G, Benetis N-P, Taousanid N. An easy uncertainty evaluation of the COD titrimetric analysis in correlation with quality control and validation data. Method applicability region. Analytical Methods. 2012;**4**:4204-4412

[12] Hussain CM, Keçili R. Wet chemical techniques in environmental analysis. In: Modern Environmental Analysis Techniques for Pollutants. 1st ed. Amsterdam, Holland: Elsevier; 2020. pp. 121-113

[13] Rouessac F, Rouessac A. Chemical Analysis: Modern Instrumentation Methods and Techniques. 2nd ed. New Jersey, USA: Wiley; 2007

[14] Ghorani-Azam A, Riahi-Zanjani B, Balali-Mood M. Effects of air pollution on human health and practical measures for prevention in Iran. Journal of Research in Medical Sciences : The Official Journal of Isfahan University of Medical Sciences. 2016;**21**:1-12. DOI: 10.4103/1735-1995.189646

[15] Alcock R, Bashkin V, Bisson M, Brecher RW, Bree LV, Chrast R, et al. Health Risks of Persistent Organic Pollutants from Long-range Transboundary Air Pollution. Netherland: World Health Organization, Europe; 2003. p. 252

[16] Ilyas M, Ahmad W, Khan H, Yousaf S, Yasir M, Khan A. Environmental and health impacts of industrial wastewater effluents in Pakistan: A review. Reviews on

Environmental Health. 2019;**34**(2): 171-186. DOI: 10.1515/reveh-2018-0078

[17] Akpor OB, Muchie M. Environmental and public health implications of wastewater quality. African Journal of Biotechnology. 2011; **10**(13):2379-2387. DOI: 10.5897/AJB10.1797

[18] Akpor OB, Ohiobor GO, Olaol TD. Heavy metal pollutants in wastewater effluents: Sources, effects and remediation. Advances in Bioscience and Bioengineering. 2014;**2**(4):37-43. DOI: 10.11648/j.abb.20140204.11

[19] Owalude SO, Tella AC. Removal of hexavalent chromium from aqueous solutions by adsorption on modified groundnut hull. Beni-suef University Journal of Basic and Applied Sciences. 2016;**5**:377-388

[20] Crini G, Lichtfouse E. Advantages and disadvantages of techniques used for wastewater treatment. Environmental Chemistry Letters. 2019;**17**:145-155

[21] Odegard H, Karlsson I. Chemical wastewater treatment: Value for money. In: Klute R, IHH H, editors. Chemical Water and Wastewater Treatment III. Berlin Heidelberg: Springer-Verlag; 1994

[22] Amit Sonune A, Ghate R. Developments in wastewater treatment methods. Desalination. 2004;**167**:55-63. DOI: 10.1016/j.desal.2004.06.113

[23] Li N, Sheng G-P, Lu Y-Z, Zeng RJ, Yu H-Q. Removal of antibiotic resistance genes from wastewater treatment plant effluent by coagulation. Water Research. 2017;**111**:204-212. DOI: 10.1016/j.watres.2017.01.010

[24] Sibiya NP, Rathilal S, Tetteh ET. Coagulation treatment of wastewater: Kinetics and natural coagulant evaluation. Molecules. 2021;**26**:698. DOI: 10.3390/molecules26030698

[25] Zhao C, Zhou J, Yan Y, Yang L, Xing G, Li H, et al. Application of coagulation/flocculation in oily wastewater treatment: A review. Science Total Environment. 2021;**765**:142795

[26] Teh CY, Mori P, Katrina B, Shak PY, Wu TY. Recent advancement of coagulation–flocculation and its application in wastewater treatment. Industrial and Engineering Chemistry Research. 2016;**55**(16):4363-4389. DOI: 10.1021/acs.iecr.5b04703

[27] Yousef R, Qiblawey H, El-Naas MH. Adsorption as a process for produced water treatment: A review. PRO. 2020;**8**: 1657. DOI: 10.3390/pr8121657

[28] Ayub S, Changani F. Adsorption process for wastewater treatment by using coconut shell introduction. International Journal of Civil, Structural, Environmental and Infrastructure Engineering Research and Development (IJCSEIERD). 2014;**4**(3):21-34

[29] Rashid R, Shafiq I, Akhter P, Iqbal MJ, Hussain M. A state-of-the-art review on wastewater treatment techniques: The effectiveness of adsorption method. Environmental Science and Pollution Research. 2021;**28**: 9050-9066. DOI: 10.1007/s11356-021-12395-x

[30] Kumar B, Smita K, Sánchez E, Carina Stael C, Cumbal L. Andean Sacha inchi (*Plukenetia volubilis L.*) shell biomass as new biosorbents for Pb^{2+} and Cu^{2+} ions. Ecological Engineering. 2016; **93**:152-158

[31] Gupta A, Sharma V, Sharma K, Kumar V, Choudhary S, Mankotia P,

et al. A review of adsorbents for heavy metal decontamination: Growing approach to wastewater treatment. Materials. 2021;**14**(16):4702

[32] Pouran SR, Aziz ARA, Daud WMAW. Review on the main advances in photo-Fenton oxidation system for recalcitrant wastewater. Journal of Industrial and Engineering Chemistry. 2015;**21**:53-69

[33] Mirzaei A, Chen Z, Haghighat F, Yerushalmi L. Removal of pharmaceuticals from water by homo/heterogonous Fenton-type processes: A review. Chemosphere. 2017;**174**:665-688

[34] Clarizia L, Russoa D, Di Somma I, Marotta R, Randreozzi R. Review Homogeneous photo-Fenton processes at near neutral pH: A review. Applied Catalysis B: Environmental. 2017;**209**: 358-371

[35] Kishimoto N, , Kitamura T, Kato M, Otsu H. Influence of chelating agents on fenton-type reaction using ferrous ion and hypochlorous acid. Jurnal of Water and Environment Technology. 2013;**11** (1):21-32

[36] Buitrago JL, Sanabria J, Gútierrez-Zapata HM, Urbano-Ceron FJ, García-Barco A, Osorio-Vargas P, et al. Photo-Fenton process at natural conditions of pH, iron, ions, and humic acids for degradation of diuron and amoxicillin. Environmental Science and Pollution Research. 2020;**27**:1608-1624

[37] Wang C, Liu H, Qu Y. Review article TiO_2-based photocatalytic process for purification of polluted water: Bridging fundamentals to applications. Hindawi Publishing Corporation. Journal of Nanomaterials. 2013;**2013**:319637

[38] Albiter E, Valenzuela MA, Alfaro S, Valverde-Aguilar G, Martínez-Pallares FM. Photocatalytic deposition of Ag nanoparticles on TiO_2: Metal precursor effect on the structural and photoactivity properties. Journal of Saudi Chemical Society. 2015;**19**(5):563-573

[39] Nigussie GY, Tesfamariam GM, Tegegne BM, Weldemichel YA, Gebreab TW, Gebrehiwot DG, Gebremichel GE. Antibacterial activity of Ag-doped TiO_2 and Ag-doped ZnO nanoparticles. Hindawi International Journal of Photoenergy. 2018;**2018**:7. Article ID 5927485

[40] Sescu AM, Favier L, Lutic D, Soto-Donoso N, Ciobanu G, Harja M. TiO_2 doped with noble metals as an efficient solution for the photodegradation of hazardous organic water pollutants at ambient conditions. Watermark. 2021; **13**:19

[41] Mohammed MKA. Sol-gel synthesis of Au-doped TiO_2 supported SWCNT nanohybrid with visible-light-driven photocatalytic for high degradation performance toward methylene blue dye. Optik. 2020;**223**:165607

[42] Assadi MHN, Hanaor DAH. The effects of copper doping on photocatalytic activity at (101) planes of anatase TiO_2: A theoretical study. Applied Surface Science. 2016;**387**: 682-689. DOI: 10.1016/j.apsusc.2016.06.178 1

[43] Tong T, Zhang J, Tian B, Chen F, He D. Preparation of Fe^{3+}-doped TiO_2 catalysts by controlled hydrolysis of titanium alkoxide and study on their photocatalytic activity for methyl orange degradation. Journal of Hazardous Materials. 2008;**155**:572-579

[44] Safari M, Talebi R, Rostami MH, Nikazar M, Dadvar M. Synthesis of iron-doped TiO_2 for degradation of reactive Orange16. Journal of Environmental

Health Science and Engineering. 2014; **12**:19

[45] Mahy JG, Cerfontaine V, Poelman D, Devred F, Gaigneaux EM, Heinrichs B, et al. Highly efficient low-temperature N-doped TiO_2 catalysts for visible light photocatalytic applications. Materials. 2018;**11**:584. DOI: 10.3390/ma11040584

[46] Lin Y-H, Hsueh H-T, Chan C-W, Chua H. TiO_2-S The visible light-driven photodegradation of dimethyl sulfide on S-doped TiO_2: Characterization, kinetics, and reaction pathways. Applied Catalysis B: Environmental. 2016;**199**:1-10

[47] Hua L, Yin Z, Cao S. Recent advances in synthesis and applications of carbon-doped TiO_2 nanomaterials. Catalysts. 2020;**10**:1431

[48] Kumar B. Green synthesis of gold, silver, and iron nanoparticles for the degradation of organic pollutants in wastewater. Journal of Composite Science. 2021;**5**:219

[49] Kumar B. Graphene- and graphene oxide-bounded metal nanocomposite for remediation of organic pollutants. In: Carbon-Based Material for Environmental Protection and Remediation. London: IntechOpen; 2020. p. 19

[50] Vizuete KS, Kumar B, Vaca AV, Debut A, Cumbal L. Mortiño (*Vaccinium floribundum Kunth*) berry assisted green synthesis and photocatalytic performance of Silver–Graphene nanocomposite. Journal of Photochemistry and Photobiology A: Chemistry. 2016;**329**:273-279

Chapter 2

Fundamental Principles to Address Green Chemistry and Green Engineering for Sustainable Future

Nikhat Farhana, Mohammed Gulzar Ahmed, Mohammed Asif Iqbal, Natasha Naval Aggarwal, Prajitha Biju, Ashwini Somayaji, Abdul Rahamanulla, Nishmitha Gretta D'Souza, Sudhina Makuttan, Tahreen Taj, Abdullah Khan and Roshan Sayeed

Abstract

The background of green chemistry represents the dramatic module of a new millennium, the substantiable chemical process steam for evaluation in designing phase to incorporate the principles of GC (Green Chemistry) in 1990s. there has been a tremendous success in developing a new product and process which are more compatible with biological, zoological and botanical perspective to illuminate the sustainability goal, this chapter represents the simplified way to lookout different approach adopted in GC-research, the methodology enhance the chemical process economics, concomitant which deduct the environmental burden. This review merely focusing on eco-friendly protocol which replace the traditional method of synthesis followed in chemistry to synthesize lifesaving drugs, with prevention outgoing waste from industries. GC and chemical engineering or green engineering (GE) should produce eco-friendly chemical process for drug design which likely to be spread rapidly in next few decades. This chapter explains in-depth and compact with detailed glimpse of environment friendly-protocol and principle bridging continent and scientific discipline to create new solution.

Keywords: green Chemistry, sustainability, biological, zoological, botanical, eco-friendly, industries, drug-design, environmental, scientific

1. Introduction

Green chemistry approach is an advanced field of science and it attracted the scientist and researcher since decade, it opens the enormous gateways to modified version of known synthetic reaction in a newer way with associated enlarged potentiality accompanied by sustainability [1]. Revolutionary measures against solvent based synthetic reaction were invented in green synthesis roots, whereas organic solvents were replaced with non-organic medias in synthetic reaction to overcome the volatility and corrosiveness of hazardous nature of organic solvent to preserve the environmental greenness. Chemistry and engineering working in the area of green chemistry are proliferating, and quite justifiably [2], asked to explain why their experimental reactions, process or product is actually sustainable or green, it was an on-going debate on the topic, often the assert of sustainable or green reactions cut-throat disagreement. In this context, the 12-Principles of green chemistry become widely accepted set of criteria for high-speed screening towards "Greenness" for comparing the environmental acceptability of two rival process. The 12-Principle are not sufficient to elaborate the whole concept of green chemistry (**Figure 1**) mostly relevant to important concept in regards to environmental impact; for example, fundamental process or product need to be monitored for its lifecycle and heat recovery from exothermic or endothermic reaction. For this reason, some scientist proposes the 12-principles of green engineering (**Figure 2**).

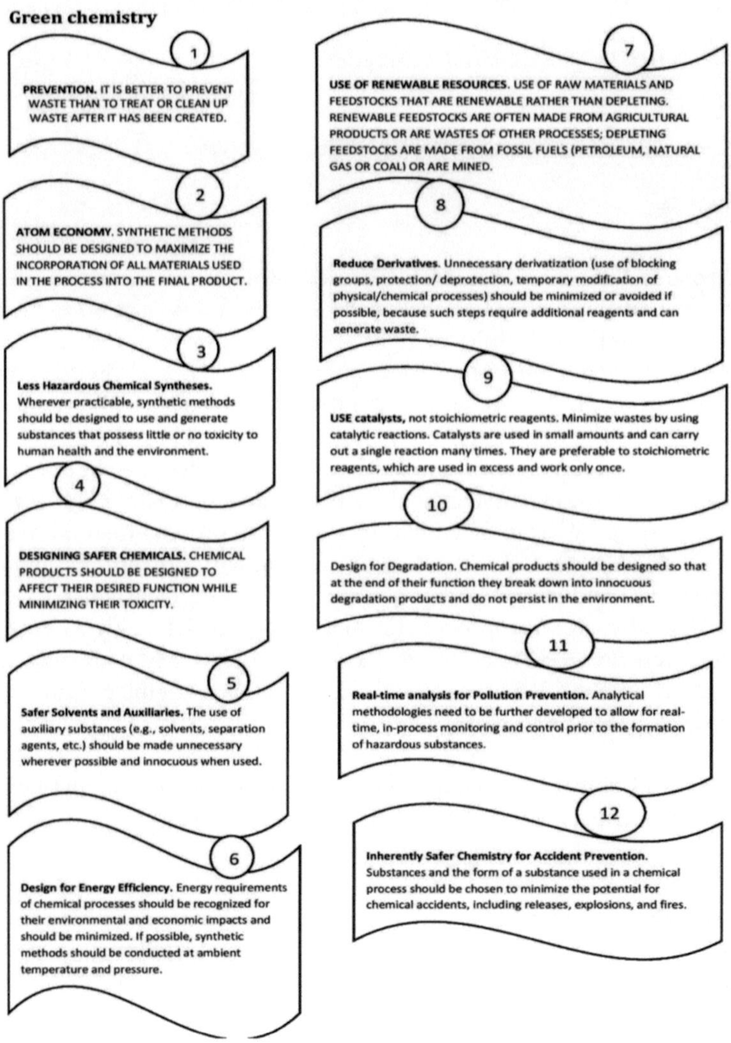

Figure 1.
Classical principle used to followed in green chemistry for synthesis.

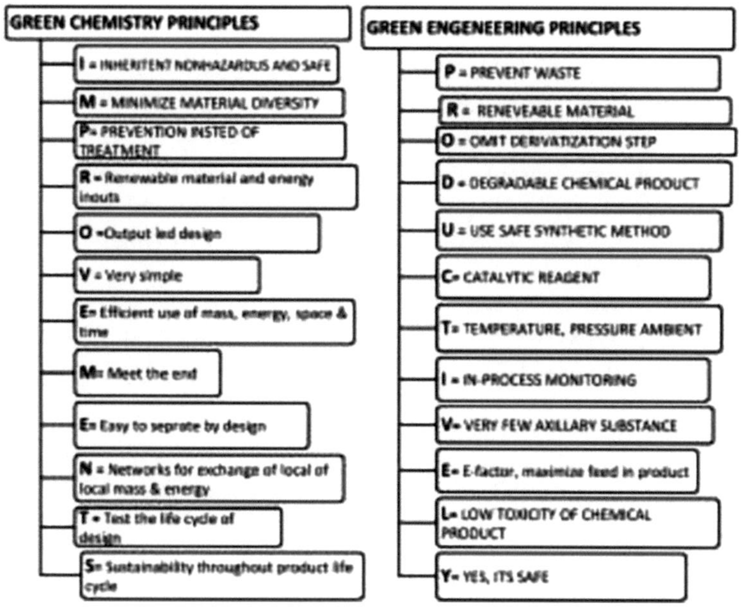

Figure 2.
Principles of green chemistry and green engineering.

In-depth scripture of both principles was, well articulated till date, here we representing the linking key which opens the roadmap, to get it, into the principle of green chemistry and green engineering, without lengthening the discussion, we are discussing the bullet points to crystal clear the importance of green technology which is lost in debate while digging-out the available repositories. Hence forth we are proposing a condensed version of green chemistry and engineering, fitting onto single sentence and incorporated mnemonic PRODUCTIVELY (green chemistry) and IMPOVEMENTS (Green engineering), we believe these two condensed forms are sufficient to manage the entitled topic of this chapter, this chapter re-present both the topic in form of 24-principle for sustainable green technological chemistry [3] (**Figure 2**). With this combined set we encourage the reader to reciprocate the sustainable future in new way called IMPROVEMENTS PRODUCTIVELY.

2. Methodology

This work speaks about the technical branch imbedded, environmental friendly and cost-effective utilization of resources that minimize or even exclude the production of harmful bi-product in designing and manufacturing of product, which will ultimately increase the yield of product [4].

2.1 Green chemistry application and synthetic methodology

With reciprocating to above principles some examples related to synthetic application, utilization, minimization of bi-product, invention, design, application of chemical process and products in academic laboratories and industries based on

Environmental Protection Agencies (EPA) & Organization of economic co-operation and development (OECD) [5].

2.1.1 Synthesis of polymer

Synthesis of polycaprolactone. Poly (Ɛ-caprolactone) PCL is important biodegradable and biocompatible synthetic polymer used in prosthesis and controlled drug release matrix for active substance, this polymer obtained by ring polymerization reaction (**Figures 3–10**) [6].

- **Synthesis of polyurethane**: Novel synthetic routes to polyurethane production without using diisocyanate

- Synthesis of series of polyhydroxy-urethane

- Synthesis from commercially available ethylene carbonate

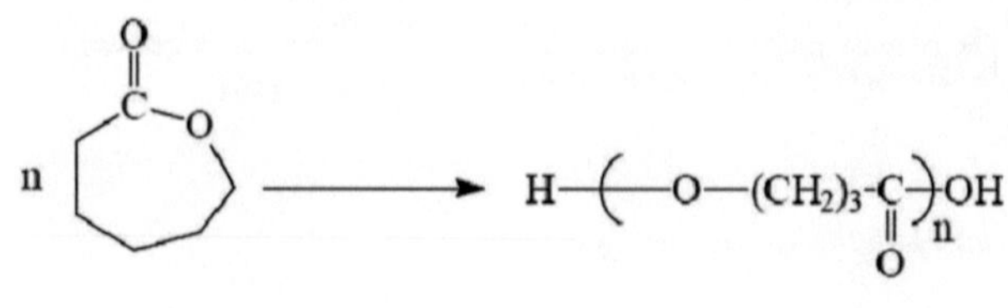

Figure 3.
Polymerization of PCL.

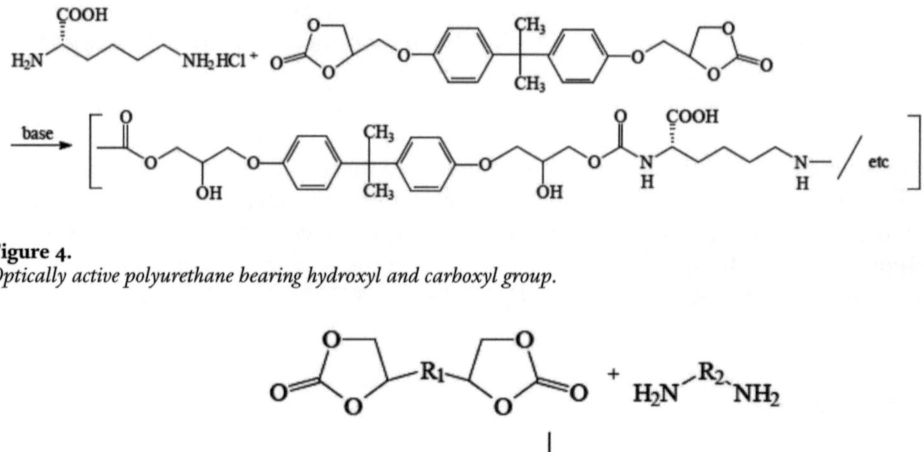

Figure 4.
Optically active polyurethane bearing hydroxyl and carboxyl group.

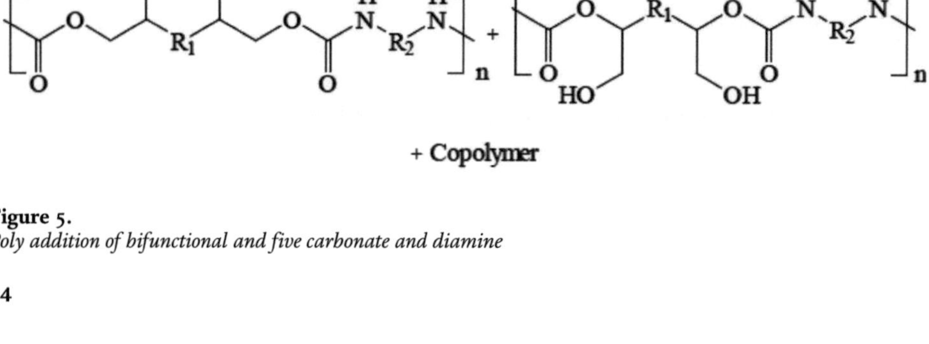

Figure 5.
Poly addition of bifunctional and five carbonate and diamine

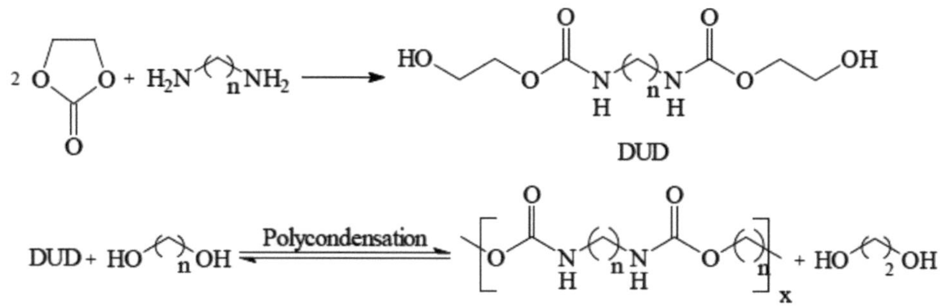

Figure 6.
Polyurethane from ethyl carbonate.

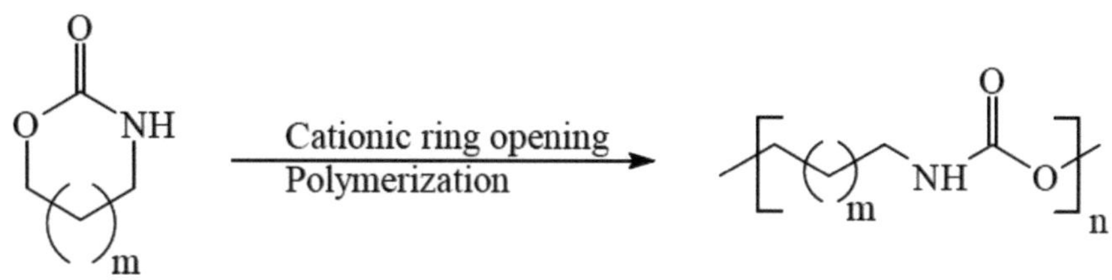

Figure 7.
Cationic ring opening polymerization reaction.

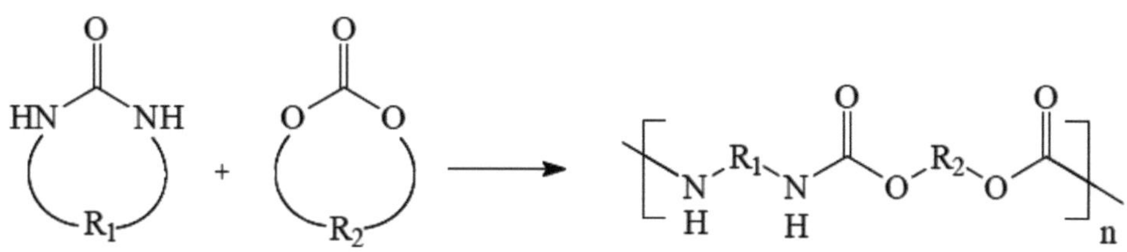

Figure 8.
New rout to polyurethane containing urethane bond.

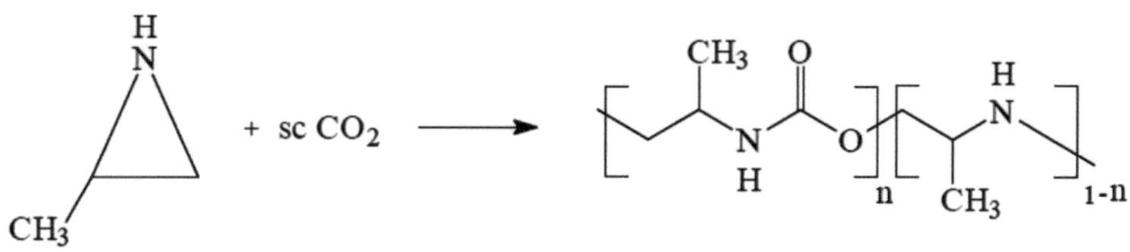

Figure 9.
Sensitive phase transformation reaction in water.

- Synthesis of Alternative route to synthesis poly (tri methylene urethane) and poly (tetra methylene urethane)
- Synthesis of 2,2-dimethyltrimethylene carbonate
- Synthesis of thermo-responsive polyurethane

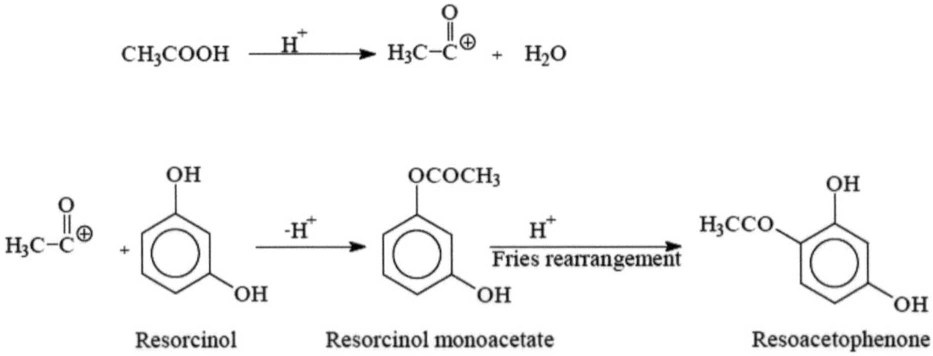

Figure 10.
Dihydroxy acetophenone from resorcinol and acetic acid.

- Synthesis of Reso-acetophenone (2,4-dihydroxyacetophenone)
- Bromination

2.1.2 Alternative methods to classical synthesis

Less hazardous chemical synthetic reactions in chemistry to prevent accident, safeguard the environment and prevention of by-product, not only to the public but also to laboratory-worker [7]. How to use safe alternative methods in process material in process some examples depicted in the **Table 1** represents the protocol for halogenation where as sustainable method safe, ecofriendly, less time consuming with high yield compare to classical methods, **Tables 2–6** represents the green synthetic protocol which already replaced with classical methods.

Where as the **Table 6** represents the modern method of drug synthesis by using Microwave irradiation technique.

2.2 Green engineering and its application

Green engineering becomes a sustainable and training technology in construction sector [14], its values are influential and all-inclusive, its very broad field to meet the demand for more sustainable process, the innovation includes; Smaller continuous flow reactor, new scale reactor and designer, new separation technique which decreases the amount of necessary solvent and energy, Process designing requires no separation, an integrated system approach to maximize heat transfer. Green chemistry design of inherently safe reaction and use of nonhazardous materials also contribute to the use and application of green engineering principle. With this safe process, the risk management does not have to be engineered to limit the exposure or potential for accidents [15, 16].

Tables: Sustainable synthetic schemes with application and replacement of catalysts [17, 18].

3. Conclusion and future perspectives

The rapid industrialization and urbanization in present scenario are only to achieve growth and development around the globe. The natural creations unfortunately

DOI: http://dx.doi.org/10.5772/intechopen.104717

SN	Name of the reaction	Sustainable synthetic reaction with replacement of carcinogenic reagents & solvents with application	
1	**Halogenation**		
	Electrophilic substitution reaction	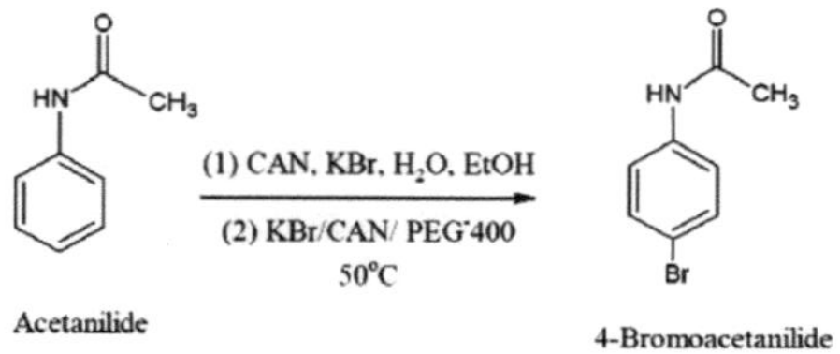	**Application:** The aromatic bromo compounds can act as potent anti-tumor anti-bacterial anti-fungal anti-neoplastic anti-viral and anti-oxidizing agent and important compound for Grignard reagents. Classical Catalyst Br2:
		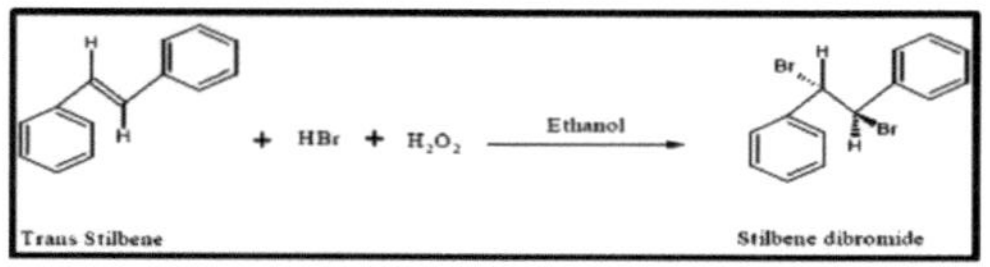	**Application:** Manufacturing of dye. Classical Catalyst: Liq Br2 in CH2Cl2 Catalyst replaced: HBr & H2O **Application: Trans stilbene.** Classical Catalyst: Liq Br2 in CH2Cl2 Catalyst replaced: NaBr & NaOBr
2	**Benzoin condensation**	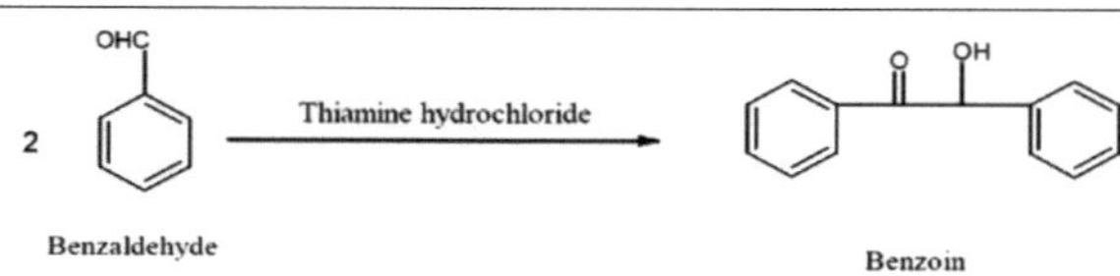	**Application:** Anti-inflammatory gemicidal. Classical Catalyst: NaCN (Poisonous) Catalyst replaced: Thiamine HCl

Table 1.
Halogenation and condensation reaction [8].

2	Nitration	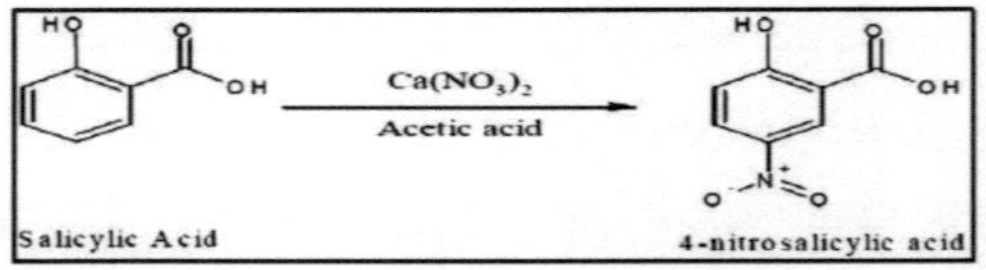	Application: **Anti-inflammatory germicidal.** **Classical Catalyst: NaNo3 & Conc H2SO4** **Catalyst replaced:** [Ca (NO3)]2 + CH3COOH **Regioselective product.**
3	**Ionic liquids & Green solvent** The molten TBAB can act both solvent and catalyst.	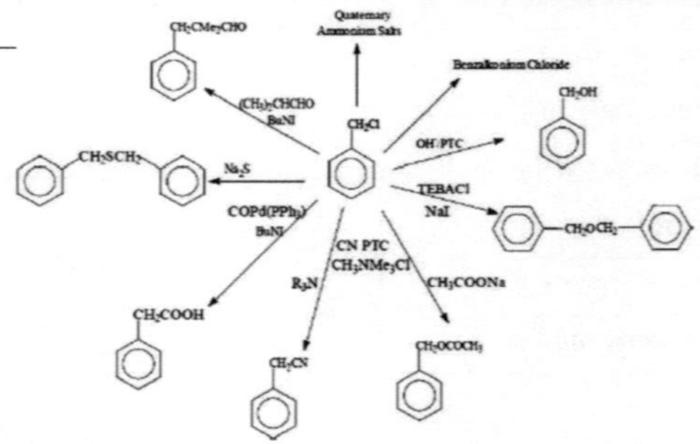	**Application:** Holds Unique array of Physio-chemical properties, such as High electrical conductivity, High thermal stability, Lowe nucleophilicity and capability of providing weekly coordinating or non-coordinating environment, for organic inorganic organometallic & polymeric compound synthesis. This solvent can be removed and recycled
	Quaternary ammonium compound essential for higher reactivity (Terephthalic Acid) TPA to Poly amide (tetrabutylammonium bromide) TBAB	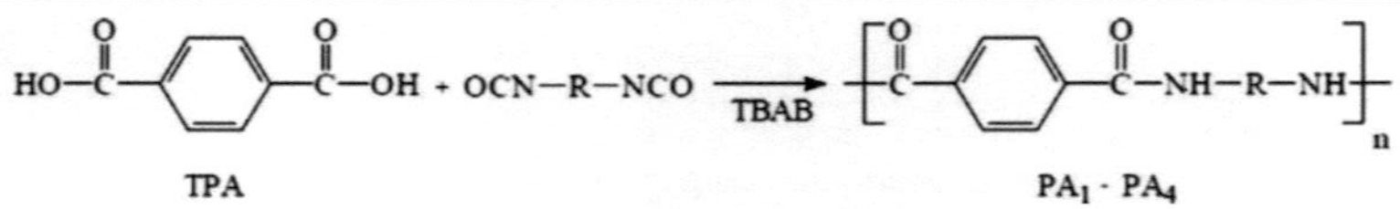	

Table 2.
Nitration ionic liquid and green solvent for future reference [8, 9].

4	Safe solvent and auxiliaries/solvent free reactivity **Good process to follow Grindstone Chemistry. Absence of solvent** Schem-5: **Aldo condensation.** Schem-6: **Diels's alder reaction**	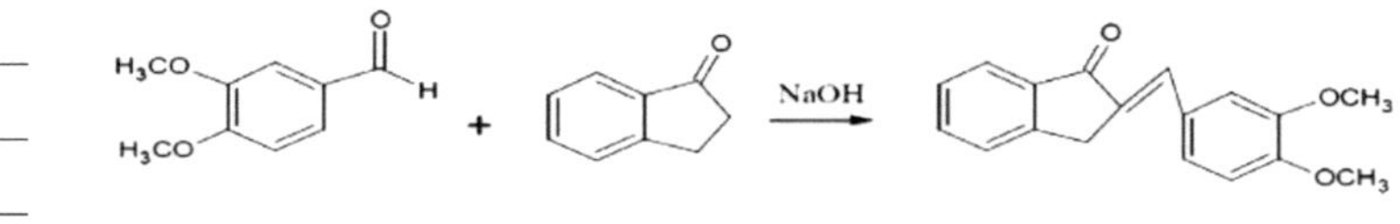 3,4-Dimethoxybenzaldehyde 1-Indanone Aldol **(Scheme 5)** 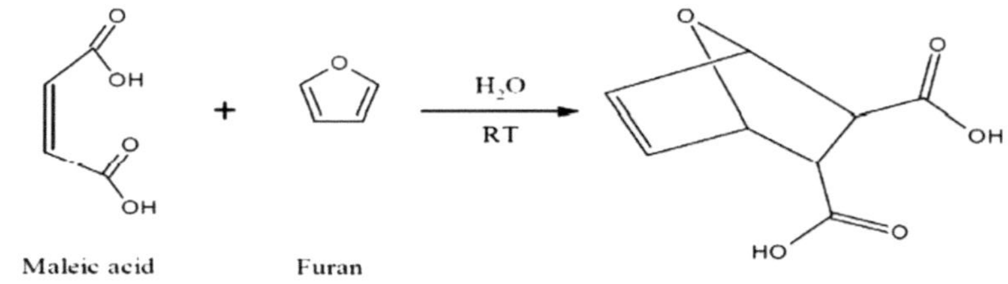 Maleic acid Furan Adduct **(Scheme 6)**
5	**Preparation of carbonyl compound with oxime**	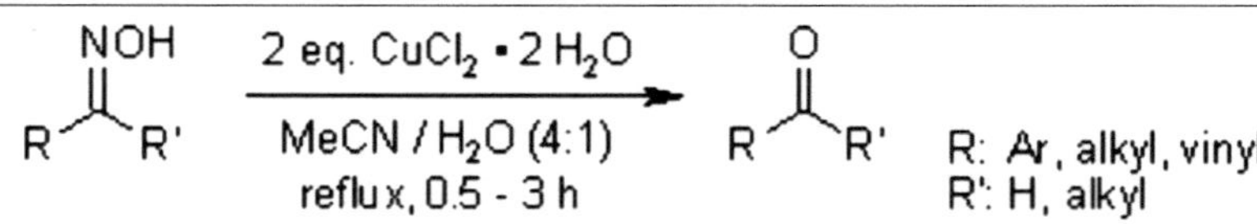

Table 3.
Safe solvent auxiliary solvent/solvent free reactivity [10].

6	Preparation oxime using Green chemistry	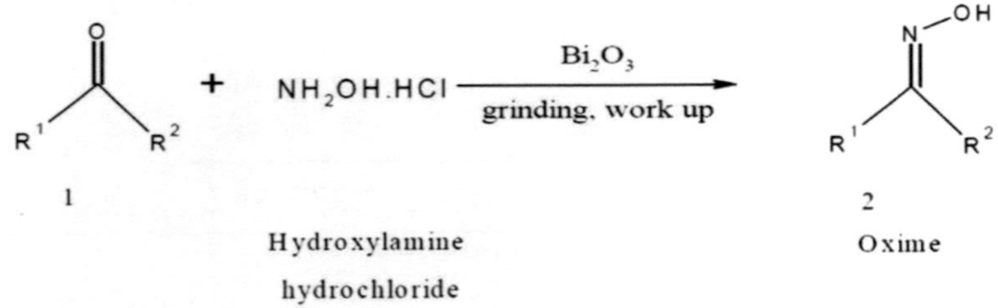 R^1= Ar, Aliphatic, R^2= Ar, Aliphatic, H, R^1, R^2= Cycloalkyl, R^1=Heteroyclic, R^2= H **(Scheme 7)**
9	**Base catalyzed Aldo condensation**	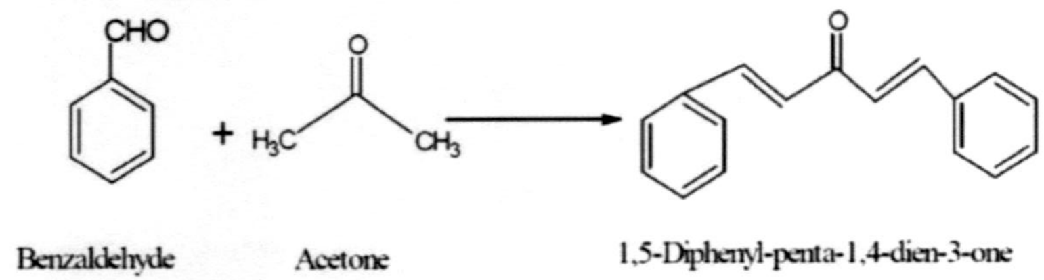 **(Scheme 8)**

Table 4.
Protocol to synthesis oxime and basic Aldo condensation by utilizing sustainable catalysis [11].

DOI: http://dx.doi.org/10.5772/intechopen.104717

10	Cannizaro's reaction	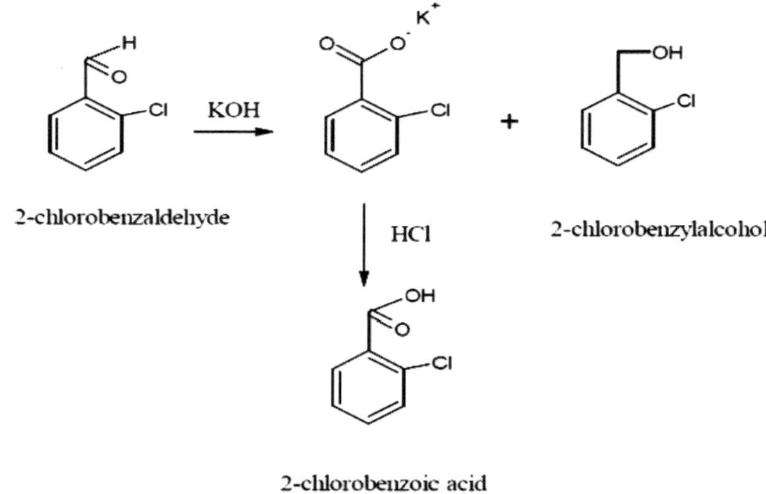
11	**Use of catalytical reagent.** Synthesis of 7-hydroxy-4-methyl cumarone in presence of resorcinol	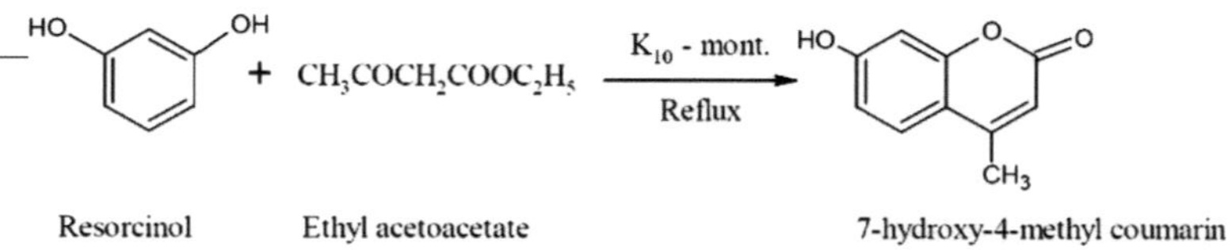

Table 5.
Green protocol for Cannizaro's reaction and use of catalytic reagents [10].

12	Benzylic acid rearrangement 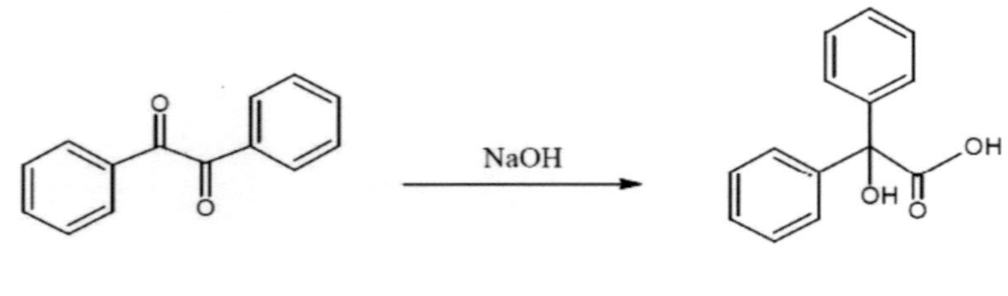 Benzil Benzilic Acid
13	**Synthesis of Dihydro-pyrimidinone** 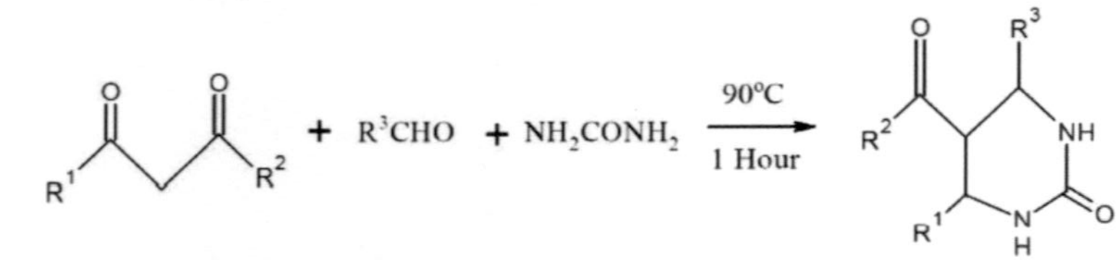 Ethyl acetoacetate Benzaldehyde Urea Dihydropyrimidinone R^1=CH_3 R^2=C_2H_5 R^3=C_6H_5
	Microwave assisted synthesis. $R^1 = CH_3$ $R^2 = C_2H_5$ $R^3 = C_6H_5$
14	The microwave-assisted synthesis is unique and simple technique which offers fast and efficent processing of material with reproducibility, this interaction leads to the rotation and vibrational motion of polar molicule and generates heat within the material, this methodology enhance the reaction rate, higher yield, greater selectivity, and economics for the synthesis of large number of organic compound, provided the momentum for many chemists to switch from conventional heating method to microwave assisted
15	**Ultrasonic synthesis.** Utilization of sonicatore to carryout the sustainable reaction with ultrasound rediation (20 kHz-10 MHz).

Table 6.
Green protocol for rearrangement and condensation [12] of drug molecules and microwave and ultrasonic synthetic reactions [13].

ignored by today's generation to achieve goals. Whereas three essential pillars (soil, water & air) need to be sustainable to lead successful happy present life without compromising future. Where as green chemistry and green engineering emerged out necessity at this junction, to safeguard and substance of all leaving kingdom. It is our responsibility to maintain ecological balance.

The standpoint of this review chapter to show the possible improvement done, by improving energy efficiency, through elimination of unnecessary energy used and saving of 30% of cost on energy, which makes green technology of utmost important, the adaptation of vista of green chemistry with technological implementation will help to achieve conservation of resources, above all, research and construction, goes hand in hand, towards the sustainable development by adopting the 24principle of green synthesis, for sustainable future.

As author believe that it could start by creating awareness by implementing the portion of green chemistry and technology in academic to sensitize the students about the need of today without compromising tomorrow by following the slogan **GO GREEN WITH CHIMISTRY & TECHNOLOGY.**

Acknowledgements

Author immensely grateful to the management of Yenepoya Pharmacy college and Research center, Yenepoya University Deralkatte Mangaluru, and I extent me sincere gratitude to IntechOpen for giving me the opportunity to become a part of this novelty work. Last but not least my heartfelt thanks to Principle & Vice-Principle of Yenepoya Pharmacy college and Research center providing me the timely suggestions cooperation and support to carry out the novelty work. I never deny the support and love of my children and family for providing me the quality time to draft this chapter.

Conflict of interest

The author declared no conflict of interest.

Notes/Thanks/Other declarations

Thanks to publishing house for giving me the opportunity to contribute.

Synonyms and abbreviations

GC	Green chemistry
GE	Green Engineering
EPA	Environmental Protection Agencies
OECD	Organization of economic co-operation and development
PCL	Polycaprolactone

Author details

Nikhat Farhana[1*], Mohammed Gulzar Ahmed[1], Mohammed Asif Iqbal[1], Natasha Naval Aggarwal[1], Prajitha Biju[1], Ashwini Somayaji[1], Abdul Rahamanulla[1], Nishmitha Gretta D'Souza[1], Sudhina Makuttan[1], Tahreen Taj[1], Abdullah Khan[2] and Roshan Sayeed[3]

1 Department of Pharmaceutical Chemistry, Yenepoya Pharmacy College and Research Center, Yenepoya (Deemed To Be University), Mangalore, India

2 Centre of Excellence for Pharmaceutical Sciences (CoEPS), School of Pharmacy, KPJ Healthcare University College, Nilai, Malaysia

3 Department of Pharmacognosy, Deccan School of Pharmacy, Hyderabad, India

*Address all correspondence to: nikhatfarhana@yenepoya.edu.in

References

[1] Letters GC. Greenness of chemical reactions – Limitations of mass metrics. Green Chemistry Letters and Reviews. 2013;**6**:8253

[2] Ribeiro MGTC, Costa DA, Machado AASC, Ribeiro MGTC, Costa DA, Adélio ASC. "Green star": A holistic green Chemistry metric for evaluation of teaching laboratory experiments. Green Chemistry Letters and Reviews. 2010;**3**:8253

[3] Anastas PT. Green Chemistry-Green Engineering. Chemical & Engineering News. 2011;**89**(26):62-65

[4] Bhandari M, Raj S. Practical approach to green chemistry. International Journal of Pharmacy and Pharmaceutical Sciences. 2017;**9**:10:4-26

[5] Anastas PT, Beach ES. Green chemistry : The emergence of a transformative framework. Green Chemistry Letters and Reviews. 2008;**1**:8253

[6] Manahan SE. Green chemistry for a sustainable future. In: Fundamentals of Environmental Chemistry. Boca Raton: CRC Press LLC; 2001

[7] Anastas PT. Green engineering and sustainability. Environmental Science & Technology. 2003;**13**(1):423

[8] Dichiarante V, Ravelli D, Albini A. Green chemistry: State of the art through an analysis of the literature. Green Chemistry Letters and Reviews. 2010;**3**: 8253

[9] Kumar B, Smita K, Cumbal L, Debut A. Biogenic synthesis of iron oxide nanoparticles for 2-arylbenzimidazole fabrication. Journal of Saudi Chemical Society. 2014;**18**(4): 364-369. DOI: 10.1016/j.jscs.2014.01.003

[10] Byrne FP, Jin S, Paggiola G, Petchey THM, Clark JH, Farmer TJ, et al. Tools and techniques for solvent selection : Green solvent selection guides. Sustain Chem Process. 2016:1-24

[11] Beach ES, Cui Z, Anastas PT. Green Chemistry : A design framework for sustainability. Energy & Environmental Science. 2009;**2**:1038-1049

[12] Veleva VR, Jr BWC, Todorova S, Mehta NH, Padia KB. Benchmarking green chemistry adoption by the Indian pharmaceutical supply chain. Green Chemistry Letters and Reviews. 2018;**11**:8253

[13] Science E. Green Chemistry : A framework for a sustainable future. Organometallics. 2021;**40**:1801-1805

[14] Aboginije A, Aigbavboa C, Aghimien DO. The application of "green technology". In: The Modern Day Construction Projects-A Review. 2019

[15] Sheet F. Sustainable/Green Chemistry and Chemical Technology. European Commision; 2007. pp. 1-14

[16] Nikhat F, Satyanarayana D, Shastry C, Thouheed A, Moid A. A review on development of retro-synthetic methodology using natural molecules. Asian Journal of Research in Chemistry. 2015;**8**(4):285

[17] Cue BW, Zhang J. Green process chemistry in the pharmaceutical industry. Green Chemistry Letters and Reviews. 2009;**2**:8253

[18] Kumar B, Smita K, Kumar B, Cumbal L. Ultrasound promoted and SiO2/CCl3COOH mediated synthesis of 2-aryl-1-arylmethyl-1H-benzimidazole derivatives in aqueous media: An eco-friendly approach. Journal of Chemical Sciences. 2014;**126**(6):1831-1840

Section 2

Green Organic Synthesis

Chapter 3

Green Synthesis of Chalcone Derivatives Using Chalcones as Precursor

Surbhi Dhadda, Prakash Giri Goswami and Himanshu Sharma

Abstract

Recently, the use of green methodologies like sonication, use of ionic liquids, etc. attracted the attention of researchers in the field of organic synthesis as they have advantages such as mild reaction conditions, environmentally benign procedures, etc. Herein, this chapter highlights some recyclable ionic liquids (ILs) catalyzed ring closure reactions of chalcones to obtain several heterocyclic rings viz.; pyrazoles, pyrans, pyrimidines under ultrasonification. These reactions have very important features i.e., short routine, high yields, being environmentally friendly, high functional group tolerance, formation of a single product, high atom economy, high yielding, no need for column purification, etc. The various synthesized compounds were prepared in optimized reaction conditions in good to efficient yields. Analytical and spectral (FTIR, ^{1}H, and ^{13}C NMR) techniques were employed for the structural elucidation of the synthesized compounds. The ionic liquids used in the synthesis are recycled and reused several times.

Keywords: Chalcones, green synthesis, ionic liquid, ring closure reactions, sonication

1. Introduction

In recent years, the emphasis of science and technology has shifted more toward environmental benign and sustainable resources and progress. Green Chemistry is paramount concept in chemistry for sustainability, which is the implementation of a set of principles that minimize or get rid of the utilization or generation of hazardous substances in the design, manufacture, and applications of chemical products [1]. Presently, Sonochemistry is a simplistic pathway for a huge variety of syntheses in organic chemistry. Hence, significant features of the ultrasound approach compared with traditional methods are in higher yields, milder conditions, lesser reaction times, improved reaction rates, formation of purer products, easier manipulation and a role in waste minimization and energy protection [2–5].

Multicomponent reactions [6] leading to facinating heterocyclic scaffolds must appear as Potent tools for delivering the molecular diversity required in combinatorial approaches for the synthesis of bioactive compounds and producing varied chemical libraries of drug-like molecules for biological screening [7, 8]. Chalcones, or 1,3-diphenyl-2-propen-1-ones, are commonly occurring heterocyclic ring systems and are important structural motifs found in many natural products and pharmaceuticals. It is also known as benzalacetophenone and benzylidene acetophenone. Chalcones are one of the most important classes of flavonoids [9, 10]. Further ring closure reactions of Chalcones can be used to obtain various heterocyclic rings viz.; Pyrazoles, Pyrans, Cyanopyridines, isoxazoles and pyrimidines having different hetero-cyclic ring systems and multiple derivatives can be synthesized using chalcones [11–15].

The increased environmental concerns needed the replacement of present methods with new more sustainable processes which used the ionic liquids in place of organic catalysts and solvents [16–25]. Ionic Liquids (ILs), as a class of molten salts, are composed entirely of ions and their melting point is around or below 100°C [26–36]. Due to short reaction times, mild reaction conditions, better yields, easy recyclability thermally stable, non-flammable character with negligible vapor pressure, adjustable miscibility with organic substrates and tunable solvating ability ionic liquids (ILs) have attracted the attention of organic chemists [37–49]. Furthermore, unique physiochemical properties that make them potential candidates for many applications in pharmaceuticals, industry and academia [50–52].

There are several varieties of ionic liquids being studied, out of them a few Simple functionalized ILs have created unparalleled fascination as they display some benefits for certain base-catalyzed processes, like easy recycling and better catalytic performance [53]. The environmentally benign basic ionic liquids are used as reaction media as well as catalysts in the development of multicomponent reactions (MCRs). Among all such basic ionic ILs [DBU][OAc] has shown the desired results. Some of the key benefits that can be highlighted for utilization of this IL as catalyst are, the desired product obtained without any further purification and the recyclability of the catalyst was found to be up to 5 cycles. The investigation of alternatives with the help of ionic liquids to conventional organic solvents is a developing research area due to increased environmental concerns.

Herein, we are especially interested in developing the potential use of efficient, simple methodology for the ring closure reactions of chalcones using [DBU][OAc] as ionic liquids as a solvent and catalyst. Chalcones can be used to obtain various heterocyclic rings through ring closure reactions (**Figure 1**).

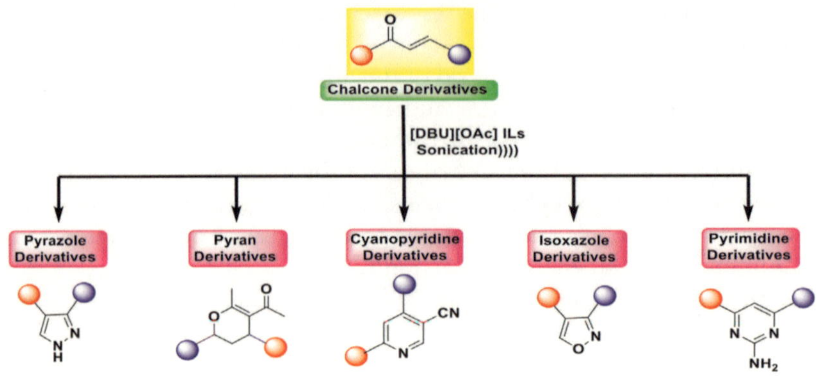

Figure 1.
General scheme of ring closure reactions of chalcones.

2. Experimental section

2.1 Materials and methods

Melting points were recorded in open glass capillary tube using Gallenkamp melting point apparatus and are uncorrected. Checked by Thin layer chromatography (TLC) was applied to check the purity of synthesized compounds and Spots were visualized by irradiation with UV lights (254 nm) or by staining with iodine vapors. The Fourier-transform infrared (FT-IR) spectra were recorded on SHIMADZU 8400S FT-IR spectrophotometer and wave number is given in cm^{-1}. The ^1H NMR spectra and ^{13}C NMR (by broad band proton decoupling technique) were recorded on JEOL AL spectrometer in $CDCl_3$/DMSO-d_6 solvents at 400 and 100 MHz and chemical shift were measured in δ ppm relative to TMS as an internal standard. The Mass (HRMS) spectra were recorded on JEOL SX 102/DA-600 using Argon/Xenon gas. The elemental analysis (C, H and N) were performed using vario-III analyzer at CDRI Lucknow.

2.2 General procedure for preparation of DBU based ionic liquid

According to the reported literature [DBUH][OAc] ILs [54] and [DBUH][Cl] ILs [55] were synthesized by the reaction of 2,3,4,6,7,8,9,10-octahydropyrimido[1,2-a] azepine (DBU) and acetic acid or hydrochloric acid, respectively.

2.3 General procedure for preparation of chalcones (3a-c)

Chalcones were synthesized according to the reported procedure with minor modification (**Figure 2**), the synthesized products were characterized by ^1H NMR, and physical data and compared with those reported in literature [54].

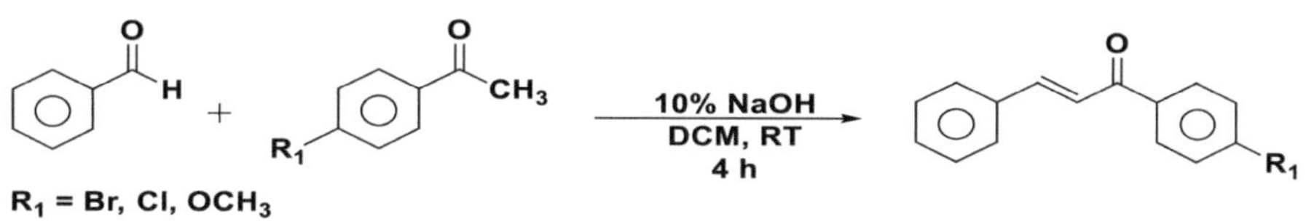

Figure 2.
General procedure of synthesis of chalcones (3a-c).

2.4 Model reaction for preparation of pyrazole derivative (4a)

Chalcone derivative (1 mmol) and methylhydrazine (1 mmol) were ultrasonicated catalyzed by [DBUH][OAc] (5 ml) at 50°C for about 4 h (**Figure 3**). The crude product was refrigerated overnight. The precipitate formed was filtered off and crystallized from ethanol yielding yellow crystals of the product (4a).

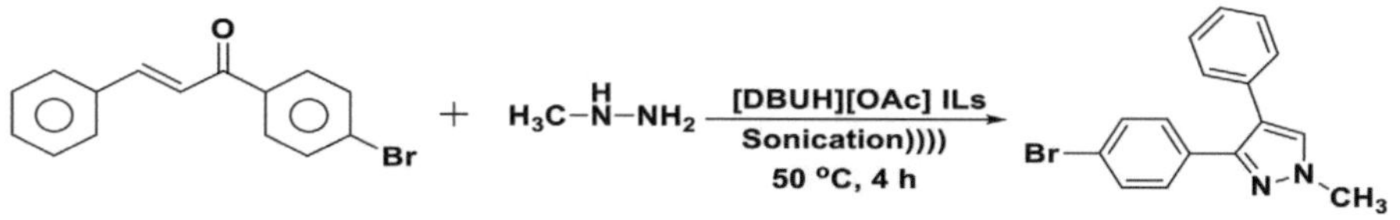

Figure 3.
Model reaction for preparation of pyrazole derivative (4a).

2.4.1 Spectroscopic data of (4a)

^{1}H NMR (400 MHz, DMSO-d$_6$) δ 8.81, 7.94, 7.59, 7.47, 7.44, 7.38, 3.96; ^{13}C NMR (100.15 MHz, DMSO-d$_6$) δ 145.51, 133.32, 131.99, 130.37, 129.66, 128.69, 128.34, 126.78, 125.67, 123.12, 122.13, 40.57; HRMS; *m/z* 312.04 (M+); $C_{16}H_{13}BrN_2$: calcd. C, 61.36; H, 4.18; N, 8.94; found C, 61.34; H, 4.20; N, 8.97.

2.5 Model reaction for preparation of pyran derivative (5a)

Chalcone derivative (1 mmol) mixed with *α,β*-diketone (1 mmol) was ultrasonicated in [DBUH][OAc] (5 mL) for about 45 minutes. The mixture was heated to 60°C for 2 h to complete the reaction which was monitored by TLC. The organic layer was extracted with ethyl acetate, washed with water and then dried over Na_2SO_4 which was followed by filtration and concentration. The crude was recrystallized from ethyl acetate and hexane mixture to give pure product (5a). The catalyst remained in the aqueous phase was reused in other reactions (**Figure 4**).

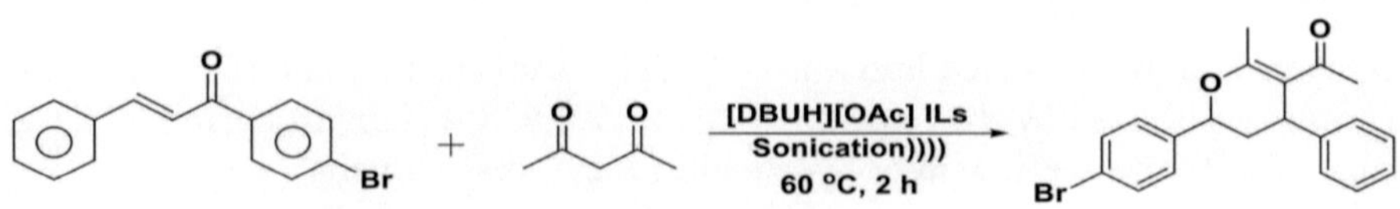

Figure 4.
Model reaction for preparation of pyran derivative (5a).

2.5.1 Spectroscopic data of (5a)

^{1}H NMR (400 MHz, DMSO-d$_6$) δ 7.45, 7.32, 7.28, 7.25, 7.01, 5.27, 3.60, 2.28, 2.21, 2.12, 2.05; ^{13}C NMR (100.15 MHz, DMSO-d$_6$) δ 195.02, 164.92, 142.45, 138.87, 131.92, 129.14, 128.97, 128.27, 128.12, 122.84, 107.13, 76.87, 35.62, 35.14, 27.61, 16.76; HRMS; *m/z* 370.01 (M+); $C_{20}H_{19}BrO_2$: calcd. C, 64.72; H, 5.17; found C, 64.75; H, 5.21.

2.6 Model reaction for preparation of cyanopyridine derivative (6a)

A mixture of chalcone derivative (2 mmol) with malononitrile (2 mmol) in 5 mL of [DBUH][OAc] was ultrasonicated at atmospheric pressure at 65°C for 3 h (**Figure** 5). After completion of the reaction, the mixture was cooled to room temperature and the organic layer was concentrated. The pure product was obtained by column chromatography (n-hexane:ethyl acetate = 80:20) to afford the preferred product (6a).

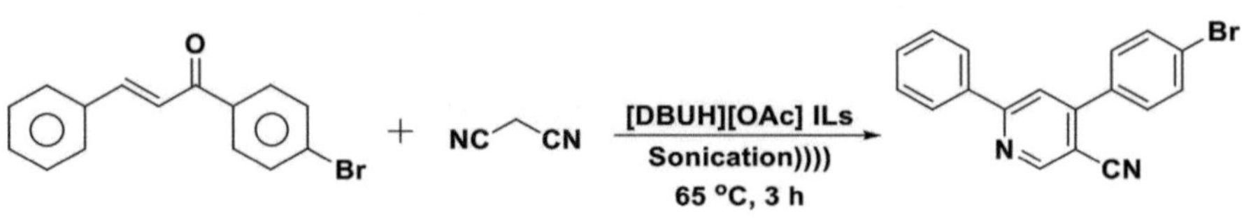

Figure 5.
Model reaction for preparation of cyanopyridine derivative (6a).

2.6.1 Spectroscopic data of (6a)

^{1}H NMR (400 MHz, DMSO-d_6) δ 9.21, 8.40, 7.95, 7.62, 7.51, 7.44, 7.46; ^{13}C NMR (100.15 MHz, DMSO-d_6) δ 160.21, 152.79, 151.01, 138.51, 138.45, 131.34, 130.03, 129.67, 128.99, 127.91, 121.91, 120.71, 117.22, 110.19; HRMS; *m/z* 334.06 (M+); $C_{18}H_{11}BrN_2$: calcd. C, 64.52; H, 3.35; N, 8.37; found C, 64.49; H, 3.33; N, 8.36.

2.7 Model reaction for preparation of isoxazole derivative (7a)

Chalcone derivative (1 mmol) was ultrasonicated with hydroxylamine hydrochloride (1 mmol) in catalytic influence of [DBUH][OAc] ILs (5 mL) at 70°C for 1 h (**Figure 6**). The formation of product was monitored by TLC. Isoxazole derivative was obtained by keeping the reaction mixture on ice bath, then the desired product was isolated, washed with water, and dried (7a).

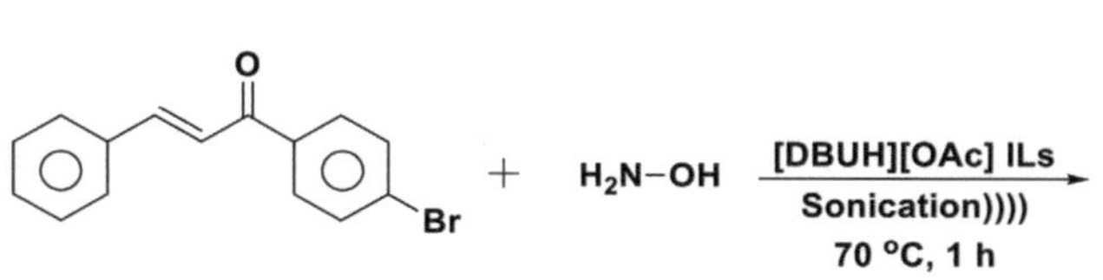

Figure 6.
Model reaction for preparation of isoxazole derivative (7a).

2.7.1 Spectroscopic data of (7a)

^{1}H NMR (400 MHz, DMSO-d_6) δ 8.66, 7.69, 7.57, 7.50, 7.41, 7.32; ^{13}C NMR (100.15 MHz, DMSO-d_6) δ 170.26, 154.95, 131.82, 130.67, 128.61, 128.30, 128.07, 127.55, 126.73, 125.41, 116.74; HRMS; *m/z* 299.04 (M+); $C_{15}H_{10}BrNO$: calcd. C, 60.01; H, 3.37; N, 4.69; found C, 60.03; H, 3.39; N, 4.71.

2.8 Model reaction for preparation of pyrimidine derivative (8a)

To the mixture of chalcone derivative (1 mmol), guanidine hydrochloride (2 mmol) was added with [DBUH][OAc] ILs was heated under ultrasonication for 2 h at 55°C. The completion of the reaction was checked by TLC (**Figure 7**). The reaction mixture poured into ice water and formed product was filtered and recrystallized from ethanol (8a).

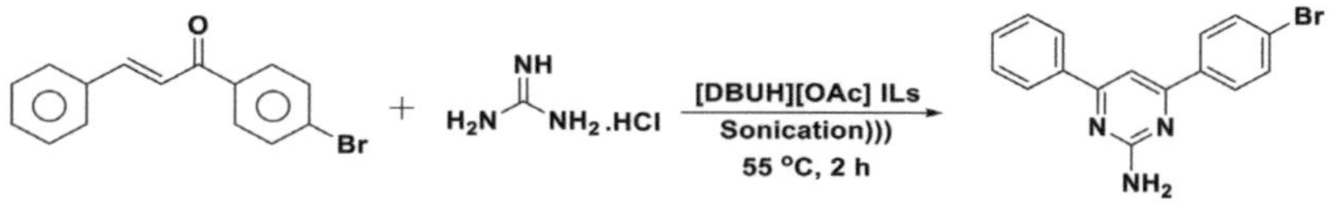

Figure 7.
Model reaction for preparation of pyrimidine derivative (8a).

2.8.1 Spectroscopic data of (8a)

^{1}H NMR (400 MHz, DMSO) δ 7.77, 7.64, 7.47, 7.41, 7.15, 2.29; ^{13}C NMR (100.15 MHz, DMSO-d_6) δ 160.02, 158.70, 138.09, 136.17, 132.01, 131.05, 130.28,

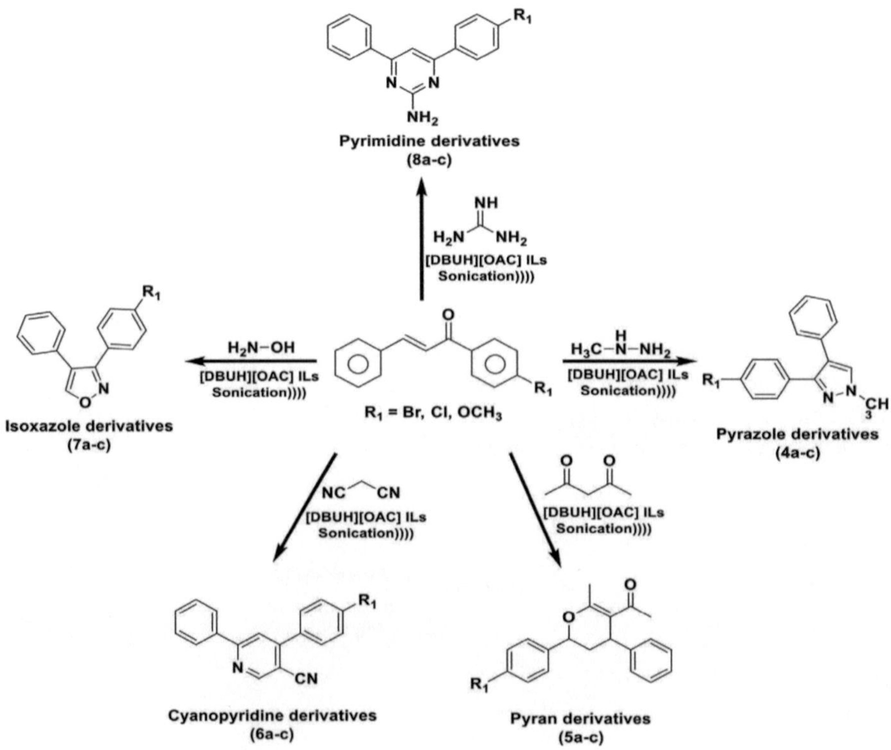

Figure 8.
General representation of preparation of chalcone derivatives (4-8a-c).

129.20, 128.35, 125.47, 112.89; HRMS; *m/z* 325.05 (M+); $C_{16}H_{12}BrN_3$: calcd. C, 58.93; H, 3.72; N, 12.90; found C, 58.97; H, 3.70; N, 12.91 (see **Figure 8**).

3. Results and discussion

In this chapter, the ring closure reaction of chalcone derivatives in the presence of basic ionic liquid [DBUH]OAc to afford the several derivatives like pyrazoles, pyrans, pyrimidines, isoxazoles, and cyanopyridines. Different catalytic systems were used to optimize the reaction conditions on the set of model reactions.

3.1 Optimization of reaction conditions

The reaction conditions were optimized on the respective model reactions, further these optimized reaction conditions were used to produce corresponding derivatives of chalcones (**Table 1**).

We have carried out the synthesis of a number of chalcone derivatives (4-8a,b,c) under different reaction conditions. The optimized conditions for all the ring closure reactions of chalcones involved use of [DBUH]OAc ILs as catalyst under sonication for appropriate time at adequate temperature (**Table 2**, Entry 6).

S. No.	Reactant 1	Reactant 2	Product	% Yield
1.	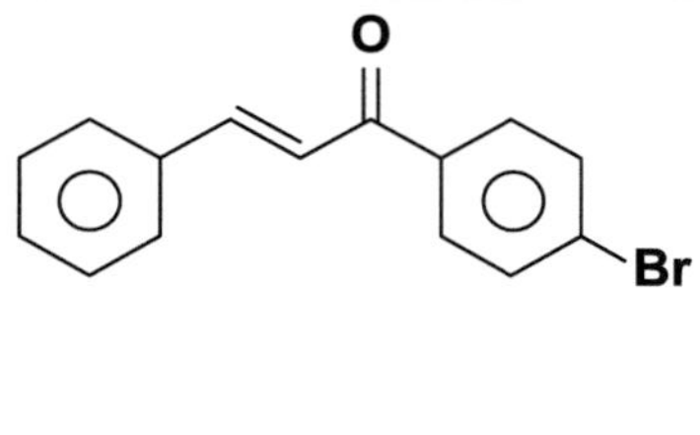	$H_3C-NH-NH_2$	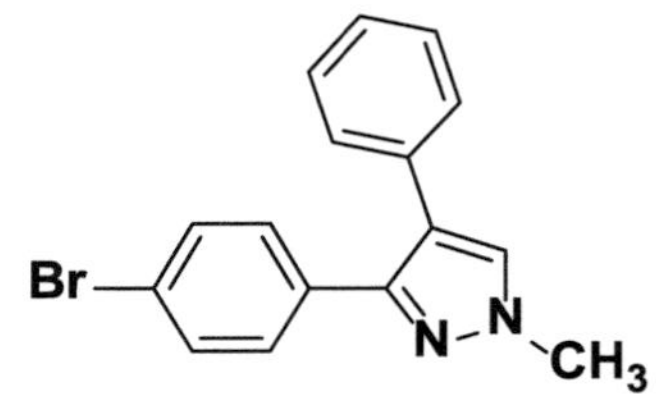 (4a)	97
2.	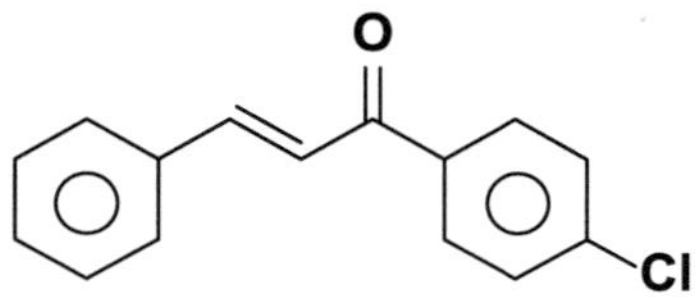	$H_3C-NH-NH_2$	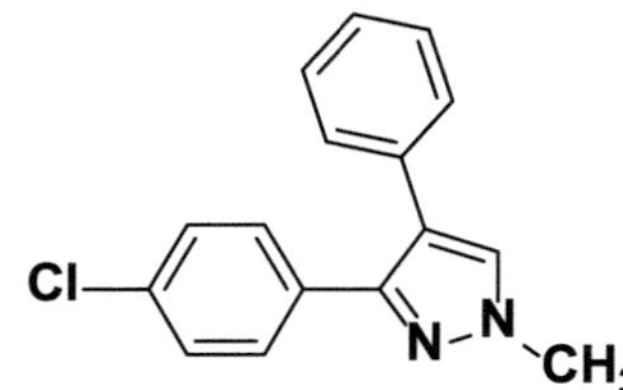 (4b)	94
3.	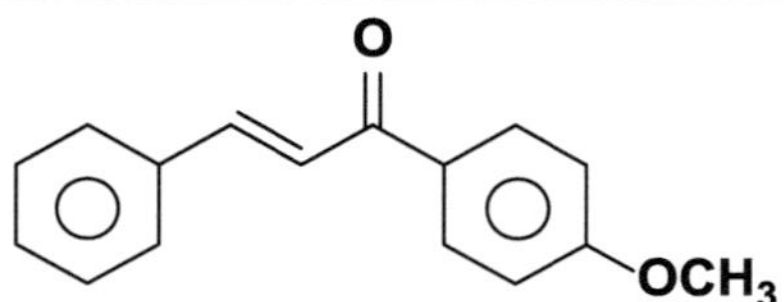	$H_3C-NH-NH_2$	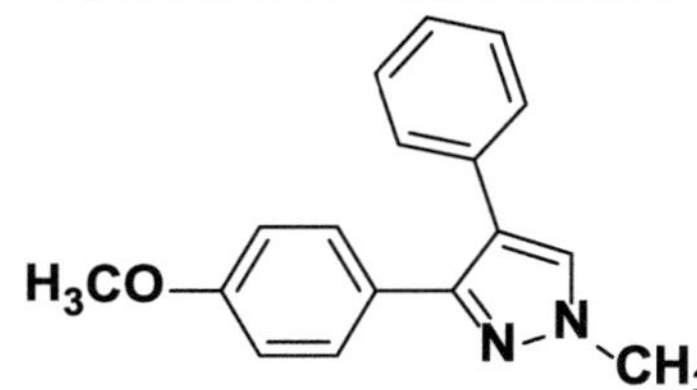 (4c)	98
4.	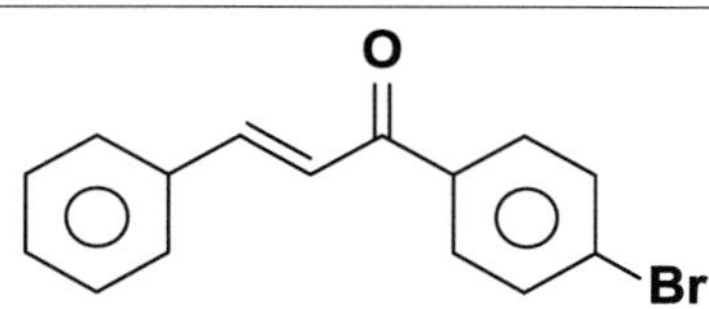	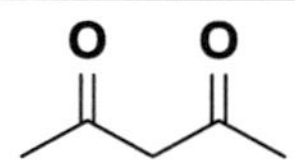	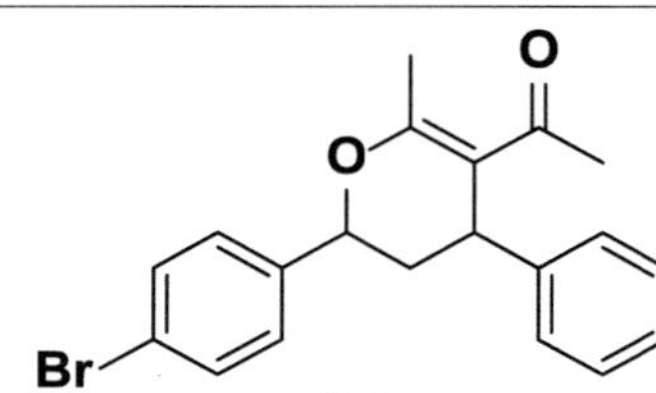 (5a)	95

S. No.	Reactant 1	Reactant 2	Product	% Yield
5.	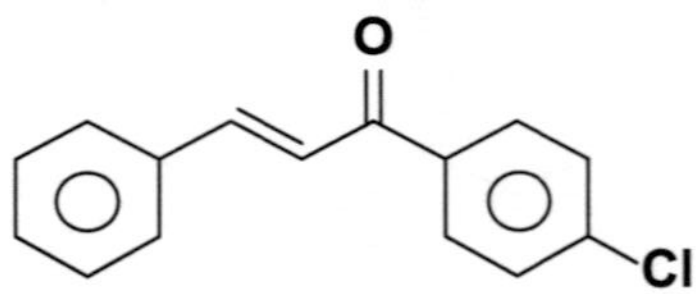	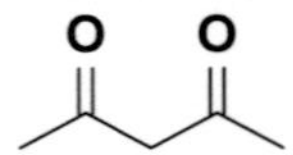	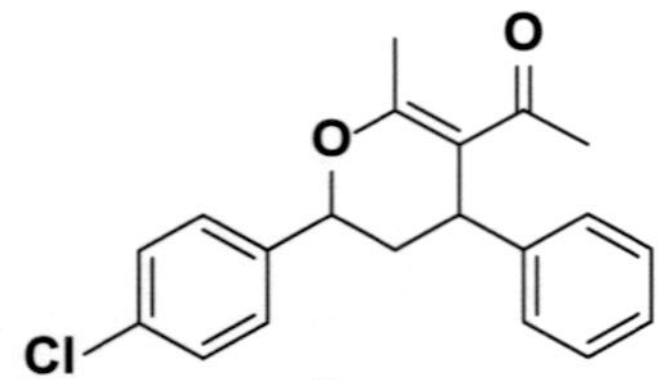 (5b)	92
6.	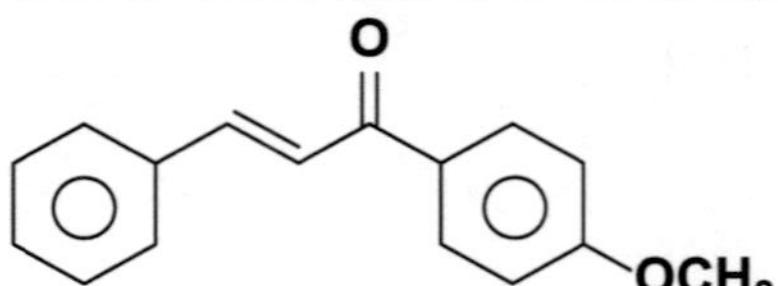	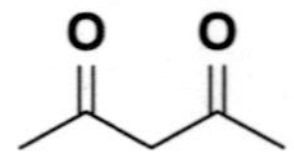	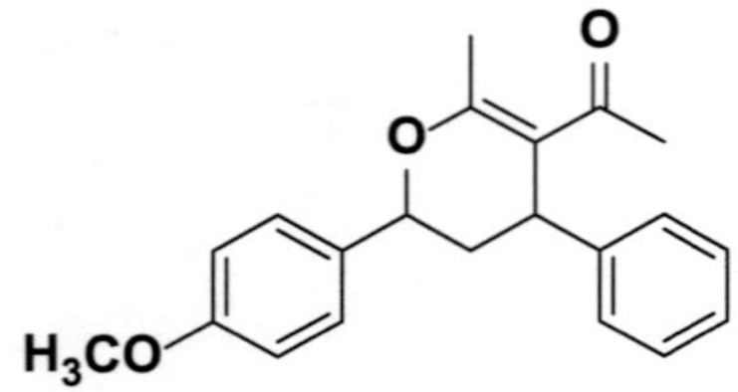 (5c)	96
7.	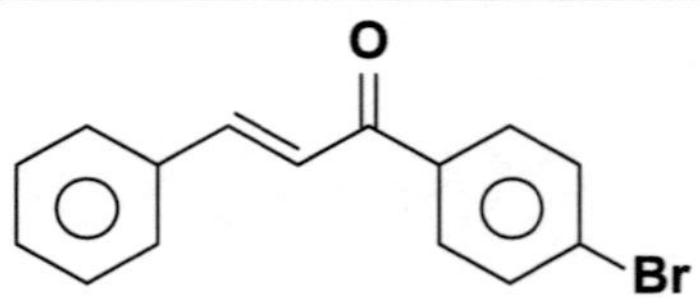		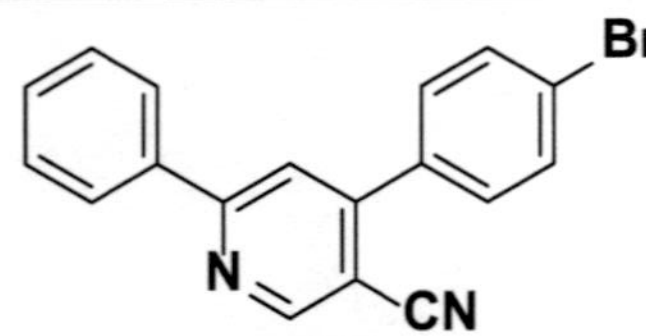 (6a)	93
8.	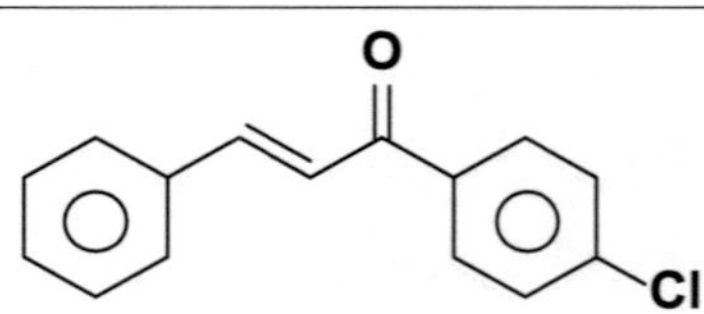	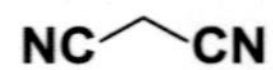	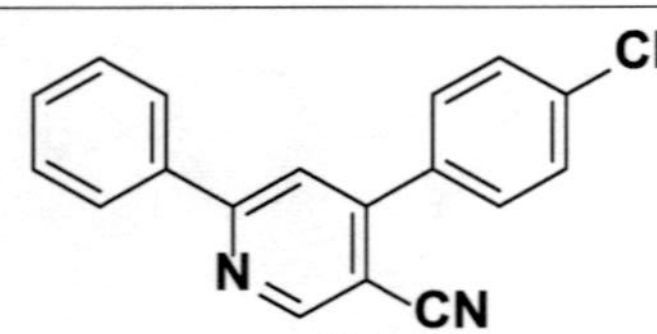 (6b)	90

S. No.	Reactant 1	Reactant 2	Product	% Yield
9.	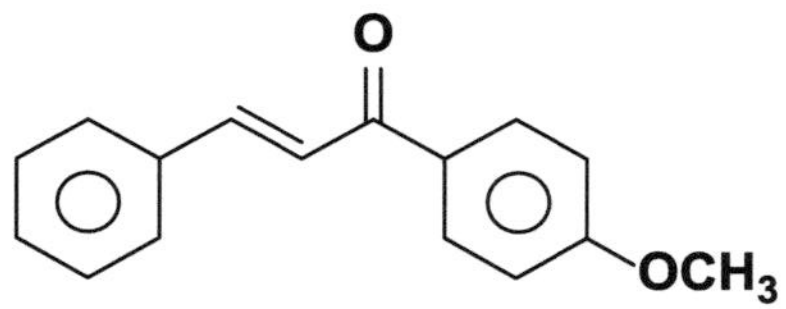	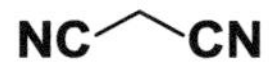	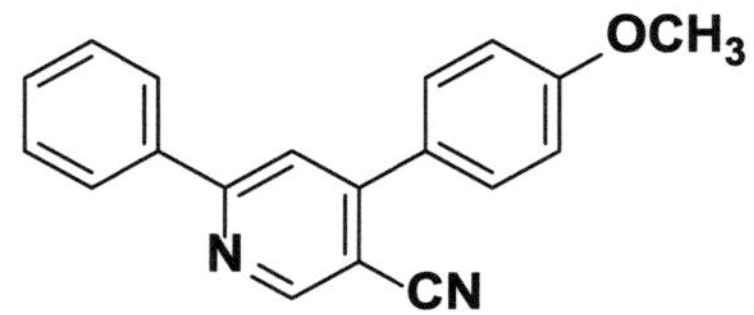 (6c)	95
10.	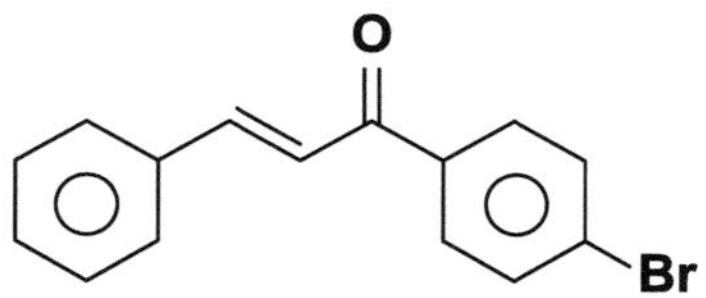	H_2N-OH	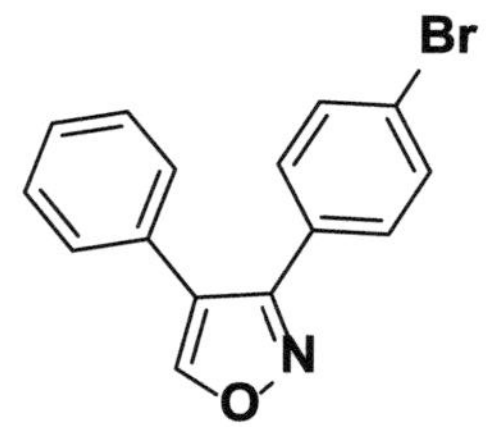 (7a)	94
11.	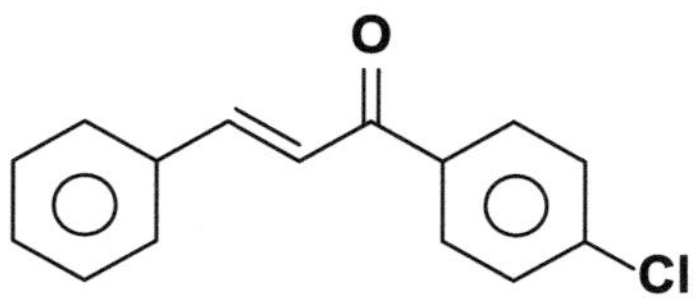	H_2N-OH	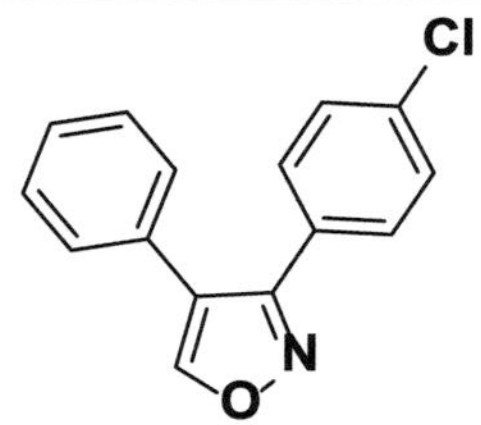 (7b)	93
12.	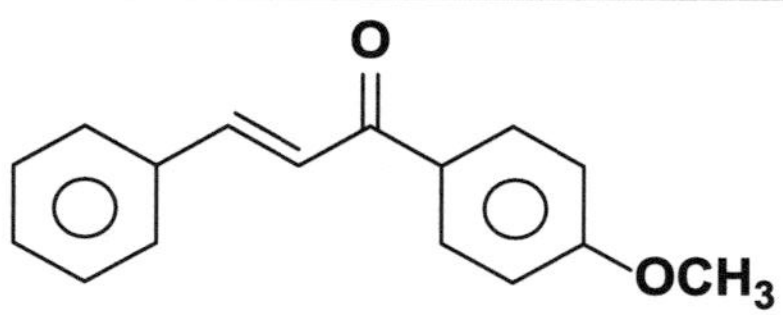	H_2N-OH	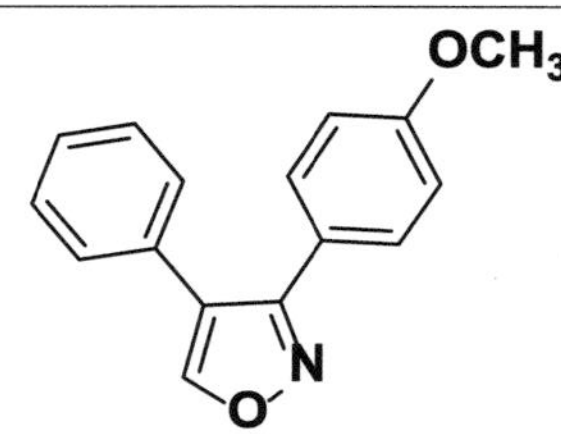 (7c)	95

S. No.	Reactant 1	Reactant 2	Product	% Yield
13.	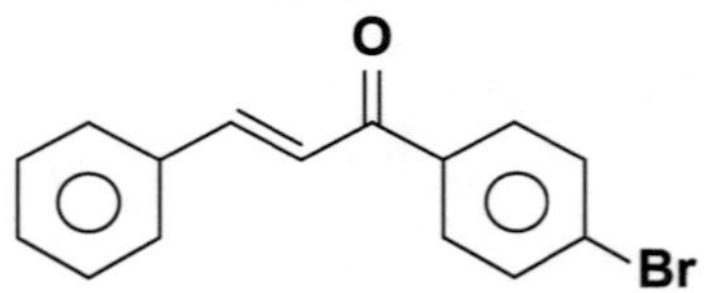	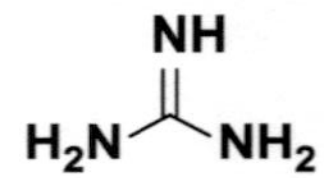	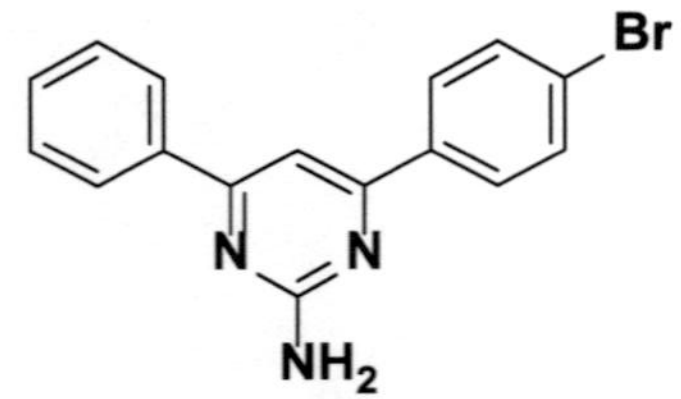 (8a)	96
14.	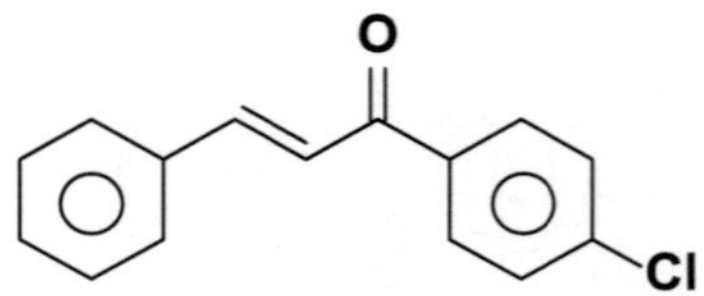	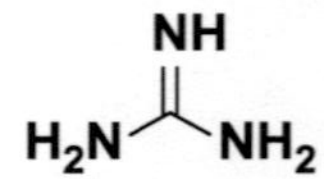	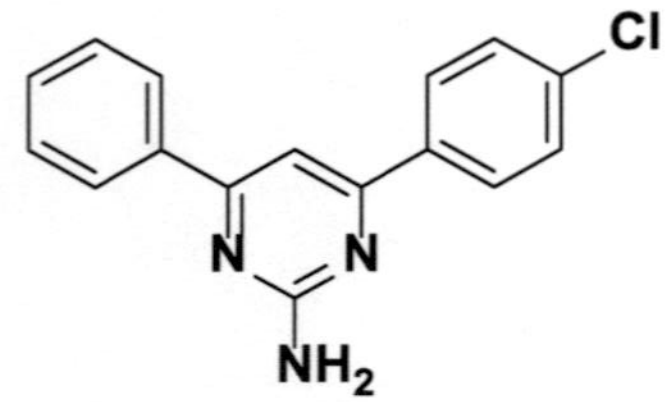 (8b)	92
15.	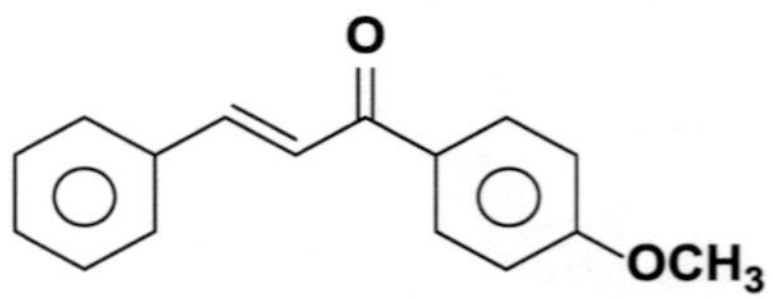	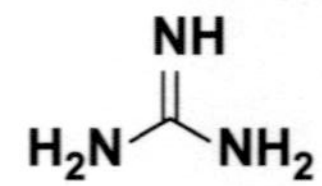	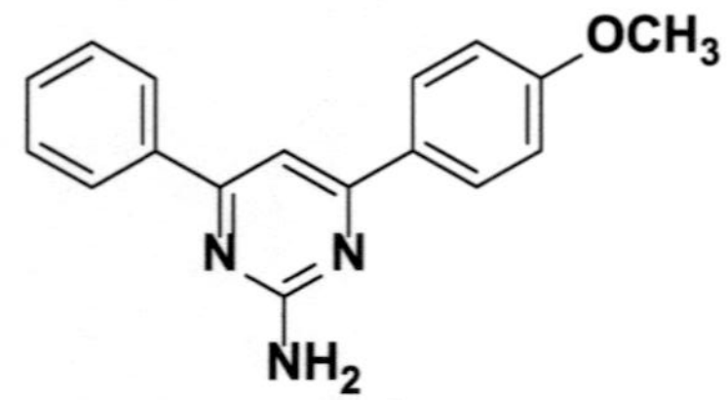 (8c)	96

Table 1.
% yield of desired products (4-8a,b,c).

DOI: http://dx.doi.org/10.5772/intechopen.103959

S. No.	Catalyst / Solvent	Reaction Condition	Pyrazole Derivative (4a)		Pyran Derivative (5a)		Cyanopyridine Derivative (6a)		Isoxazole Derivative (7a)		Pyrimidine Derivative (8a)	
			Temp (°C)	% Yield	Temp (°C)	% Yield	Temp (°C)	% Yield	Temp (°C)	% Yield	Temp (°C)	% Yield
1.	No catalyst / DCM	Reflux	120	<5	120	<5	120	<5	120	<5	120	<5
2.	NaOH / DCM	Sonication	100	>15	100	>15	100	>15	100	>15	100	>15
3.	[MIM]BF_4 IL	Sonication	100	40	100	35	100	42	100	45	100	38
4.	[MIM]OH IL	Sonication	90	48	90	42	90	51	90	55	90	44
5.	[DBUH]Cl IL	Sonication	90	62	90	65	90	68	90	62	90	60
6.	[DBUH]OAc IL	Sonication	50	97	60	95	65	93	70	94	55	96

Table 2.
Optimization of reaction conditions.

3.2 Reusability of ionic liquids

The catalytic reusability of ILs was observed during optimized reaction conditions. The ILs were easily recovered as filtration after the completion of reaction. The recovered ILs were used four times without remarkable loss in activity but after that there is sudden decrease (**Figure 9**) in yield of products.

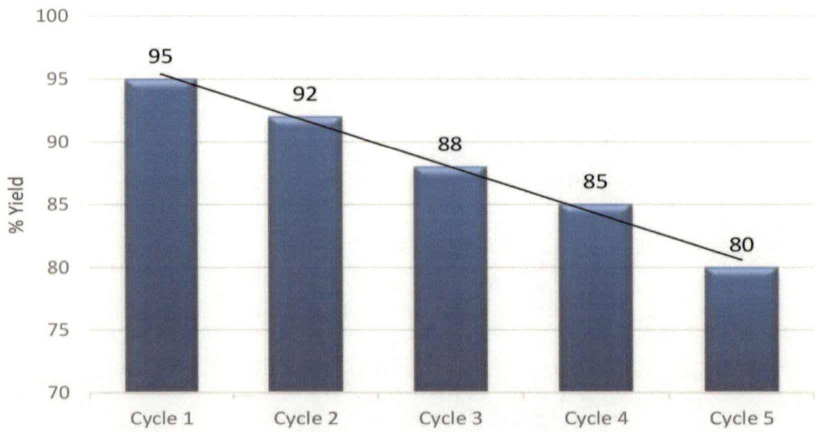

Figure 9.
Reusability and recyclability of [DBUH]OAc ILs.

4. Conclusion

In Summary, we developed a simple and efficient catalytic system that can effectively promote the conversion of chalcones into different derivatives viz.; Pyrazoles, Pyrans, Cyanopyridines, isoxazoles and pyrimidines *via* [DBU][OAc] IL catalyzed ring closure reactions under mild conditions. A series of functional ILs was screened and [DBU][OAc] was determined as the optimal catalyst. This mild and environmental friendly synthetic methodology permitted us to synthesize products in good to excellent yields. There are many merits of the used protocol like, low cost of green catalyst, operational simplicity, obtaining products in high yield, and the catalyst can be reused without any significant loss of catalytic property up to five catalytic cycles.

Author details

Surbhi Dhadda[1], Prakash Giri Goswami[2] and Himanshu Sharma[3]*

1 Department of Chemistry, Vedic Kanya P.G. College, Jaipur, Rajasthan, India

2 Department of Chemistry, University of Rajasthan, Jaipur, Rajasthan, India

3 Department of Chemistry, Mohanlal Sukhadiya University, UCOS, Microwave Chemistry Lab, Udaipur, Rajasthan, India

*Address all correspondence to: himanshu@mlsu.ac.in

References

[1] Hua Y, Zou Y, Wu H, Shi D. A facile and efficient ultrasound-assisted synthesis of novel dispiroheterocycles through 1, 3-dipolar cycloaddition reactions. Ultrasonics Sonochemistry. 2012;**19**:264-269

[2] Zbancioc G, Florea O, Jones PG, Mangalagiu II. An efficient and selective way to new highly functionalized coronands or spiro derivatives using ultrasonic irradiation. Ultrasonics Sonochemistry. 2012;**19**:399-403

[3] Wang SY, Ji SJ, Loh TP. The Michael addition of indole to α,β-unsaturated ketones catalyzed by iodine at room temperature. Synlett. 2003;**15**:2377-2379

[4] Dandia H, Singh R, Bhaskaran S. Ultrasound promoted greener synthesis of spiro[indole-3,5′-[1,3]oxathiolanes] in water. Ultrasonics Sonochemistry. 2010; **17**:399-402

[5] Nair V, Rajesh C, Vinod AU, Bindu S, Sreekanth AR, Mathen JS, et al. Strategies for heterocyclic construction via novel multicomponent reactions based on isocyanides and nucleophilic carbenes. Accounts of Chemical Research. 2003;**36**:899-907

[6] Dömling A. Recent advances in isocyanide-based multicomponent chemistry. Current Opinion in Chemical Biology. 2002;**6**(3):306-313

[7] Babu AS, Raghunathan R. Ultrasonic assisted-silica mediated [3+ 2] cycloaddition of azomethine ylides—A facile multicomponent one-pot synthesis of novel dispiroheterocycles. Tetrahedron Letters. 2007;**48**(38): 6809-6813

[8] Ni L, Meng CQ, Sikorski JA. Recent advances in therapeutic chalcones. Expert Opinion on Therapeutic Patents. 2004;**14**(12):1669-1691

[9] Sahu NK, Balbhadra SS, Choudhary J, Kohli DV. Exploring pharmacological significance of chalcone scaffold: A review. Current Medicinal Chemistry. 2012;**19**:209-225

[10] El-Hashah MA, El-Kady M, Saiyed MA, Elaswy AA. Arylidene derivatives as synthons heterocyclic synthesis. Egyptian Journal of Chemistry. 1985;**27**:715

[11] Crawley LS, Fanshawe WJ. Neighboring group participation in cyclodehydration. A regiospecific isoxazole synthesis. Journal of Heterocyclic Chemistry. 1977;**14**(3): 531-534

[12] Taylor EC, Morrison RW Jr. An unusual molecular rearrangement of an N-aminopyrimidine. The Journal of Organic Chemistry. 1967;**32**(8):2379-2382

[13] Utale PS, Raghuwanshi PB, Doshi AG. Synthesis of some new 1-Carboxamido-3-(substituted-2-hydroxy phenyl)-5-aryl-[delta] 2-pyrazolines. Asian Journal of Chemistry. 1998;**10**(3): 597-599

[14] Kidwai M, Misra P. Ring closure reactions of chalcones using microwave technology. Synthetic Communications. 1999;**29**(18):3237-3250

[15] Le ZG, Chen ZC, Hu Y, Zheng QG. Organic reactions in ionic liquids: A simple and highly regioselective N-substitution of pyrrole. Synthesis. 2004; **12**:1951-1954

[16] Nara SJ, Naik PU, Harjani JR, Salunkhe MM. Potential of ionic liquids in greener methodologies involving

biocatalysis and other synthetically important transformations. Indian Journal of Chemistry. 2006;**45B**: 2257-2269

[17] Catal KN. Rev. synthesis of five-membered N-heterocycles fused with other heterocycles. Catalysis Reviews. 2015;**57**(1):1-78

[18] Kaur N, Kishore D, Kaur N, Kishore D. Microwave-assisted synthesis of six-membered O-heterocycles. Synthetic Communications. 2014; **44**(21):3047-3081

[19] Kaur N, Dwivedi J, Kishore D, Kaur N, Dwivedi J, Kishore D. Solid-phase synthesis of nitrogen-containing five-membered heterocycles. Synthetic Communications. 2014;**44**(12): 1671-1729

[20] Nair V, Vellalath S, Poonoth M, Suresh E, Viji S. N-heterocyclic carbene catalyzed reaction of enals and diaryl-1, 2 diones via homoenolate: Synthesis of 4, 5, 5-trisubstituted γ-butyrolactones. Synthesis. 2007;**20**:3195-3200

[21] Potewar TM, Siddiqui SA, Lahoti RJ, Srinivasan KV. Efficient and rapid synthesis of 1-substituted-1H-1, 2, 3, 4-tetrazoles in the acidic ionic liquid 1-n-butylimidazolium tetrafluoroborate. Tetrahedron Letters. 2007;**48**(10): 1721-1724

[22] Xu JM, Qian C, Liu BK, Wu Q, Lin XF. A fast and highly efficient protocol for Michael addition of N-heterocycles to α, β-unsaturated compound using basic ionic liquid [bmIm] OH as catalyst and green solvent. Tetrahedron. 2007;**63**(4): 986-990

[23] Hutka M, Toma S. Hydrogen-transfer reduction of aromatic ketones in basic ionic liquids. Monatshefte für Chemie-Chemical Monthly. 2009; **140**(10):1189-1194

[24] Syamala M. Recent progress in three-component reactions. An update. Organic Preparations and Procedures International. 2009;**41**(1):1-68

[25] Wasserscheid P, Keim W. Ionic liquids—New "solutions" for transition metal catalysis. Angewandte Chemie International Edition. 2000;**39**:3772-3789

[26] Frizzo CP, Tier AZ, Bender CR, Gindri IM, Villetti MA, Zanatta N, et al. Structural and Physical Aspects of Ionic Liquid Aggregates in Solution. InIonic Liquids-Current State of the Art Rijeka. London, UK: InTech; 2015. pp. 161-198

[27] Berthod A, Ruiz-Angel MJ, Carda-Broch S. Recent advances on ionic liquid uses in separation techniques. Journal of Chromatography A. 2018;**1559**:2-16

[28] Javed MN, Muhammad S, Hashmi IA, Bari A, Musharraf SG, Ali FI. Newly designed pyridine and piperidine based ionic liquids: Aggregation behavior in ESI-MS and catalytic activity in CC bond formation reactions. Journal of Molecular Liquids. 2018;**272**: 84-91

[29] Kaur N. Environmentally benign synthesis of five-membered 1, 3-N, N-heterocycles by microwave irradiation. Synthetic Communications. 2015;**45**(8): 909-943

[30] Kaur N. Advances in microwave-assisted synthesis for five-membered N-heterocycle synthesis. Synthetic Communications. 2015;**45**(4): 432-457

[31] Kaur N. Microwave-assisted synthesis of five-membered S-heterocycles. Journal of the Iranian Chemical Society. 2014;**11**(2):523-564

[32] Kaur N. Review on the synthesis of six-membered N, N-heterocycles by microwave irradiation. Synthetic Communications. 2015;**45**(10):1145-1182

[33] Kaur N. Greener and expeditious synthesis of fused six-membered N, N-heterocycles using microwave irradiation. Synthetic Communications. 2015;**45**(13):1493-1519

[34] Kaur N. Applications of microwaves in the synthesis of polycyclic six-membered N. N-heterocycles. Synthetic Communications. 2015;**45**(14):1599-1631

[35] Kaur N. Synthesis of five-membered N, N, N-and N, N, N, N-heterocyclic compounds: Applications of microwaves. Synthetic Communications. 2015;**45**(15):1711-1742

[36] Bao Q, Qiao K, Tomida D, Yokoyama C. Preparation of 5-hydroymethylfurfural by dehydration of fructose in the presence of acidic ionic liquid. Catalysis Communications. 2008; **9**(6):1383-1388

[37] Shen J, Wang H, Liu H, Sun Y, Liu Z. Brønsted acidic ionic liquids as dual catalyst and solvent for environmentally friendly synthesis of chalcone. Journal of Molecular Catalysis A: Chemical. 2008; **280**(1–2):24-28

[38] Wang W, Shao L, Cheng W, Yang J, He M. Brønsted acidic ionic liquids as novel catalysts for Prins reaction. Catalysis Communications. 2008;**9**(3): 337-341

[39] Kaur N. Role of microwaves in the synthesis of fused five-membered heterocycles with three N-heteroatoms. Synthetic Communications. 2015;**45**(4): 403-431

[40] Kaur N. Recent impact of microwave-assisted synthesis on benzo derivatives of five-membered N-heterocycles. Synthetic Communications. 2015;**45**(5):539-568

[41] Kaur N, Kishore D. Microwave-assisted synthesis of seven-and higher-membered N-heterocycles. Synthetic Communications. 2014;**44**(18): 2577-2614

[42] Kaur N, Kishore D. Microwave-assisted synthesis of six-membered S-heterocycles. Synthetic Communications. 2014;**44**(18): 2615-2644

[43] Kaur N, Kishore D. Microwave-assisted synthesis of seven-and higher-membered O-heterocycles. Synthetic Communications. 2014;**44**(19): 2739-2755

[44] Luo S, Mi X, Zhang L, Liu S, Xu H, Cheng JP. Functionalized ionic liquids catalyzed direct aldol reactions. Tetrahedron. 2007;**63**(9):1923-1930

[45] Carvalho PJ, Álvarez VH, Marrucho IM, Aznar M, Coutinho JA. High pressure phase behavior of carbon dioxide in 1-butyl-3-methylimidazolium bis (trifluoromethylsulfonyl) imide and 1-butyl-3- methylimidazolium dicyanamide ionic liquids. The Journal of Supercritical Fluids. 2009;**50**(2):105-111

[46] Holbrey JD, Reichert WM, Reddy R, Rogers R. Ionic Liquids as Green Solvents: Progress and Prospects, ACS Symposium Series. Washington, DC: American Chemical Society; 2003. pp. 121-133

[47] Cevasco G, Chiappe C. Are ionic liquids a proper solution to current environmental challenges. Green Chemistry. 2014;**16**:2375-2385

[48] Yang Z, Pan W. Ionic liquids: Green solvents for nonaqueous biocatalysis.

Enzyme and Microbial Technology. 2005;**37**(1):19-28

[49] Mecerreyes D. Polymeric ionic liquids: Broadening the properties and applications of polyelectrolytes. Progress in Polymer Science. 2011;**36**(12): 1629-1648

[50] Mishra N, Arora P, Kumar B, Mishra LC, Bhattacharya A, Awasthi SK, et al. Synthesis of novel substituted 1, 3-diaryl propenone derivatives and their antimalarial activity in vitro. European Journal of Medicinal Chemistry. 2008; **43**(7):1530-1535

[51] Awasthi SK, Mishra N, Kumar B, Sharma M, Bhattacharya A, Mishra LC, et al. Potent antimalarial activity of newly synthesized substituted chalcone analogs in vitro. Medicinal Chemistry Research. 2009;**18**(6):407-420

[52] Hajipour AR, Rafiee F. Basic ionic liquids. A short review. Journal of the Iranian Chemical Society. 2009;**6**(4): 647-678

[53] Ying AG, Liu L, Wu GF, Chen G, Chen XZ, Ye WD. Aza-Michael addition of aliphatic or aromatic amines to α, β-unsaturated compounds catalyzed by a DBU-derived ionic liquid under solvent-free conditions. Tetrahedron Letters. 2009;**50**(14):1653-1657

[54] Tolstikova LL, Shainyan BA. Ionic liquids on the basis of 2, 3, 4, 6, 7, 8, 9, 10-octahydropyrimido-[1, 2-a] azepine (1, 8-diazabicyclo [5.4. 0] undec-7-ene). Russian Journal of Organic Chemistry. 2006;**42**:1068-1074

[55] Li J, Zhuang R, Qian Y. Synthesis of novel Chalcone derivatives by organic catalysis. Materials Physics and Chemistry. 2019;**1**(1):1-564

Chapter 4

Mechanochemistry in Organocatalysis: A Green and Sustainable Route toward the Synthesis of Bioactive Heterocycles

Biplob Borah and L. Raju Chowhan

Abstract

Considering the great prevalence of heterocyclic compounds in the core structure of numerous natural products, synthetic drug candidates, active pharmaceutical ingredients, and also in optoelectronic materials; tremendous efforts have been dedicated toward their synthesis and functionalization. But, the exploitation of hazardous, volatile organic solvents and toxic reagents caused disadvantageous effects on the atom economy and eco-friendly nature of the chemical transformation. Therefore, developing chemical processes providing easy access to complex target molecules by avoiding the utilization of hazardous solvents and reagents for making our environment toxic-free is of increasing significance for chemists in both academia and industry. The synergic combination of the features of mechanochemical activation as alternative energy input with the efficiency associated with small organic molecules that can catalyze chemical reactions is predominantly relevant to fulfill the goal of green and sustainable chemistry. This chapter is dedicated to providing a critical overview on the application of mechanochemical techniques for the synthesis of five- and six-membered heterocycles, as well as complex-fused heterocycles and spiro-heterocycles under organocatalytic conditions.

Keywords: mechanochemistry, bioactive heterocycles, organocatalysis, ball-milling, grinding method

1. Introduction

Heterocyclic compounds comprise a broad range of structural motifs ubiquitously found in the architecture of numerous natural products and active pharmaceutical ingredients [1–3]. They are frequently existed in the markedly available drug candidates, fine chemicals and play a fundamental role in medicinal chemistry as a consequence of their outstanding biological activities, such as anticancer, antibacterial, anti-HIV, antidiabetic, antimalarial [4–8]. Furthermore, they are considered as significant fragments in many optoelectronic materials, such as laser dyes, fluorescent whiteners, organic light-emitting diodes (OLEDs), polymers, optical recording, organic solar cells, organic semiconductors, fluorescent probes, fluorescent activity, and sensitizers for dye-sensitized solar cells [9–12].

Owing to these above-mentioned properties and broad chemical landscape, the construction and functionalization of molecules featuring heterocyclic framework as the key ingredients have attracted much more attention in synthetic organic chemistry [13–17]. However, the utilization of volatile organic solvents in a chemical process often results in the formation of chemical waste on both laboratory and industrial scales. This chemical waste was supposed to be one of the main sources of environmental pollution. Therefore, the design and development of a synthetic chemical route that leads to the expedient and rapid synthesis of diverse and highly functionalized heterocyclic scaffolds by avoiding or reducing the utilization of volatile organic solvents, toxic reagents, and hazardous chemicals to make our environment green and sustainable is highly desired and has emerged as a key challenge of modern synthetic organic chemistry. Furthermore, exploitation of energy in a chemical process either for heating or for cooling leads to an undesirable effect on the living environment.

To address many of these problems, mechanochemical methods, including ball-milling and grinding *via* a mortar and pestle have recently received a considerable and steadily increasing interest that has proved to be an excellent alternative and highly feasible environmentally benign energy inputs for organic synthesis. Concurrent to ultrasonic sonochemistry, solar light- and microwave-assisted chemistry, the introduction of mechanochemistry as an attractive and eco-friendly activation method has made rapid strides to be considered as the method of choice for organic synthesis, as it avoids the use of hazardous solvents and is less energy-consuming. The mechanical energy produced by grinding or milling of two solids or a solid and a liquid material breaks the order of the crystalline structure and makes close contact between the starting materials on a molecular scale, thereby producing the desired products. Mechanochemistry allows chemical transformation to be carried out in solvent-free conditions and makes them energy efficient by reducing high-temperature conditions to ambient temperature. Similar to ultrasound- and microwave-assisted organic synthesis, as well as solar light-induced organic synthesis which are associated with not only the enhancements of the reaction rates but also, involves in reducing the reaction times; the avoidance of toxic organic solvents, reduced reaction time, improved safety, less energy consumption, simple workup, and improved yields make mechanochemical method incredibly advantageous economically and ecologically favorable procedure in green chemistry [18–23].

But, again the occurrence of transition-metal-catalyst(s) in chemical processes even at the lowest level communicates the unfavorable effects on the atom economy and sustainability of the transformation. Notwithstanding, transition metal catalyst(s) has been successfully employed in the synthesis of valuable structural building blocks [24–26]; their occurrences in the chemical process caused serious effects because of their highly toxic nature, and the requirements of high cost for the preparation of catalytic system. Apart from these, the removal of transition-metal-catalyst(s) from the chemical transformation which is predominately needed in the pharmaceutical industry is not so easy and as a consequence, there will be high chances for contamination of the final compounds. Interestingly, the development of a synthetic chemical route for the construction of structural scaffolds with high atom- and step-economy which utilized alternative materials that are not only environmentally benign but also found to be in large scale in anywhere with minimum cost by reducing or circumventing the exploitation of transition metal catalyst(s), additives, supportive ligands, and toxic reagents to make a pollution-free environment are highly desired. For this purpose, the application of small organic molecules described as organocatalysts, in organic transformations have provided a new alternative route for the efficient synthesis of complex molecular structure in terms of synthetic efficiency and from the green chemistry viewpoint. The unique ability to accomplish chemical transformation through different activation modes, avoidance of expensive catalysts and metal catalyst(s), high stability, ready availability and easy recoverability, lower activation energy, high efficiency, as well as with an immediate reduction in the toxicity and reaction costs makes organocatalysis a highly advantageous and considerable approach in synthetic organic chemistry [13, 27–30]. These advantages of organocatalysis can contribute to many of the requirements of green and sustainable chemistry.

Considering the versatile applicability of mechanochemical activation in organic synthesis and the significant contribution of organocatalysis in organic transformation, here we provide a critical overview on the organocatalytic expedient synthesis of different types of highly functionalized five- and six-membered heterocycles as well as complex-fused heterocycles and spiro-heterocycles by using mechanochemical techniques, including ball-milling and grinding with mortar and pestle. The mechanochemical activation in organic reactions is well-reviewed by many researchers [19–23, 31–35] and we hope, the present chapter would be helpful for researchers working in these fields.

2. Mechanochemical organocatalytic reactions for the synthesis of five-membered heterocycles

2.1 Synthesis of five-membered heterocycles containing one-heteroatom

2.1.1 Synthesis of pyrroles

The five-membered nitrogen-containing heterocycle, pyrroles and its derivatives are well-established building blocks of many naturally occurring and synthetic drug molecules [13]. The most commonly applied method for the synthesis of pyrroles realizes the Paal-Knorr method that involves the reaction of 1,4-dicarbonyl compounds and primary amines or ammonia.

In 2016, Akelis et al. [36] developed a simple, facile, and highly efficient mechanochemical method for the synthesis of a variety of substituted pyrroles **3** from the reaction of 1,4-dicarbonyl compound **1** and amines **2** utilizing 1 mol% of citric acid as the organocatalyst under ball-milling condition for 15–30 minutes in the absence of solvent (**Figure 1**). This reaction offers the corresponding pyrroles **3** in 08–84% yields and the products were obtained in a very short reaction time. Encouraged by this result, they further extended their methodology for the desymmetrization of amines or to access bis(pyrroles) **6** by using several aryl or aliphatic diamines **4** and 1,4-dicarbonyl compound **1** as the reactants under the same reaction condition. The formation of mono-pyrroles **5** *that is* desymmetrization of amines and bis(pyrroles) **6** depends on the reactant diketones **1** and diamines **4**.

Another ball-milling approach for the synthesis of 3,4-disubstituted pyrroles has been accomplished by Bolm et al. [37] in 2021 (**Figure 2**). Under the influences of organic base DBU (1,8-Diazabicyclo[5.4.0]undec-7-ene), the desired products **9** derived from the reaction of enones **8** with TosMIC (toluenesulfonylmethyl isocyanide) **7** at milling frequency of 35 Hz for 1 hour in solvent-free condition has been obtained in moderate to good yield (34–92% yield). A vast array of

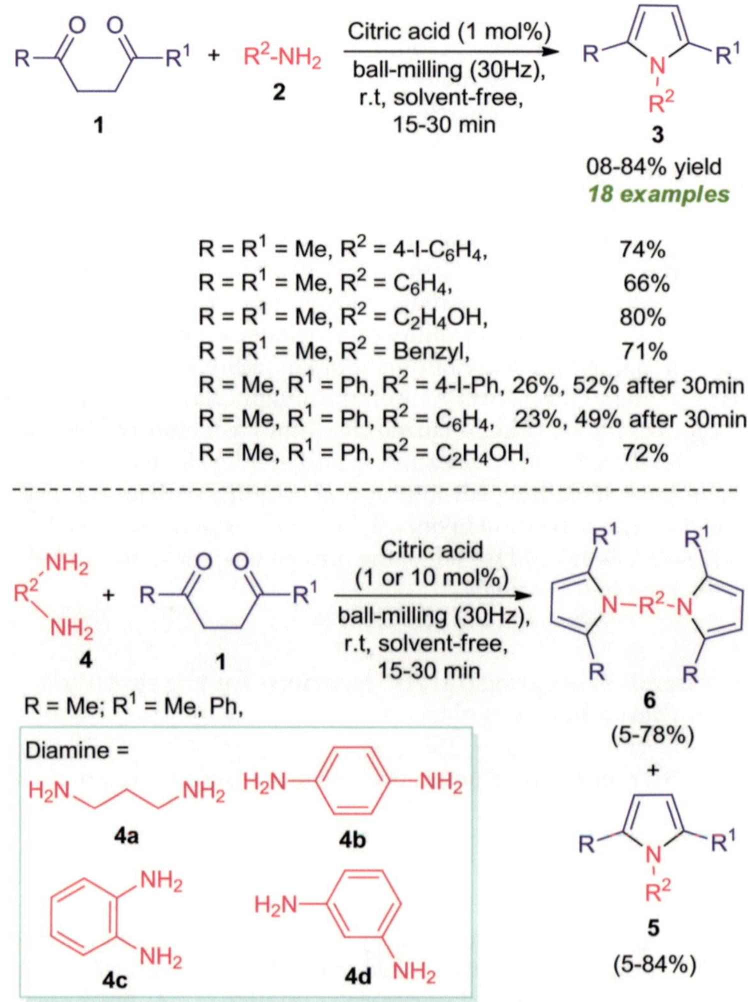

Figure 1.
Synthesis of pyrroles assisted by ball-mill under the organocatalytic condition.

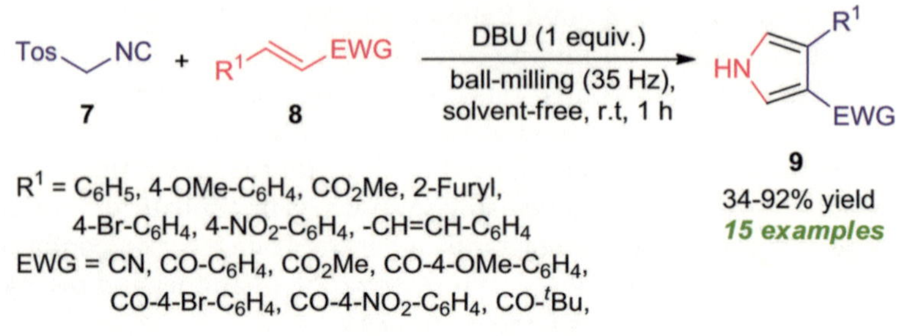

Figure 2.
DBU catalyzed van Leusen pyrrole synthesis under mechanochemical activation.

electron-withdrawing and electron-donating groups was found to be well worked under this standard condition. The tolerability of a broad functional group, simple operational procedure, short reaction time, is some of the key features of this strategy.

2.1.2 Synthesis of furans

A simple but highly attractive one-pot procedure for the synthesis of *trans*-2,3-di-substituted 2,3-dihydrofurans in a diastereoselective manner by employing mechanochemical techniques as a powerful green energy source under solvent-free conditions was developed by Chuang and Chen (**Figure 3**) [38]. With the help of piperidine as the organocatalyst, the desired products **13** were accomplished *via* grinding of several 1,3-dicarbonyl compounds **10**, aldehydes **11**, and *N*-phenacyl pyridinium bromides **12** in a mortar and pestle for 1–1.5 hours at room temperature have been achieved in 51–79% yield. This reaction was also possible to carry out in conventional solution conditions, however, the mechanochemical method was found to be very superior in terms of green chemistry point of view and synthetic efficiency. The reaction condition was found to be tolerable to a variety of 1,3-dicarbonyl compounds and also aryl aldehydes comprising different electron-withdrawing and electron-donating substituents.

2.1.3 Synthesis of thiophenes

The Gewald method which involves the reaction of ketones, α-methylene carbonyl compounds, activated nitriles, and elemental sulfur is a well-established approach for the synthesis of 2-amino thiophenes. In this regard, Mack et al. [39] reported a Gewald reaction of acetophenone **14**, ethyl cyanoacetate **15**, and sulfur **16** under the ball-milling conditions with a stainless-steel ball (1/8th inch) at 18 Hz for the synthesis of thiophenes **17** by using morpholine as the organocatalyst in solvent-free condition at 130°C. By applying this method, a total of six compounds were synthesized in a 14–53% yield. Although solvent-free, as well as metal-free, waste-free short reaction time, makes the advantages of this protocol, however, the low-substrate scopes mark a limitation of this procedure and call for further developments otherwise outstanding developments (**Figure 4**).

2.2 Synthesis of five-membered heterocycles containing two-heteroatoms

2.2.1 Synthesis of pyrazoles

As nitrogen-containing heterocycle, pyrazole and their derivatives have a significant role in the field of medicinal chemistry and material sciences. As a consequence, substantial efforts have been dedicated to their synthesis [16]. In line with this, a highly efficient one-pot mechanochemical method for the synthesis of a series of

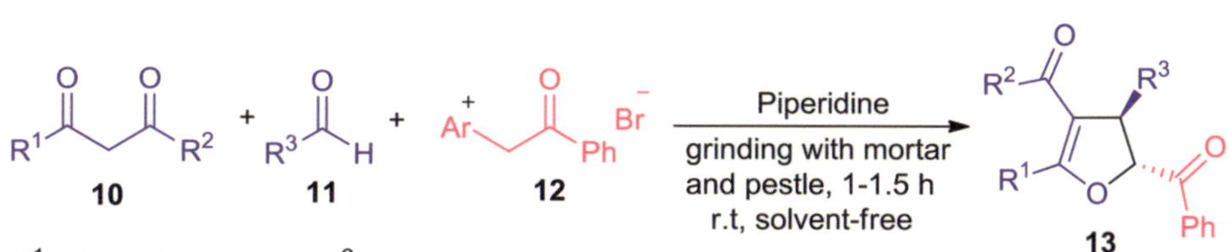

R^1 = CH_3, C_6H_5, C_2H_5; R^2 = CH_3, OC_2H_5, OCH_3;
R^3 = C_6H_5, 4-Me-C_6H_4, 4-MeO-C_6H_4, 4-Br-C_6H_4, 4-Cl-C_6H_4, 4-(MeO_2C)-C_6H_4
Ar = Py, 4-Me-Py, 4-Bn-Py, 3,5-(Me)$_2$-Py, 3-(EtO_2C)-Py

Figure 3.
Secondary amine catalyzed grinding assisted one-pot three-component diastereoselective synthesis of dihydrofurans.

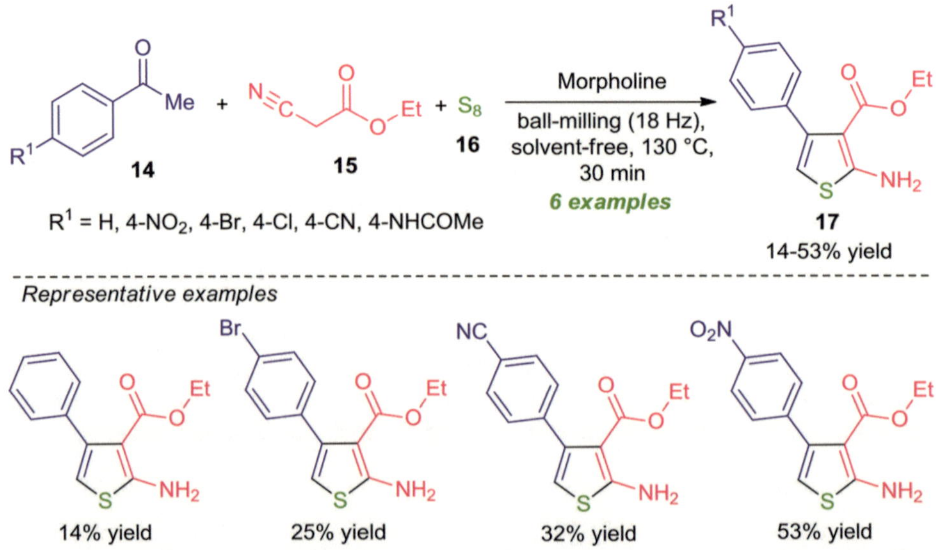

Figure 4.
Organocatalytic mechanochemical one-pot synthesis of thiophene derivatives.

sulfur-containing pyrazole derivatives **21** under solvent-free conditions has been developed by Saeed and Channar (**Figure 5**) [40]. With the help of a mortar and pestle, the authors manually ground the readily available 3-chloro-2,4-pentanedione **18**, hydrazine **19**, and thiophenol **20** by using piperidine as the base organocatalyst at room temperature for 10–17 minutes. This three-component reaction afforded the corresponding products **21** in 72–88% yields. Broad functionality, short reaction time, mild reaction condition, metal-free, operational simplicity are some of the salient features of this protocol.

2.2.2 Synthesis of thiazoles

Edrees, Gomha et al. [41] demonstrated the successful application of mechanochemical techniques in the synthesis of a library of highly functionalized thiazole derivatives bearing pyrazole core in their structure (**Figure 6**). By using DABCO (1,4-Diazabicyclo[2.2.2]octane) as the catalyst, the treatment of pyrazole-1-carbothioamide **22** with α-haloketones **23** or **26** under grinding with mortar and

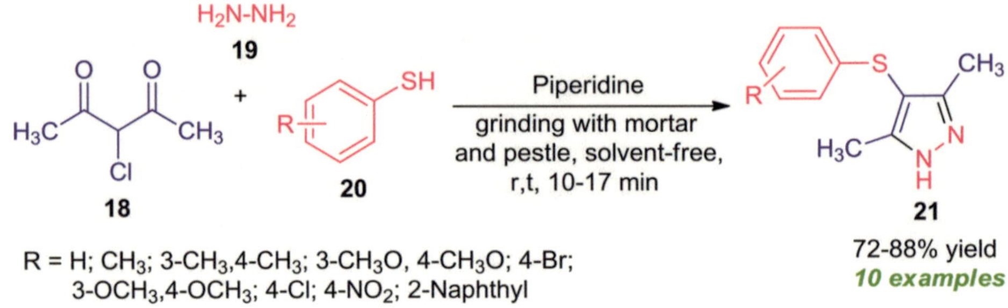

Figure 5.
Piperidine catalyzed one-pot three-component grinding assisted synthesis of pyrazoles.

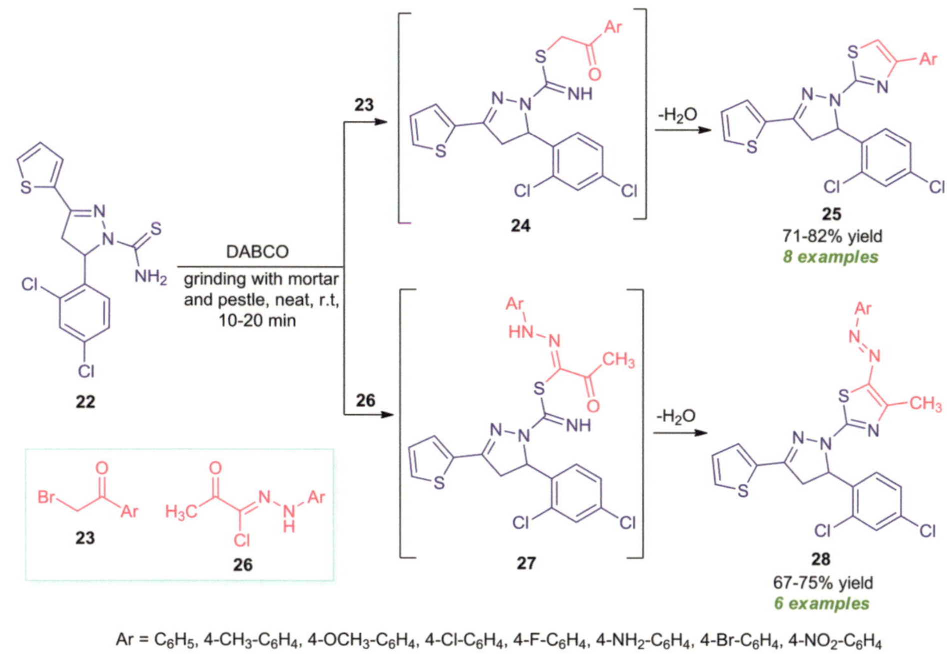

Figure 6.
Grinding assisted one-pot synthesis of diverse thiazole derivatives bearing pyrazole moiety.

pestle at room temperature was found to proceed in solvent-free condition to form the desired products **25** and **28** in moderate to good yields, respectively. In both cases, the reaction was completed through the initial formation of intermediate **24** and **27** which undergo cyclization and dehydration to afford the final products. A wide variety of electron-withdrawing and electron-donating substituents present on the aryl ring of the α-haloketones **23** or **26** were found to be well worked under this reaction condition.

A very simple and straightforward grinding-assisted method to access benzo-fused thiazole derivatives under organocatalytic conditions was disclosed by Agarwal and Gandhi (**Figure 7**) [42]. In this context, they manually grind the readily available 2-aminobenzenethiol **29** and several aldehydes **11** by using urea nitrate as the organocatalyst in solvent-free conditions at room temperature. By applying this operationally simple methodology, a total of six benzo-fused thiazole products **30** were synthesized in excellent yield within a very short reaction time.

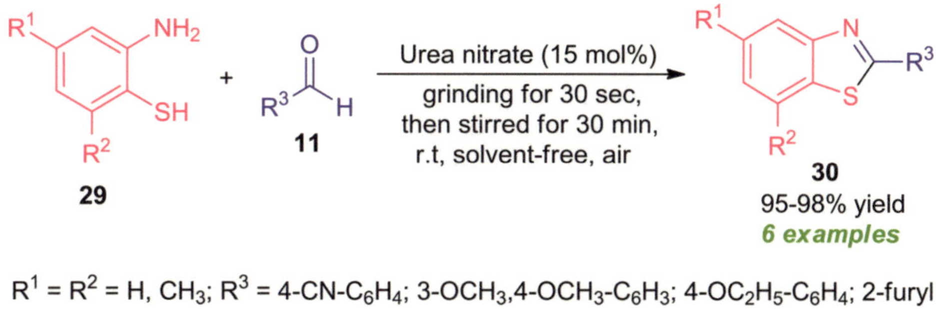

Figure 7.
Grinding-assisted construction of benzo-fused thiazoles under organocatalysis.

2.2.3 Synthesis of imidazoles

Rajitha et al. [43] disclosed the utilization of grinding techniques for the condensation reaction of benzo[*c*][1,2,5]thiadiazole-4,5-diamine **31** with different substituted aldehydes **32** under the influences of cellulose sulfuric acid as the organocatalyst in solvent-free condition at room temperature to access a variety of benzo-fused imidazoles **33** in moderate to excellent yield. The reaction was found to be well tolerated for both aryls and heteroaryl-substituted aldehydes. The mild reaction condition, short reaction time, use of recyclable and reusable catalyst, environmentally as well as eco-friendly benign, simple work-up procedure, wide substrate scopes are some of the advantages of this protocol (**Figure 8**).

2.3 Synthesis of five-membered heterocycles containing three-heteroatoms

2.3.1 Synthesis of oxadiazoles

Kategaonkar [44] developed an environmentally benign procedure for the construction of oxadiazole derivatives **36** from the condensation reaction of 1*H*-indazole-3-carboxylic acid hydrazide **34** and aromatic acids **35** by applying the high catalytic activity of the cellulose sulfuric acid as the organocatalyst under grinding condition by using a simple mortar and pestle in solvent-free condition at room temperature. The reaction was completed within only 5–10 minutes to afford the desired products **36** in good to excellent yield. To broaden the substrate scopes, a variety of aromatic acids bearing electron-rich and electron-poor substituents on the aryl ring were subjected to 1*H*-indazole-3-carboxylic acid hydrazide under the optimized reaction condition and all are found to be efficiently worked by this mechanochemical reaction (**Figure 9**).

2.3.2 Synthesis of thiadiazoles

Thiadiazoles are well-established five-membered heterocyclic compounds with three heteroatoms including two nitrogen atoms and one sulfur in their structure. They are known to be an important skeleton in medicinal and synthetic chemistry due to their wide prolific pharmacological profile. Considering their importance, Aziem et al. [45] introduced the grinding process as an eco-friendly and environmentally friendly chemical technology for the organocatalytic synthesis of various 1,3,4-thiadiazole derivatives **39** comprising benzofuran moiety in their structure. With the help of a mortar and pestle, the author's manually ground benzofuran-bearing hydrazinecarbodithioate **37** and hydrazonoyl halides **38** in presence of TEA (triethylamine) as the catalyst under the solvent-free condition at room temperature which eventually

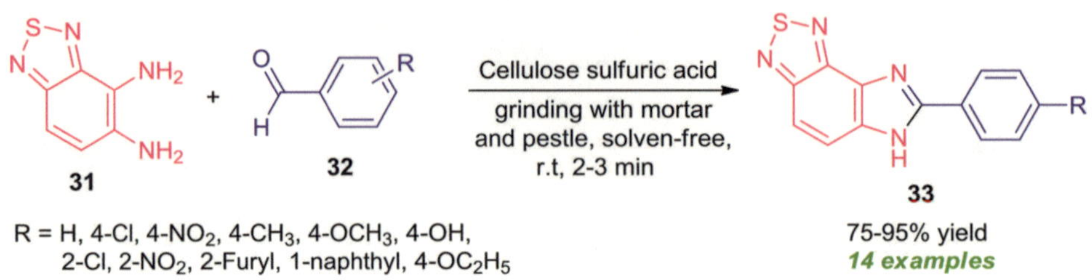

Figure 8.
Organocatalytic mechanochemical synthesis of benzo-fused imidazoles.

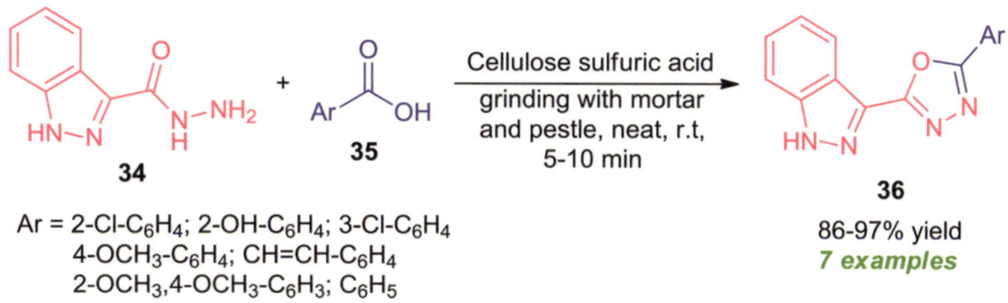

Figure 9.
CSA catalyzed rapid access to oxadiazoles 36 by means of mechanochemical activation.

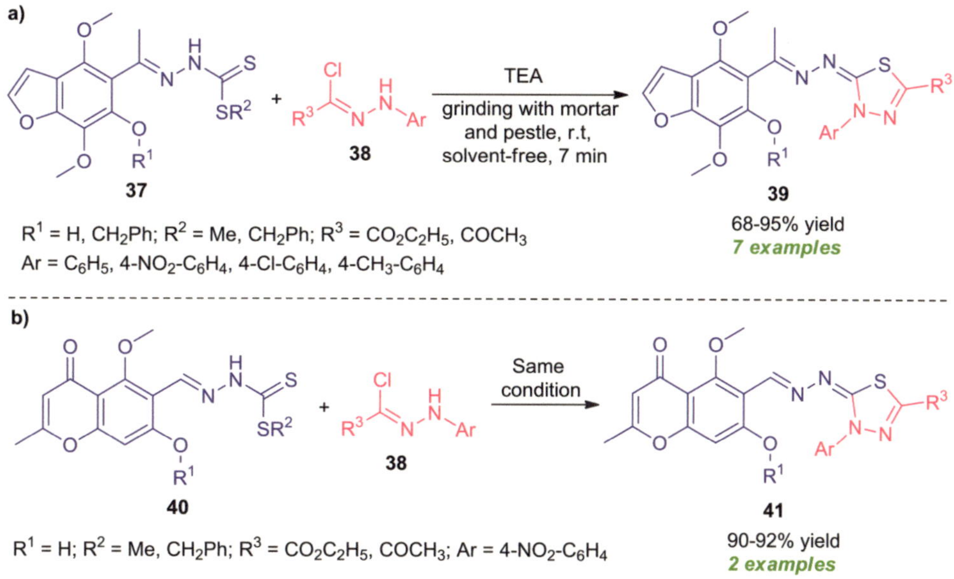

Figure 10.
Tertiary amine catalyzed grinding assisted synthesis of thiadiazoles comprising benzofuran moiety.

leads to the formation of the desired products **39** in moderate to excellent yield within a very short reaction time (**Figure 10a**). Enlightened by this result, they extended their protocol to synthesize another type of thiadiazole derivatives containing chromone moiety in their structure. The same reaction condition was found to be tolerable to hydrazine-carbodithioate of type **40** and hydrazonoyl halides **38** to deliver the corresponding 1,3,4-thiadiazole derivatives **41** in good yields (**Figure 10b**).

3. Mechanochemical organocatalytic reactions for the synthesis of six-membered heterocycles

3.1 Synthesis of six-membered heterocycles containing one-heteroatom

3.1.1 Synthesis of pyridines

Synthesis of a vast array of 1,4-dihydropyridine derivatives under mechanochemical activation has been achieved by Sarada et al. [46]. With the help of a mortar and pestle, grinding of readily available aldehydes **11**, 2 equivalent of ethyl acetoacetate **42**, and ammonium acetate **43** in presence of 20 mol% of chlorosulfonic acid (CSA) as the organocatalyst under the solvent-free condition at room temperature results in the formation of the corresponding products **44** in moderate to good yields within a very short reaction time. A wide variety of aliphatic and substituted aryl aldehydes were found to be well tolerated by this methodology (**Figure 11a**). Similar to this, the application of ball-milling techniques as an alternative energy input for the one-pot synthesis of 1,4-dihydropyridines was demonstrated by Ghafuri et al. [47]. By using imidazole dicarboxylic acid (IDCA) as the organocatalyst, the desired products **46**, derived from several substituted aldehydes **11**, β-ketoesters **45**, and ammonium acetate **43** were obtained in good to excellent yield under solvent-free conditions. Short reaction time, energy efficiency, mild reaction conditions are some of the advantages of this protocol (**Figure 11b**).

3.1.2 Synthesis of quinolines

A highly efficient and environmentally benign approach for the synthesis of polysubstituted quinolines *via* Friedländer reaction under ball-milling conditions was developed by Javanshir et al. [48]. For this purpose, authors performed a solvent-free two-component reaction of 2-aminoaryl ketones **47** and a variety of active methylene compounds **48** under the influences of 30 mol% of *p*-TSA (*p*-toluene sulfonic acid). Initial optimization for the reaction condition of this reaction in presence of different catalytic systems, such as chitosan, cyanuric chloride, *p*-TSA, MCM-41 suggested the utilization of *p*-TSA as the best catalytic system under solvent-free conditions. A total of twelve quinoline products **49** were synthesized in poor to excellent yield by this method within a very few minutes (**Figure 12**).

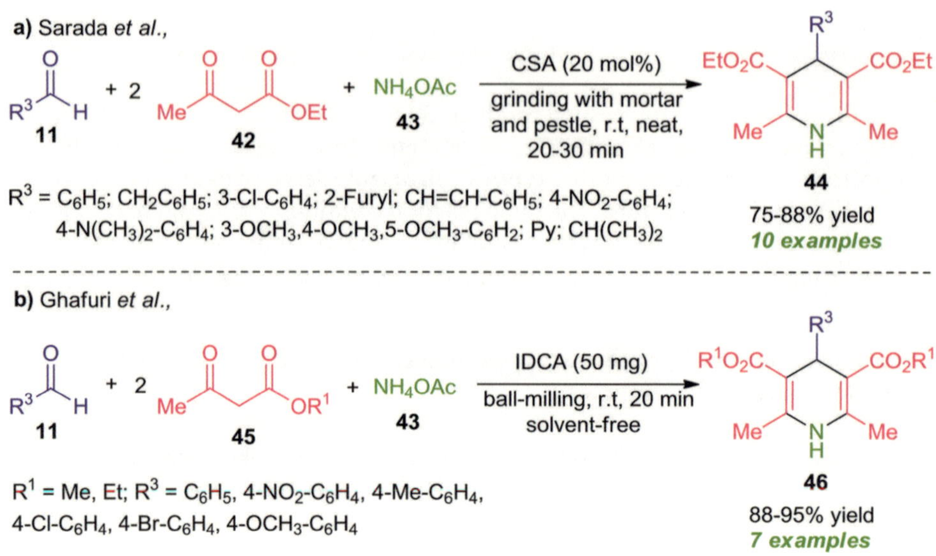

Figure 11.
One-pot mechanochemical synthesis of 1,4-dihydropyridines under organocatalysis.

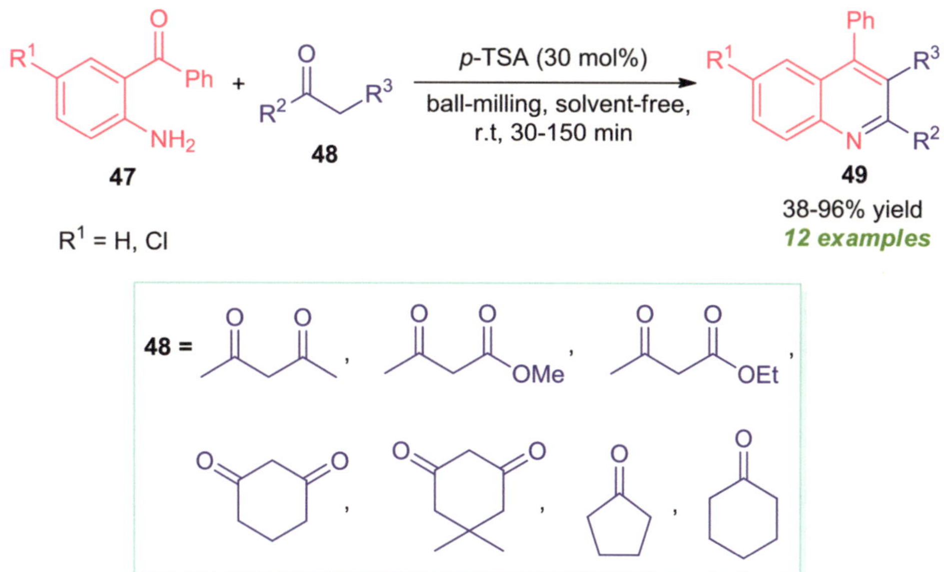

Figure 12.
Organocatalytic mechanochemical assisted synthesis of quinolines.

3.1.3 Synthesis of pyrans

The synthesis of 4*H*-Pyran core and their derivatives have attracted tremendous attention over the last decades due to their great prevalence in natural product chemistry, and medicinal chemistry [49]. To develop a rapid, facile, and simple method inconsistent with the green chemistry principle, the group of Naimi-Jamal [50] introduced ball-milling techniques in combination with the organocatalytic system as a perfect chemical process for the synthesis of a vast array of 2-amino-4*H*-pyrans. The synthesis involves the one-pot three-component reaction of aldehydes **50**, malononitrile **51**, and ethyl acetoacetate **42** in presence of piperazine as the organocatalyst under the ball-milling condition at frequency 20 Hz–25 Hz for 20–90 minutes. Noticeably, the reaction was carried out at room temperature in absence of solvent, and the corresponding products **52** were achieved in good to excellent yields (**Figure 13a**). Subsequent to this report, Dekamin and Eslami [51] disclosed the utilization of potassium phthalimide (POPI) as the metal-free organocatalyst for the mechanochemical one-pot three-component synthesis of 2-amino-4*H*-pyrans from the reaction of aldehydes **50**, malononitrile **51**, and ethyl acetoacetate **42** under the solvent-free condition at ambient temperature. The reaction required only 5 mol% of catalyst to efficiently form the desired products **52** in 90–98% yields (**Figure 13b**).

3.2 Synthesis of six-membered heterocycles containing two-heteroatoms

3.2.1 Synthesis of quinoxalines

Cellulose sulfuric acid was applied as an efficient metal-free organocatalytic system for the solid-state construction of highly functionalized quinoxaline derivatives by Rajitha et al. (**Figure 14**) [52]. By using a mortar and pestle, grinding of substituted 3-bromoacetyl coumarins **53** and thiadiazole-substituted diamine **31** under the

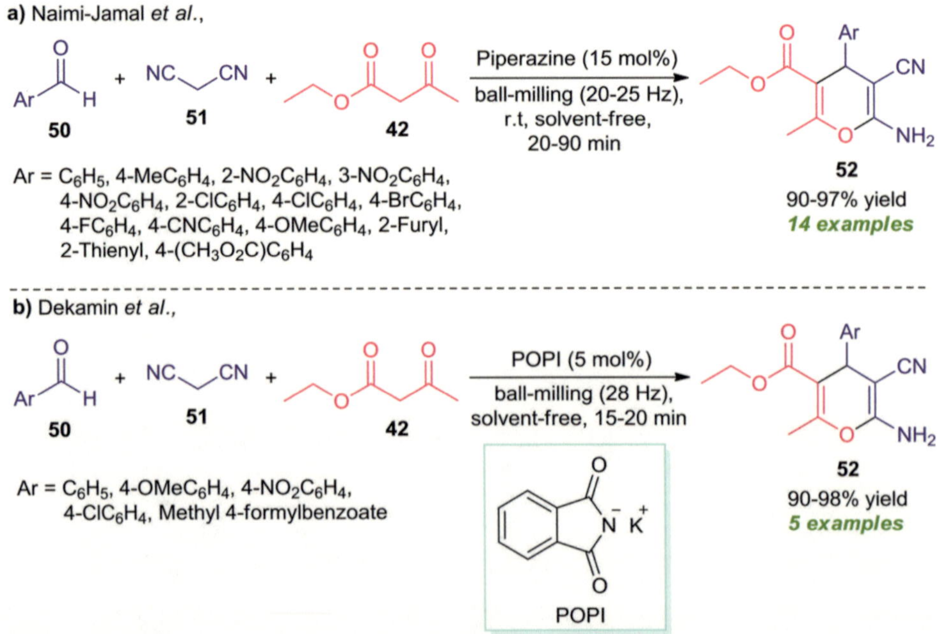

Figure 13.
Organocatalytic three-component synthesis of 4H-pyrans under ball-milling condition.

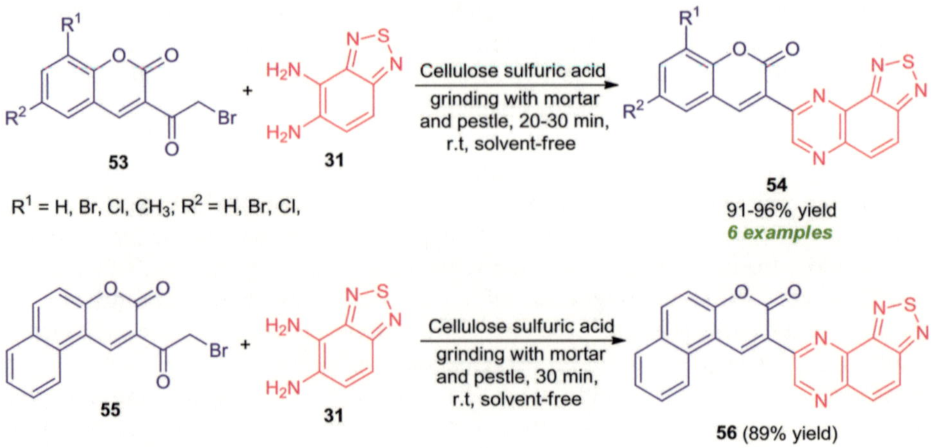

Figure 14.
Cellulose sulfuric acid-catalyzed mechanochemical synthesis of quinoxalines.

solvent-free condition at room temperature for 20–30 minutes results in the formation of coumarin-substituted quinoxaline derivatives **54** in good to excellent yields. Different substitutions by electron-rich and electron-poor groups on the coumarin ring smoothly underwent the reaction and had no detrimental effect on the product yields. Similarly, other types of substituted coumarins **55** were found to be well worked under the same reaction condition to provide the desired product **56** in quantitative yield. Environmentally benign conditions, inexpensive, good yield of the products, exploitation of recyclable and reusable catalysts are some key features of this protocol.

3.2.2 Synthesis of pyrimidines

Barman et al. [53] disclosed a metal-free highly convenient one-pot approach under mechanochemistry for the synthesis of 3,4-dihydropyrimidines. With the help of a mortar and pestle, the author's grind substituted 1,3-diketones **10**, aldehydes **11**, and urea or thiourea **57** under the influences of L-tyrosine as the organocatalyst in solvent-free condition at room temperature which eventually led to the desired products **58** in 81–91% yields. Initially, the reaction was performed in presence of different catalytic systems, such as L-proline, glycine, L-serine, L-tyrosine, camphorsulphonic acid, and different reaction mediums, such as ethanol, microwave, solvent-free; among them, the solvent-free grinding method using L-tyrosine as the catalyst was found to be the best condition for this reaction. Not only the aryl aldehydes possessing different electron-poor and electron-rich groups but also heteroaryl aldehydes were efficiently underwent the reaction under this condition (**Figure 15**).

3.2.3 Synthesis of quinazolinones

Quinazolinones and their derivatives are well-established heterocycles commonly encountered in many natural products and synthetic drug candidates. To realize their importance, a rapid mechanochemical assisted one-pot methodology for the synthesis of diverse quinazolinone derivatives has been developed by Shingare et al. (**Figure 16**) [54]. By introducing 10 mol% of vitamin B_1, also known as thiamine hydrochloride as the organocatalyst, the solid-state treatment of anthranilic acid **59**, triethyl orthoformate **60**, and various amines **61** under the grinding condition at room temperature afforded the corresponding quinazolinone products **62** in good to excellent yields. A wide variety of aryl amines bearing electron-withdrawing as well as electron-donating substituents smoothly worked well by this environmentally benign protocol. The effectiveness of the methodology was demonstrated by recycling and reusing the catalyst up to five consecutive reactions without affecting the significant outcome of the protocol.

Another achievement for the synthesis of different types of quinazolinones **64** and **66** has been accomplished by Saha et al. [55] by employing ball-milling techniques as an environmentally benign energy source (**Figure 17**). With the help of 10 mol% of *p*-TSA (*p*-toluene sulfonic acid) as the Brønsted acid catalyst, the solvent-free reaction of anthranilamide **63** and aldehydes **11** under mechanochemical grinding conditions delivers the corresponding quinazolinone products **64** in moderate to excellent yield within 3–15 minutes. While the reaction of anthranilamide **63** with

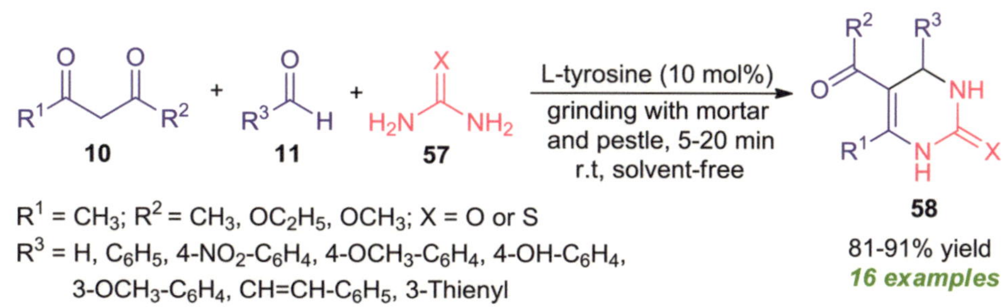

Figure 15.
L-tyrosine catalyzed three-component synthesis of 3,4-dihydropyrimidine.

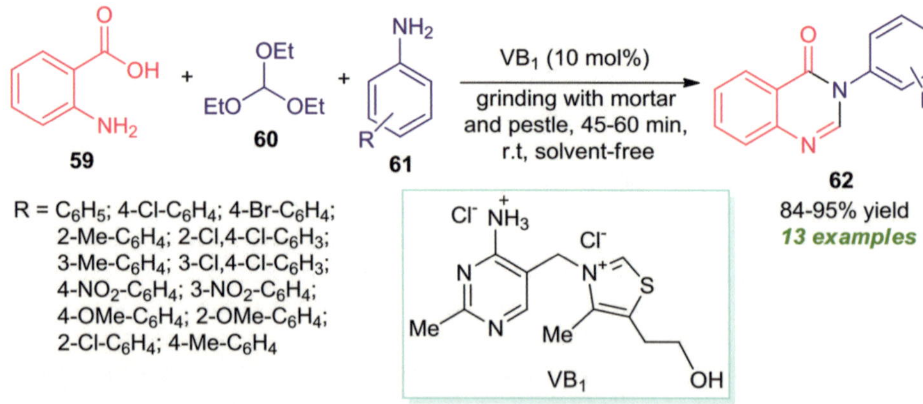

Figure 16.
Thiamine hydrochloride catalyzed three-component synthesis of 4-(3H)-quinazolinone by grindstone technique.

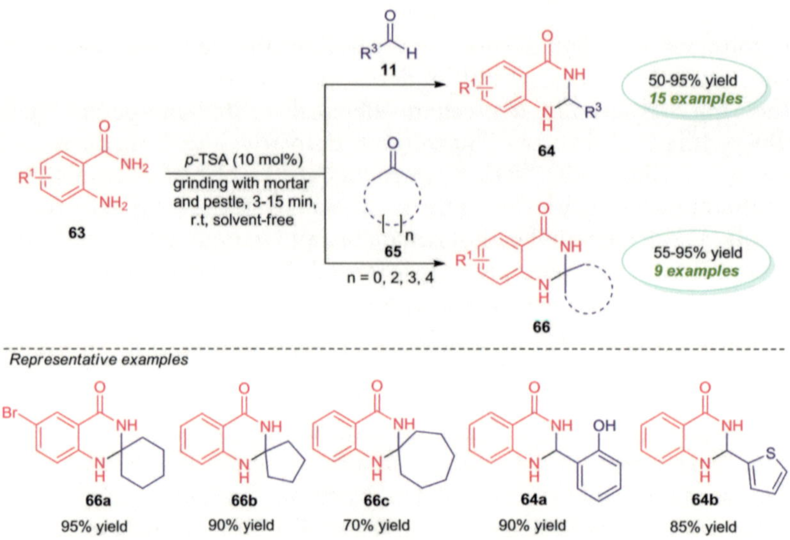

Figure 17.
p-TSA catalyzed grinding assisted synthesis of diverse quinazolinone derivatives.

different carbonyl compounds **65** under the same reaction condition afforded the quinazolinone products of type **66** in moderate to excellent yield in only 5 minutes. Broad functional group tolerances, mild reaction conditions, solvent-free, waste-free, metal-free are some of the key advantages of this protocol. The practicality of the protocol was established by performing gram scale synthesis in quantitative yield.

4. Mechanochemical organocatalytic reactions for the synthesis of complex-fused poly-heterocycles

4.1 Synthesis of indazolo[2,1-b]phthalazine

An efficient eco- and environmentally friendly approach for the synthesis of complex-fused heterocycle, namely indazolo[2,1-*b*]phthalazine under mechanochemical method was developed by Wang et al. (**Figure 18**) [56]. By using 3 mol% of *p*-TSA (*p*-toluene sulfonic acid) as the catalyst, the three-component grinding-assisted reaction between phthalhydrazide **67**, dimedone **68**, and aldehydes **50** was found to proceed under the solvent-free condition at room temperature to deliver the desired products **69** in 83–92% yields. The utilization of grinding as a green energy source, metal-free, simple work-up procedure, mild reaction condition, low catalyst loading, broad functionality is some of the salient features of this protocol.

4.2 Synthesis of naphtho[2,3-b]thiophenes

A domino one-pot mechanochemical route toward the synthesis of naphtho-fused thiophene heterocycle was developed by the research group of Singh (**Figure 19**) [57]. By utilizing DMAP (4-Dimethylaminopyridine) as the metal-free catalyst, the oxidative [3 + 2] heteroannulation of 1,4-naphthoquinone **70** and α-enolicdithioesters/β-oxothioamides **71** under mechanochemical grinding with a mortar and pestle in solvent-free condition at room temperature afforded the corresponding naphtho[2,3-*b*] thiophene products **72** in moderate to excellent yield within a very short reaction time. It is interesting to note that, this reaction does not require any co-catalyst and an activator which marks the advantages of this protocol. To broaden the substrate scopes, a variety

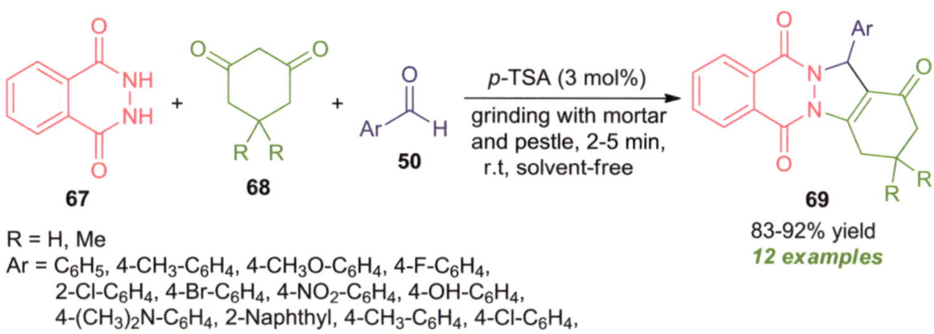

Figure 18.
Brønsted acid-catalyzed three-component one-pot synthesis of indazolo[2,1-b]phthalazine by grinding method.

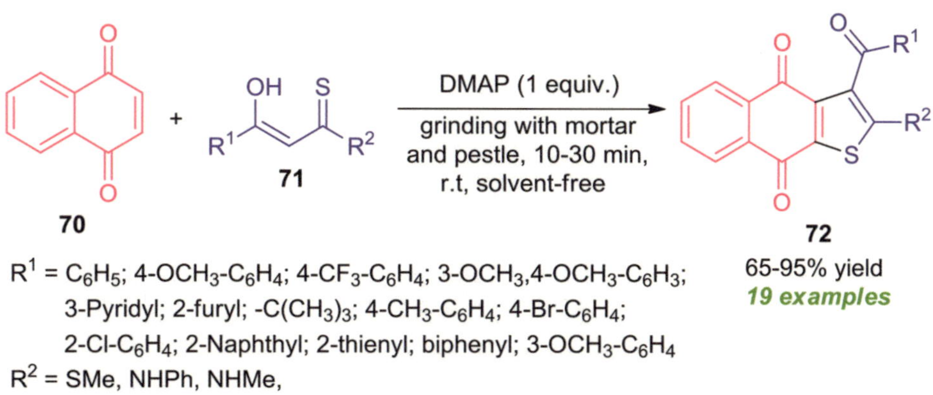

Figure 19.
DMAP catalyzed grinding assisted domino thienannulation to access naphtho[2,3-b]thiophenes.

of aryl, as well as heteroaryl-substituted α-enolicdithioesters/β-oxothioamides, was subjected to the reaction under the optimized condition and all are found to be efficiently compatible by this method.

4.3 Synthesis of pyrano[4,3-b]pyrans

Khaligh et al. [58] demonstrated the successful application of ball-milling techniques in the multicomponent reaction of substituted aldehydes **50**, malononitrile **51**, and 4-hydroxy-6-methyl-2-pyrone **73** in presence of 10 mol% of 1,4-piperazinediethane-sulfonic acid (PIPES) as the organocatalyst under solvent-free condition (**Figure 20**). This reaction offers a library of pyrano[4,3-*b*]pyran derivatives **74** in good to excellent yield after 30 minutes. The electronic effects of the substitution by different electron-withdrawing as well as electron-donating substituents on the aryl ring of aldehydes were examined and all are found to be well tolerated by these reaction conditions.

4.4 Synthesis of pyrano[2,3-c]pyrazoles

A grinding assisted one-pot multicomponent approach for the rapid construction of pyrano[2,3-*c*]pyrazoles from the four-component reaction of aldehydes **50**, malononitrile **51**, acetylene dicarboxylate **75**, and hydrazine hydrate **76** in presence of L-proline as the organocatalyst in solvent-free condition was developed by Padmini et al. (**Figure 21**) [59]. Noticeably, only 10 mol% of L-proline was found to be effective for catalyzing this reaction and a total of 13 compounds were synthesized in moderate to good yield. Not only the aryl aldehydes possessing various electronic groups on different positions but also heteroaryl aldehydes smoothly undergo this reaction. The operational simplicity, mild, green reaction medium, short reaction time, and wide-substrate scopes are some key features of this approach.

4.5 Synthesis of triazolo[1,5-a]pyrimidine

Khaligh and Mihankhah [60] reported the exploitation of ball-milling techniques as a powerful alternative energy source in the three-component reaction of amino-substituted triazoles **78**, aldehydes **50**, and ethyl cyanoacetate **15** by using poly-melamine-formaldehyde (mPMF) as the nitrogen-rich porous organocatalyst under the solvent-free condition at room temperature. This solid-state reaction provides a library of triazolo[1,5-*a*]pyrimidines **79** in moderate to excellent yield after

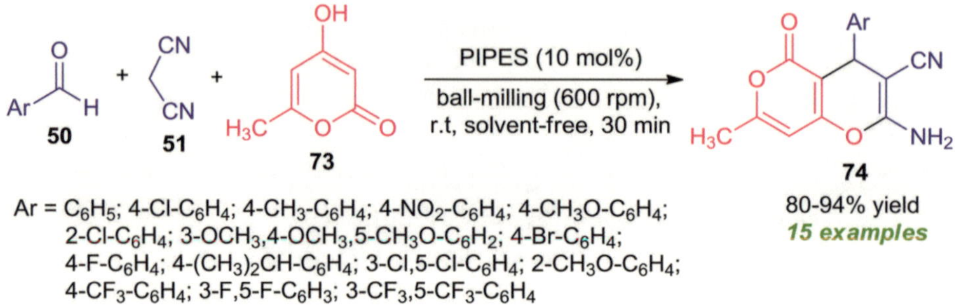

Figure 20.
Mechanochemical one-pot three-component organocatalytic synthesis of pyrano[4,3-b]pyrans.

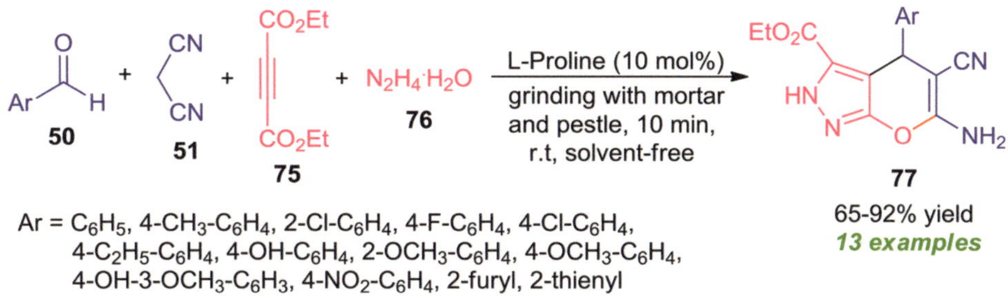

Figure 21.
L-proline catalyzed grinding assisted multicomponent synthesis of pyrano[2,3-c]pyrazoles.

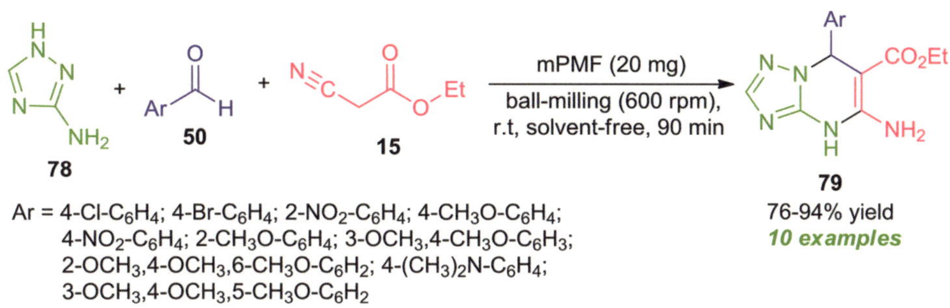

Figure 22.
Poly-melamine-formaldehyde catalyzed three-component mechanochemical synthesis of triazolo[1,5-a] pyrimidine.

90 minutes (**Figure 22**). The catalyst was found to be very effective for this reaction that could be easily recycled and reused for the next consecutive reaction without altering the reactivity and selectivity of the product. The easy work-up procedure, green reaction condition, low cost, wide abundance of the substrate scope are some advantages of this mechanochemical process.

5. Mechanochemical organocatalytic reactions for the synthesis of complex spiro-heterocycles

In the last few decades, the field of synthetic organic chemistry has witnessed outstanding developments in the synthesis of spiro-heterocycles especially spiro-oxindoles due to their outstanding reactivity and prolific pharmacological activity as well as their utilization as important building blocks for the synthesis of natural products types molecules as well as medicinally privileged heterocycles [61, 62].

Considering their importance in accordance with the significant application of mechanochemistry in synthetic organic chemistry, the research group of Bazgir [63] synthesized a series of spiro[diindenopyridine-indoline]triones **82** from the three-component reaction of various amines **61**, 2 equivalent of 1,3-indandione **80** and substituted isatins **81** by using *p*-TSA (*p*-toluene sulfonic acid) as the Brønsted acid catalyst under grinding and solvent-free condition (**Figure 23**). A total of 20 compounds were synthesized in 80–91% yield within 3–4 minutes at room temperature. Similarly, treating amines **61** and 1,3-indandione **80** with acenaphthylene-1,2-dione **83** by replacing isatins was found to proceed under the same reaction condition to

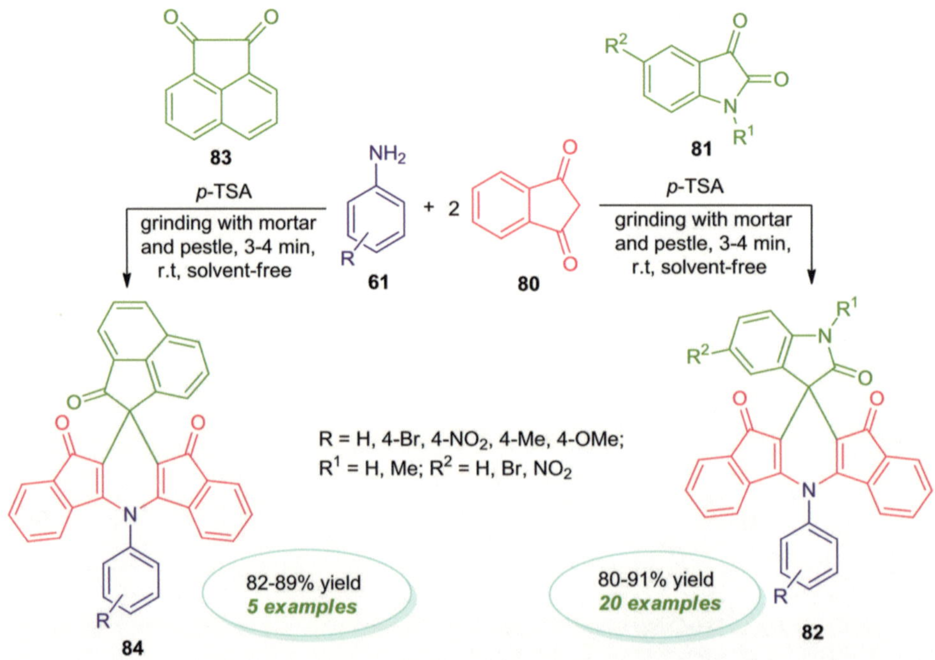

Figure 23.
Brønsted acid-catalyzed multicomponent synthesis of diverse spiro-heterocycles by means of grindstone technique.

afford the desired spiro[acenaphthylene-diindenopyridine]trione products **84** in 82–89% yields.

6. Conclusion

The frequent occurrence of heterocyclic compounds in natural, pharmaceutical, and synthetic optoelectronic materials, demands efficient methodology for their construction and selective functionalization by using them as key building blocks. But, the involvement of toxic solvents which are associated with chemical pollution often results in environmental safety concerns. Therefore, developing an alternative method to carry out organic synthesis by avoiding or minimizing the utilization of volatile organic solvents and toxic reagents by introducing environmentally benign conditions with the main focus to reduce the cost-effectiveness of the chemical transformation is highly desired. From the above observation, it is clear to conclude that the utilization of mechanochemical techniques allows all the reactions to be carried out in absence of volatile organic solvents, hazardous reagents and make them environmentally as well as eco-friendly benign. The mechanochemical techniques are found to be very efficient as compared to traditional stirring conditions from the perspective of synthetic as well as green chemistry points of view. The attractive benefits associated with mechanochemistry as a powerful alternative green energy source lead to a new frontier in the synthesis of diverse heterocyclic compounds as well as asymmetric synthesis and hope to consider as a method of choice both at the laboratory as well as in industrial level in near future.

Besides these, the development of organocatalytic reactions under mechanochemistry is set to lead the field of synthetic organic chemistry to a new height. The ability to accomplish reaction under the metal-free organocatalytic condition in the absence of solvent *via* simple grinding with mortar and pestle or milling can contribute to some of the goals of green and sustainable chemistry and are considered as highly promising routes in the synthesis of diverse and densely functionalized five-membered and six-membered heterocycles as well as complex-fused heterocycles and spiro-heterocycles.

On the other hand, some serious attention needs to be paid to broadening the substrate scopes, reducing the amount of catalyst, and developing a scalable protocol based on the industrial level.

Acknowledgements

The author thanks the Central University of Gujarat, Gandhinagar, India, and Prof. Rama Shanker Dubey, Vice-Chancellor, the Central University of Gujarat for the encouragement and continuous support. BB thanks UGC-India for the Non-NET fellowship.

Conflict of interest

"There are no conflicts to declare."

Author details

Biplob Borah and L. Raju Chowhan*
School of Applied Material Sciences, Centre for Applied Chemistry, Sector-30, Central University of Gujarat, Gandhinagar, India

*Address all correspondence to: rchowhan@cug.ac.in

References

[1] Katritzky AR. Heterocyclic chemistry: An academic subject of immense industrial importance. Chemistry of Heterocyclic Compounds. 1992;**28**:241-259. DOI: 10.1007/BF00529362

[2] Joule JA. Natural products containing nitrogen heterocycles—Some highlights 1990-2015. Advances in Heterocyclic Chemistry. 2016;**119**:81-106. DOI: 10.1016/bs.aihch.2015.10.005

[3] Cossy J, Guerinot A. Natural products containing oxygen heterocycles—Synthetic advances between 1990 and 2015. Advances in Heterocyclic Chemistry. 2016;**119**:107-142. DOI: 10.1016/bs.aihch.2016.03.002

[4] Taylor AP, Robinson RP, Fobian YM, Blakemore DC, Jones LH, Fadeyi O. Modern advances in heterocyclic chemistry in drug discovery. Organic & Biomolecular Chemistry. 2016;**14**:6611-6637. DOI: 10.1039/C6OB00936K

[5] Chugh A, Kumar A, Verma A, Kumar S, Kumar P. A review of antimalarial activity of two or three nitrogen atoms containing heterocyclic compounds. Medicinal Chemistry Research. 2020;**29**:1723-1750. DOI: 10.1007/s00044-020-02604-6

[6] Fesatidou M, Petrou A, Athina G. Heterocycle compounds with antimicrobial activity. Current Pharmaceutical Design. 2020;**26**:867-904. DOI: 10.2174/1381612826666200206093815

[7] Borah B, Chowhan LR. Recent advances in the transition-metal-free synthesis of quinoxalines. RSC Advances. 2021;**11**:37325-37353. DOI: 10.1039/D1RA06942J

[8] Ali I, Nadeem Lone M, Al-Othman AZ, Al-Warthan A, Marsin Sanagi M. Heterocyclic scaffolds: Centrality in anticancer drug development. Current Drug Targets. 2015;**16**:711-734

[9] Cooke G, Evans IR, Skabara PJ. Functional organic materials for optoelectronic applications. Journal of Materials Chemistry C. 2019;**7**:6492-6492. DOI: 10.1039/C9TC90084E

[10] Chan CY, Wong YC, Chan MY, Cheung SH, So SK, Yam VW. Bifunctional heterocyclic spiro derivatives for organic optoelectronic devices. ACS Applied Materials & Interfaces. 2016;**8**:24782-24792. DOI: 10.1021/acsami.6b09211

[11] Schramm S, Weiss D. Fluorescent heterocycles: Recent trends and new developments. Advances in Heterocyclic Chemistry. 2019;**128**:103-179. DOI: 10.1016/bs.aihch.2018.10.003

[12] Ke X, Meng L, Wan X, Sun Y, Guo Z, Wu S, et al. An oxygen heterocycle-fused fluorene based non-fullerene acceptor for high efficiency organic solar cells. Materials Chemistry Frontiers. 2020;**4**:3594-3601. DOI: 10.1039/D0QM00287A

[13] Borah B, Dwivedi KD, Chowhan LR. Recent approaches in the organocatalytic synthesis of pyrroles. RSC Advances. 2021;**11**:13585-13601. DOI: 10.1039/D1RA01690C

[14] Borah B, Dhar Dwivedi K, Chowhan LR. 4-Hydroxycoumarin: A versatile substrate for transition-metal-free multicomponent synthesis of bioactive heterocycles. Asian Journal of Organic Chemistry. 2021;**10**:3101-3126. DOI: 10.1002/ajoc.202100550

[15] Borah B, Dwivedi KD, Chowhan LR. Review on synthesis and medicinal

application of dihydropyrano [3, 2-b] pyrans and spiro-pyrano [3, 2-b] pyrans by employing the reactivity of 5-hydroxy-2-(hydroxymethyl)-4 H-pyran-4-one. Polycyclic Aromatic Compounds. 2021:1-45. DOI: 10.1080/10406638.2021.1962923

[16] Borah B, Dwivedi KD, Chowhan LR. Applications of pyrazolone in multicomponent reactions for the synthesis of dihydropyrano [2, 3-c] pyrazoles and spiro-pyrano [2, 3-c] pyrazoles in aqueous medium. Arkivoc. 2021;**part i**:273-328. DOI: 10.24820/ark.5550190.p011.481

[17] Borah B, Dwivedi KD, Chowhan LR. Recent advances in metal-and organocatalyzed asymmetric functionalization of pyrroles. Asian Journal of Organic Chemistry. 2021;**10**: 2709-2762. DOI: 10.1002/ajoc.202100427

[18] Borah B, Dwivedi KD, Kumar B, Chowhan LR. Recent advances in the microwave-and ultrasound-assisted green synthesis of coumarin-heterocycles. Arabian Journal of Chemistry. 2021;**15**: 103654. DOI: 10.1016/j.arabjc.2021.103654

[19] Chauhan P, Chimni SS. Mechanochemistry assisted asymmetric organocatalysis: A sustainable approach. Beilstein Journal of Organic Chemistry. 2012;**8**:2132-2141. DOI: 10.3762/bjoc.8.240

[20] LeonardiM,VillacampaM,MenéndezJC. Multicomponent mechanochemical synthesis. Chemical Science. 2018;**9**:2042-2064. DOI: 10.1039/C7SC05370C

[21] Claramunt RM, Lopez C, Sanz D, Elguero J. Mechano heterocyclic chemistry: Grinding and ball mills. Advances in Heterocyclic Chemistry. 2014;**112**:117-143. DOI: 10.1016/B978-0-12-800171-4.00003-2

[22] Wang GW. Mechanochemical organic synthesis. Chemical Society Reviews. 2013;**42**:7668-7700. DOI: 10.1039/C3CS35526H

[23] Baig RBN, Varma RS. Alternative energy input: Mechanochemical, microwave and ultrasound-assisted organic synthesis. Chemical Society Reviews. 2012;**41**:1559-1584. DOI: 10.1039/C1CS15204A

[24] Nakamura I, Yamamoto Y. Transition-metal-catalyzed reactions in heterocyclic synthesis. Chemical Reviews. 2004;**104**:2127-2198. DOI: 10.1021/cr020095i

[25] Cheng WM, Shang R. Transition metal-catalyzed organic reactions under visible light: Recent developments and future perspectives. ACS Catalysis. 2020;**10**:9170-9196. DOI: 10.1021/acscatal.0c01979

[26] Lee CF, Liu YC, Badsara SS. Transition-metal-catalyzed C-S bond coupling reaction. Chemistry–An Asian Journal. 2014;**9**:706-722. DOI: 10.1002/asia.201301500

[27] Bertelsen S, Jørgensen KA. Organocatalysis—After the gold rush. Chemical Society Reviews. 2009;**38**:2178-2189. DOI: 10.1039/B903816G

[28] Dalko PI, Moisan L. In the golden age of organocatalysis. Angewandte Chemie International Edition. 2004;**43**:5138-5175. DOI: 10.1002/anie.200400650

[29] Renzi P, Bella M. Non-asymmetric organocatalysis. Chemical Communications. 2012;**48**:6881-6896. DOI: 10.1039/C2CC31599H

[30] Seayad J, List B. Asymmetric organocatalysis. Organic & Biomolecular Chemistry. 2005;**3**:719-724. DOI: 10.1039/B415217B

[31] Chatterjee T, Ranu BC. Synthesis of organosulfur and related heterocycles

under mechanochemical conditions. The Journal of Organic Chemistry. 2021;**86**:13895-13910. DOI: 10.1021/acs.joc.1c01454

[32] Achar TK, Bose A, Mal P. Mechanochemical synthesis of small organic molecules. Beilstein Journal of Organic Chemistry. 2017;**13**:1907-1931. DOI: 10.3762/bjoc.13.186

[33] El-Sayed TH, Aboelnaga A, El-Atawy MA, Hagar M. Ball milling promoted N-heterocycles synthesis. Molecules. 2018;**23**:1348. DOI: 10.3390/molecules23061348

[34] Virieux D, Delogu F, Porcheddu A, García F, Colacino E. Mechanochemical rearrangements. The Journal of Organic Chemistry. 2021;**86**:13885-13894. DOI: 10.1021/acs.joc.1c01323

[35] Ould M'hamed M. Ball milling for heterocyclic compounds synthesis in green chemistry: A review. Synthetic Communications. 2015;**45**:2511-2528. DOI: 10.1080/00397911.2015.1058396

[36] Akelis L, Rousseau J, Juskenas R, Dodonova J, Rousseau C, Menuel S, et al. Greener paal–knorr pyrrole synthesis by mechanical activation. European Journal of Organic Chemistry. 2016;**1**:31-35. DOI: 10.1002/ejoc.201501223

[37] Schumacher C, Molitor C, Smid S, Truong KN, Rissanen K, Bolm C. Mechanochemical syntheses of N-containing heterocycles with TosMIC. The Journal of Organic Chemistry. 2021;**86**:14213-14222. DOI: 10.1021/acs.joc.1c01529

[38] Chuang CP, Chen KP. N-Phenacylpyridinium bromides in the one-pot synthesis of 2, 3-dihydrofurans. Tetrahedron. 2012;**68**:1401-1406. DOI: 10.1016/j.tet.2011.12.035

[39] Shearouse WC, Shumba MZ, Mack J. A solvent-free, one-step, one-pot Gewald reaction for alkyl-aryl ketones via mechanochemistry. Applied Sciences. 2014;**4**:171-179. DOI: 10.3390/app4020171

[40] Saeed A, Channar PA. A Green Mechanochemical Synthesis of New 3, 5-Dimethyl-4-(arylsulfanyl) pyrazoles. Journal of Heterocyclic Chemistry. 2017;**54**:780-783. DOI: 10.1002/jhet.2528

[41] Edrees MM, Melha SA, Saad AM, Kheder NA, Gomha SM, Muhammad ZA. Eco-friendly synthesis, characterization and biological evaluation of some novel pyrazolines containing thiazole moiety as potential anticancer and antimicrobial agents. Molecules. 2018;**23**:2970. DOI: 10.3390/molecules23112970

[42] Gandhi D, Agarwal S. Urea nitrate catalyzed synthesis of 2-arylbenzothiazoles using the grindstone technique. Heterocyclic Communications. 2018;**24**:307-310. DOI: 10.1515/hc-2018-0133

[43] Kuarm BS, Madhav JV, Rajitha B, Reddy YT, Reddy PN, Crooks PA. Cellulose sulfuric acid: Novel and efficient biodegradable and recyclable acid catalyst for the solid-state synthesis of thiadiazolo benzimidazoles. Synthetic Communications. 2011;**41**:662-669. DOI: 10.1080/00397911003632899

[44] Kategaonkar AH. Synthesis of 3-(5-aryl-[1, 3, 4] oxadiazol-2yl)-1H-indazole derivatives using cellulose sulphuric acid (CSA) as a catalyst. Journal of Biological and Chemical Chronicles. 2019;5:19-22

[45] Abdel-Aziem A, El-Sawy ER, Kirsch G. Eco-friendly synthesis of 3-Aryl-2, 3-dihydro-1, 3, 4-thiadiazoles based on benzofuran and chromone moieties. Polycyclic Aromatic Compounds. 2020:1-5. DOI: 10.1080/10406638.2020.1848890

[46] Rajeshwari M, Sammaiah B, Sumalatha D, Sarada LN. Chloro

sulphonic acid: A simple and efficient catalyst for one-pot synthesis of hantzsch 1, 4-dihydro pyridines. Indian Journal of Advances in Chemical Sciences. 2013;**1**:236-239

[47] Ghafuri H, Zand HR. Imidazole dicarboxylic acid as a new catalyst for the synthesis of 1, 4-dihydropyridines in ball-mill. In: Proceedings of the 21st International Electronic Conference on Synthetic Organic Chemistry. Basel, Switzerland: MDPI; 2017. DOI: 10.3390/ecsoc-21-04717

[48] Javanshir S, Sharifi S, Maleki A, Sohrabi B, Kiasadegh M. p-toluenesulfonic acid-catalyzed synthesis of poly-substituted quinolines via Friedländer reaction under ball-milling conditions at room temperature and theoretical study on the mechanism using a density functional theory method. Journal of Physical Organic Chemistry. 2014;**27**:589-596. DOI: 10.1002/poc.3305

[49] Dwivedi KD, Borah B, Chowhan LR. Ligand free one-pot synthesis of pyrano [2, 3-c] pyrazoles in water extract of banana peel (WEB): A green chemistry approach. Frontiers in Chemistry. 2020;7:944. DOI: 10.3389/fchem.2019.00944

[50] Amirnejad M, Naimi-Jamal MR, Tourani H, Ghafuri H. A facile solvent-free one-pot three-component method for the synthesis of 2-amino-4H-pyrans and tetrahydro-4H-chromenes at ambient temperature. Monatshefte für Chemie-Chemical Monthly. 2013;**144**:1219-1225. DOI: 10.1007/s00706-013-0938-2

[51] Dekamin MG, Eslami M. Highly efficient organocatalytic synthesis of diverse and densely functionalized 2-amino-3-cyano-4 H-pyrans under mechanochemical ball milling. Green Chemistry. 2014;**16**:4914-4921. DOI: 10.1039/C4GC00411F

[52] Kuarm BS, Crooks PA, Rajitha B. An expeditious synthesis of quinoxalines by using biodegradable cellulose sulfuric acid as a solid acid catalyst. Green Chemistry Letters and Reviews. 2013;**6**:228-232. DOI: 10.1080/17518253.2012.752041

[53] Khaskel A, Gogoi P, Barman P, Bandyopadhyay B. Grindstone chemistry: A highly efficient and green method for synthesis of 3, 4-dihydropyrimidin-2-(1 H)-ones by l-tyrosine as an organocatalyst: A combined experimental and DFT study. RSC Advances. 2014;**4**:35559-35567. DOI: 10.1039/C4RA05244G

[54] Kawade DS, Chaudhari MA, Gujar JB, Shingare MS. Thiamine hydrochloride (vitamin B1) as an efficient catalyst for the synthesis of 4-(3H)-Quinazolinone derivatives using grinding method. Iranian Journal of Catalysis. 2016;**6**:313-318

[55] Yashwantrao G, Jejurkar VP, Kshatriya R, Saha S. Solvent-free, mechanochemically scalable synthesis of 2, 3-dihydroquinazolin-4 (1H)-one using Brønsted acid catalyst. ACS Sustainable Chemistry & Engineering. 2019;7:13551-13558. DOI: 10.1021/acssuschemeng.9b03199

[56] Jia XC, Li J, Ding Y, Zhang B, Wang N, Wang YH. A simple and green protocol for 2H-indazolo [2, 1-b] phthalazine-triones using grinding method. Journal of Chemistry. 2013;**2013**:1-5. DOI: 10.1155/2013/634510

[57] Shukla G, Verma GK, Nagaraju A, Verma RK, Raghuvanshi K, Singh MS. DMAP mediated one-pot domino thienannulation: A versatile, regioselective and green mechanochemical route to naphtho [2, 3-b] thiophenes. RSC Advances. 2013;**3**:13811-13817. DOI: 10.1039/C3RA41100A

[58] Khaligh NG, Mihankhah T, Johan MR. Catalytic application of

1, 4-piperazinediethanesulfonic acid (PIPES) for the one-pot multicomponent synthesis of pyrano [4, 3-b] pyrans. Organic Preparations and Procedures International. 2020;**52**:368-373. DOI: 10.1080/00304948.2020.1761228

[59] Ambethkar S, Padmini V, Bhuvanesh N. A green and efficient protocol for the synthesis of dihydropyrano [2, 3-c] pyrazole derivatives via a one-pot, four component reaction by grinding method. Journal of Advanced Research. 2015;**6**:975-985. DOI: 10.1016/j.jare.2014.11.011

[60] Khaligh NG, Mihankhah T. Green and solid-phase synthesis of new dihydro-[1, 2, 4] triazolo [1, 5-a] pyrimidine scaffolds by using poly-melamine-formaldehyde as a nitrogen-rich porous organocatalyst. Polycyclic Aromatic Compounds. 2020:1-9. DOI: 10.1080/10406638.2020.1756357

[61] Reddy MS, Chowhan LR, Kumar NS, Ramesh P, Mukkamala SB. An expedient regio and diastereoselective synthesis of novel spiropyrrolidinylindenoquinoxalines via 1, 3-dipolar cycloaddition reaction. Tetrahedron Letters. 2018;**59**:1366-1371. DOI: 10.1016/j.tetlet.2018.02.044

[62] Singh GS, Desta ZY. Isatins as privileged molecules in design and synthesis of spiro-fused cyclic frameworks. Chemical Reviews. 2012;**112**:6104-6155. DOI: 10.1021/cr300135y

[63] Ghahremanzadeh R, Ahadi S, Shakibaei GI, Bazgir A. Grindstone chemistry: One-pot synthesis of spiro [diindenopyridine-indoline] triones and spiro [acenaphthylene-diindenopyridine] triones. Tetrahedron Letters. 2010;**51**:499-502. DOI: 10.1016/j.tetlet.2009.11.041

Chapter 5

Eco-Sustainable Catalytic System for Green Oxidation of Spirostanic Alcohols Using Hypervalent Iodine (III) Tempo-4-n-Acetoxyamine System

Joseph Cruel Sigüenza, Carla Bernal Villavicencio, María Elizabeth Canchingre, Christie Durán García and Juan E. Tacoronte

Abstract

The oxidation of the 3β-hydroxy group in the steroidal substrates obtained from naturally occurring sources, i.e., solanaceae steroidal sapogenins, is an important process in the preparation of ecdysteroid analogs. The need for selective green oxidation methodologies for steroidal alcohols (spirostenols, diosgenine, and derivatives) avoid the use of toxic Cr (VI) derivatives, without the isomerization of the double bond at 5,6 position and also without the oxidative cleavage of the spirocetal moiety is of great methodological significance. Herein, we report the oxidation of spirostanic steroidal alcohols to their carbonyl analogs using hypervalent iodine (III)/TEMPO-4-N-acetoxyamine system. The present method is simple, eco-sustainable, efficient, and high-yielding process for the oxidative transformation of secondary steroidal alcohols without any over-oxidation, isomerization of the double bond, or oxidative cleavage of spirocetalic fragment in different substrates. Therefore, this method does not involve toxic heavy metals and is expected to have wide utility in the oxidation process of these compounds.

Keywords: green chemistry, green oxidation, spirostanic substrates, TEMPO, hypervalent Iodine, ecdysteroid analogs

1. Introduction

Green or *Eco-sustainable Chemistry* is a unique, philosophically, methodologically, pedagogically, and experimentally new discipline of chemistry where the convergency of conceptual series: Structure-related Properties-Functionality-Scalability and Environmental Responsibility demonstrates the versatility and applicability of these concepts in a dynamical evolution and design of new experimental tools, techniques,

IntechOpen

technologies, and in a more advanced structural way of thinking [1]. It is helpful to chemistry and chemical engineers in the field of research, at laboratory scale, and industrial application in real-time, mainly focusing on the development and production of more eco-friendly and efficient processes and products, including scientific-technical and engineering services, which may also have great financial and strategic benefits.

Green chemistry is, generally, based on the 12 principles, which constitute the conceptual and methodological foundation of sustainable development from the point of view of environmentally benign applied chemistry, which were proposed by Anastas and Warner [2] and developed intensively by Sheldon [3, 4] and Lancaster [5]. The principles comprise theoretical and experimental instructions to implement the *green chemical eidos* related to new chemical products, new unconventional synthesis, new sources of raw materials and energy, and new processes that have been extensively described in classical today's references [6–13].

Within this frame of reference, it should be noted that 12 basic principles of green or sustainable chemical engineering have also been described [14].

Steroids constitute one of the oldest known molecular systems and have been discovered in all living systems, forming structural, compositional, and functional parts of biomembranes, defensive secretions and pheromones, and complex neurochemical and hormonal information systems [15], defining, in many points, the molecular evolutionary capacity and metabolic efficiency of living organisms [16]. In this context, several naturally occurring and synthetic steroids are important therapeutic options for a wide range of pathologies. Among them can be considered sex and corticosteroid hormones, bile acids, vitamin D derivatives, and cardiotonic steroids, which have shown unique therapeutic value for the treatment of human diseases.

Oxidation of organic substrates is one of the most important processes on an industrial and laboratory scale [17], where the application of catalytic conditions (heterogeneous or homogeneous) defines the selectivity and yield of the process [18]. One of the most important reaction in organic synthesis and industrial organic chemistry is the oxidation of alcohols (I and II) to their carbonyl derivatives (aldehydes and ketones). This transformation of functionalities, from alcohol to carbonyls, via heterogeneous or homogeneous catalysis or not, using conventional procedures or not, constitutes one of the fundamental synthetic organic reactions of applied green eco-sustainable chemistry in the fundamental fields of technological development for minimizing environmental impact, in such as pharma, agrochemistry, petrochemistry, and fine chemistry [19].

Naturally occurring steroidal derivatives, such as spirostannic alcohols, are characterized by a hydroxyl group in position 3β- in the A-ring of the cyclopentane-perhydro phenanthrene nucleus, being typical secondary alcohol, and, taking into account this consideration, most common oxidative transformation in steroid chemistry is probably the oxidation of alcohol functionality into carbonyl group [20].

A large number of oxidizing agents have been described to transform the 3β-hydroxy group into a ketone group, minimizing collateral reactions depending on the specific synthetic objectives and reaction conditions; among these, the following can be highlighted: $Cr_2O_7^{2-}$ (dichromate), CrO_3/pyridine (Collins's reagent), Pyridinium chlorochromate (PCC), Pyridinium dichromate (PDC, Cornforth reagent), Dess–Martin periodinane, Dimethylsulfoxide (DMSO)/oxalyl chloride (Swern oxidation), CrO_3/H_2SO_4/acetone (Jones oxidation), Aluminum isopropoxide/acetone (Oppenauer oxidation) [21, 22]. It should be noted that other catalytic oxidative procedures have been developed, which employ molecular oxygen, biphasic catalysis, osmium and ruthenium salts, permanganate salts impregnated on zeolitic or alumosilicate supports, organic hydroperoxides, etc. [23, 24].

In recent years an interesting increase in the application of hypervalent iodine compounds in synthetic organic chemistry has been observed. Several compounds have been recognized as useful reagents with considerable synthetic applicability [25]. Hypervalent iodine derivatives are now commonly used in organic synthesis as efficient multipurpose reagents with very low toxicity and minimal environmental problems after their utilization. The discovery of eco-sustainable catalysis via hypervalent iodine compounds and its application in micro- and meso-scale synthetic chemistry is an important achievement that has also allowed the use of green chemistry concepts [26]. In the last decade Iodo (V), Iodo (VI) reagents have been used for efficient conversion of alcohols to carbonyl compounds [27].

The oxidation of 3-cholesteryl benzoate to the corresponding 4-ketone derivative in 67% yield has been described [28], and authors reported the utilization of Iodoxybenzene ($PhIO_2$) and a catalytic amount of 2,2′-dipyridyldiselenide as a very mild and efficient catalytic system.

In this setting, and considering the main objective of the communication, should be highlighted that the reagent $PhI(OAc)_2$ in the presence of alcohol/alkali mixtures -KOH-MeOH is equivalent to PhIO-KOH-MeOH; however, $PhI(OAc)_2$ (BAIB) is commercially available whilst PhIO is not. Consequently, the reagent (Diacetoxyiodo)benzene, [$PhI(OAc)_2$], is most widely used in synthetic processes, including in green chemistry protocols. This reagent has been employed in the synthesis of the steroidal dihydroxyacetone side chain with satisfactory yields [29, 30].

Recently N-oxoammonium salts, more specifically, 2,2,6,6-tetramethylpiperidinyloxy radical (TEMPO) derivatives have been described and used as a useful reagent for the oxidation of alcohols [31]. These have been used stoichiometrically either in isolated form or generated *in situ* via acid catalyzed dismutation [32]. In order to increase the efficiency and yield of the oxidation process, different reagents and techniques have been considered for use in conjunction with oxoammonium salts such as m-CPBA [33], N-chloro succinimide, electrooxidations [34–36]; impregnation on solid support [37].

However, in spite of these relatively new mild selective oxidants, the oxidation of a hydroxy functionality in the steroidal core is a major challenge in organic synthesis particularly by employing eco-friendly oxidizing catalysts [38, 39].

Considering the aforementioned explanations, the main objective of the work is to evaluate the possibility of using the Hypervalent Iodine (III) Tempo-4-n-Acetoxyamine system in the oxidation processes of steroidal alcohols (spirostanols) in eco-sustainable green chemistry conditions.

2. Results and discussion

Brassinosteroid and ecdysteroid spirostanic oxo-analogs bearing different side-chain have shown remarkable plant growth and metamorphosis promoting activities [40]. The synthesis and study of the biological activity of spirostanic compounds with characteristic functionalities of some natural brassino- and ecdy-steroids at different rings have been reported, including some oxidative protocols for selectively oxidation of 3β-hydroxy group in the A-ring into a ketone group [41, 42].

The catalytic oxidation of 3β, 6α, 6β; 12β-hydroxy groups in steroidal substrates obtained from naturally occurring sources, (i.e., solanaceae steroidal sapogenins and phytosterol mixture isolated from sugarcane), is an important process in the preparation of ecdysteroid analogs [43, 44], blatellanosides [45],

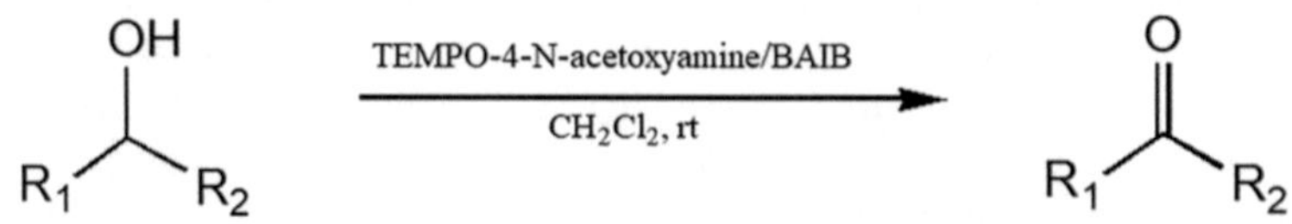

Figure 1.
Oxidation of spirostanic steroidal alcohols using hypervalent iodine (III)/ Tempo-4-N-Acetoxyamine system.

Entry	Substrate	Product	Time	Yield (%)
1	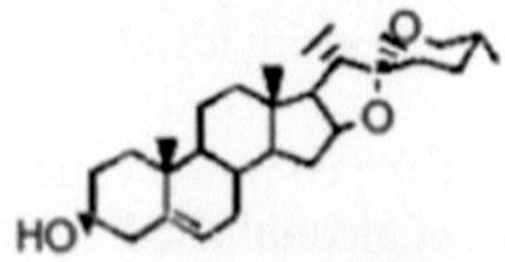	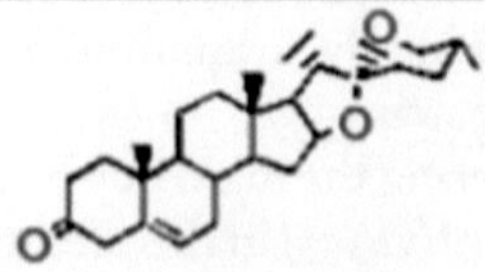	5	81
2	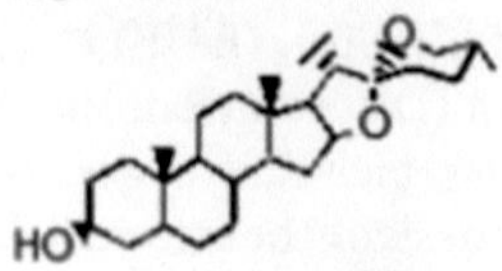	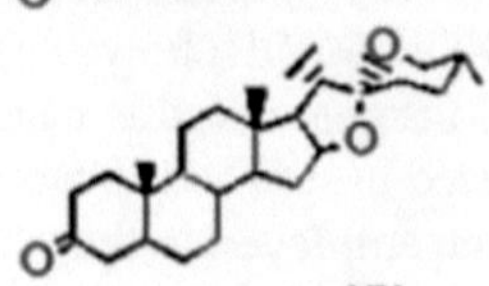	6	82
3	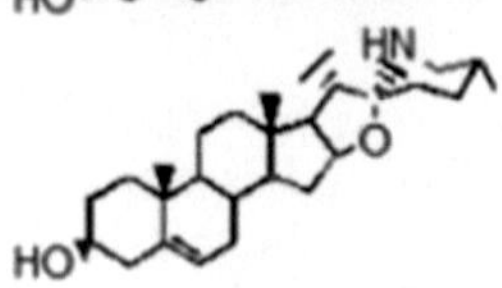	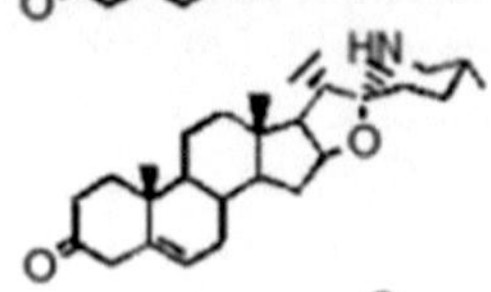	5	76
4	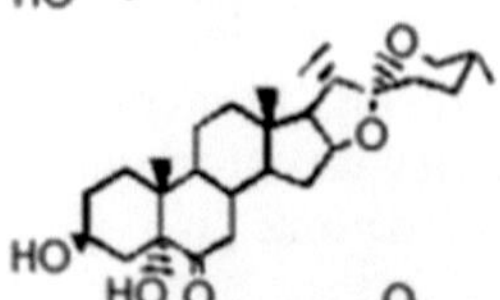	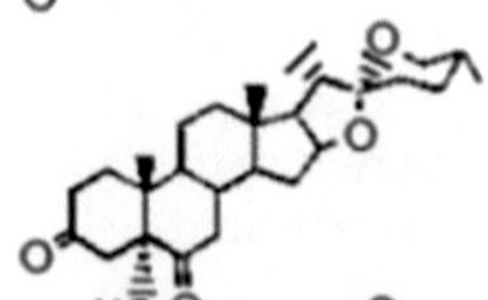	5	87
5	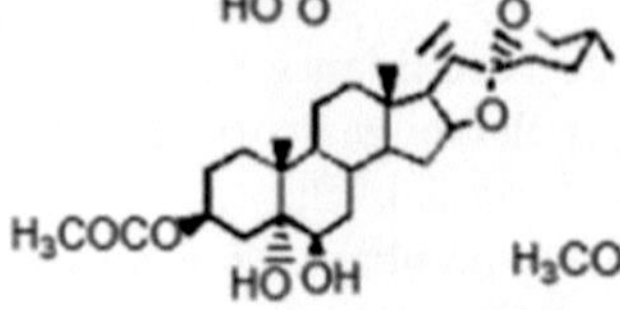	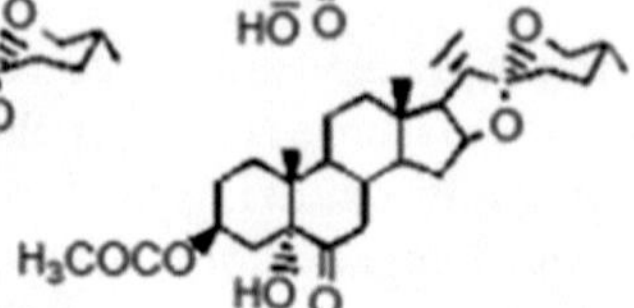	5	83
6	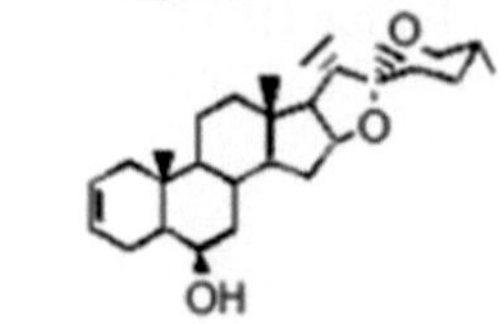	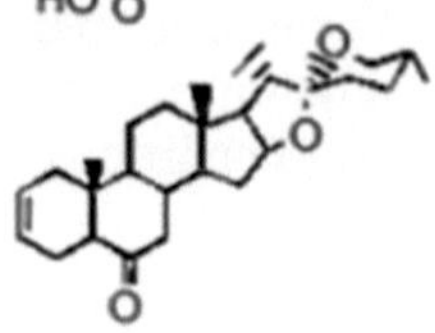	6	85
7	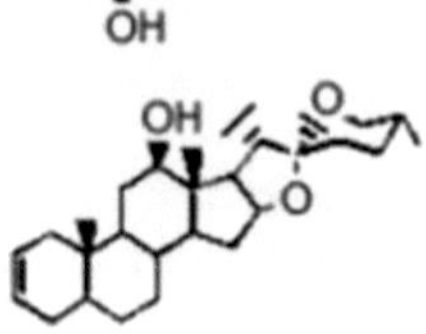	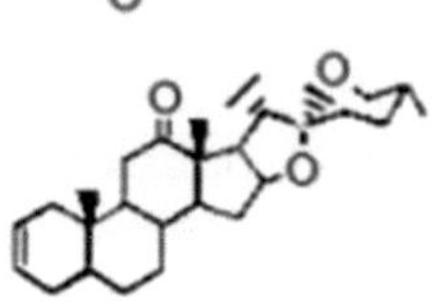	6	74

Figure 2.
Oxidation of spirostanic alcohols. The reported yields refer to isolated, chromatographically pure, or recrystallized carbonyl derivatives. All proposed structures have been confirmed by FTIR and $^1H/^{13}C$NMR analysis.

cytotoxins [46], pharmaceutical molecular templates and potential plant growth promoters [47]. The need for selective oxidation methodologies for steroidal alcohols (spirostenols, diosgenin, solasodine, hecogenin, and some polyhydroxy derivatives) without the oxidation-isomerization of the double bond at $\Delta^{5,6}$ position and also without the oxidative cleavage of the spirocetal moiety has remained unexplored.

Herein, we wish to report the oxidation of spirostanic steroidal alcohols to their carbonyl analogs using hypervalent iodine (III)/TEMPO-4-N-acetoxyamine system (**Figure 1**).

This oxidation reaction has been investigated for a large array of steroids derivatives generated from diosgenin using standard procedures and having 3β-hydroxy functionality. It is worthwhile to note, that other functionalities such as double bonds, acetoxy, and tertiary α-hydroxy groups, and spiro segments in the steroidal structure do not get affected by employing this method; the results are described in **Figure 2**. It is remarkable to note that solasodine (3, **Figure 2**) after oxidation using that catalytic system yields carbonyl derivative only without any collateral reaction in the NH functionality. The probable mechanism for this oxidation process is similar to the pathway earlier described in the literature and depicted in **Figure 3** [34, 48–51]. It appears that the role of the hypervalent iodine compound (BAIB) is to regenerate the TEMPO-4-N-acetoxyamine in the catalytic cycle. Furthermore, it is observed that in the absence of this TEMPO derivative the oxidation process does not take place. This methodology has also been extended to some non-steroidal substrates like 2-methyl-2-ene pentanol and n-heptanol for obtaining their corresponding carbonyl analogs, thus exhibiting the generality of this procedure.

$$PhI(OAc)_2 + nR\text{-}OH \longrightarrow PhI(OAc)_{2-n}(OR)_n + n\,AcOH$$

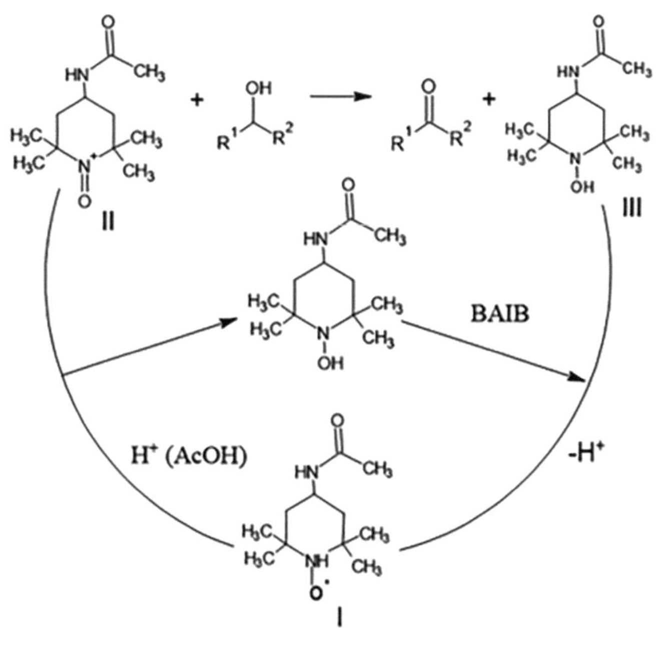

Figure 3.
Proposed mechanism for the oxidation of alcohols by BAIB/TEMPO-4-N-acetoxyamine.

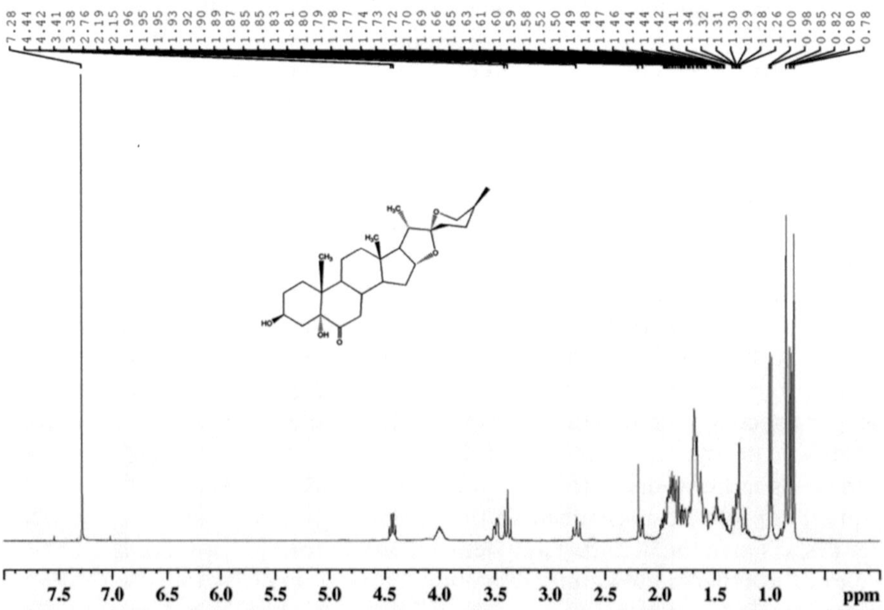

(A)

Figure 4.
1H-NMR spectrum of starting substrate ((25R)-spirost-3β,5α-dihydroxy-6-one). The typical signal of nonoxidized OH-is observed at C3-H (3.93 ppm, 1H, H-3).

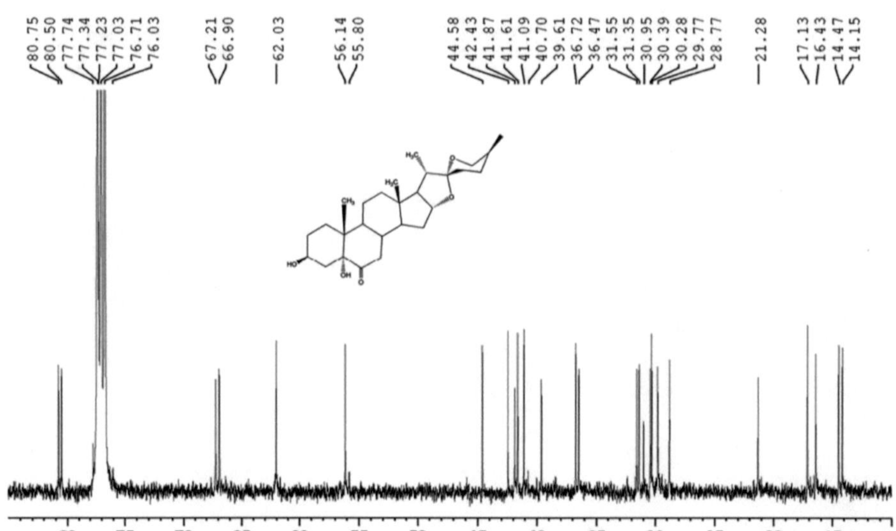

Figure 5.
^{13}C-NMR spectrum of starting substrate ((25R)-spirost-3β,5α-dihydroxy-6-one). The typical signal of non-oxidized C3-OH-is observed at C3 in 67.2 ppm.

In the case of model reactions, obtaining (25R)-Spirost-5α-hydroxy-3,6-dione (4, **Figure 2**) to corroborate the conversion of the hydroxy functional group in position 3 to its carbonyl derivative (ketone), the chemical shift values (ppm)

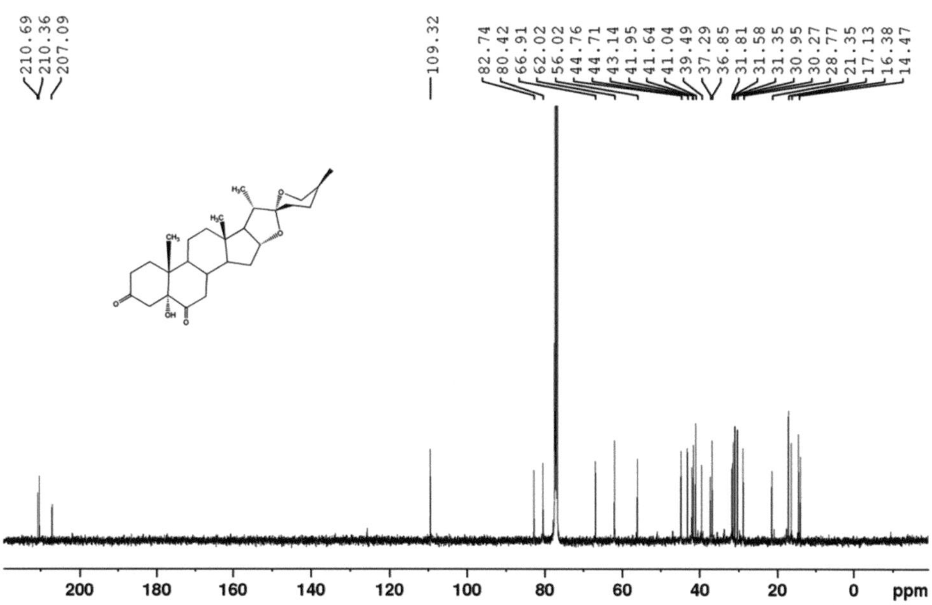

Figure 6.
*^{13}C-NMR spectrum of obtained keto-spirosteroid (25R)-spirost-5α-hydroxy-3,6-dione (4, **Figure** 2). The typical signal of oxidized C3-OH-is observed at C3 in 210.36 ppm.*

were used. In the starting substrate ((25R)-Spirost-3β,5α-dihydroxy-6-one) a typical signal (^{1}H-NMR) is observed at C3-H (3.93 ppm, 1H, H-3); and the ^{13}C signal at carbon 3 (67.2 ppm) are related to the presence of a hydroxyl substituent at this position C-3.

During the oxidation process using the hypervalent iodine (III)/TEMPO-4-N-acetoxyamine system yields the keto derivative at C-3.

In the ^{1}H nmr spectrum, a variation in the magnitude of the chemical shift values (ppm) is observed. In C-3 the signal (3.93 ppm, 1H, H-3, 67.2 ppm, C-OH) disappears and a new signal in ^{13}C nmr spectrum at C-3 (210.36 ppm) is observed. This variation of the magnitude of the signal towards lower fields (from 67.2 ppm to 210.2 ppm) corroborates the oxidation of the OH- in C-3 to a carbonyl derivative (ketone). The spectra are represented in **Figures 4–6**.

It should be noted that the use of this catalytic system, hypervalent iodine (III)/TEMPO-4-N-acetoxyamine, does not generate any dehydration product, as observed when conventional oxidants are used, and, also, it is not detected deformation of minor derivative 4-en-3,6-dione.

3. Experimental section

All chemicals and solvents were obtained from commercial sources and used without further purification. ^{1}H-NMR/^{13}C-NMR (Brucker AC-250 instrument, Germany, at 25°C, using tetramethyl silane as internal reference; magnitude in ppm); FTIR (Philips Analytical FTIR PU-9600, KBr) was used to identify the compounds. Melting points were determined on VEEGO digital automatic melting points apparatus and are uncorrected.

3.1 General procedure for the oxidation of spirostanic steroidal alcohols to their carbonyl analogs

In a typical procedure, a solution of steroidal alcohol (1 mmol) in dichloromethane (3 mL), BAIB (1.1 mmol) and TEMPO-4-N-acetoxyamine (0.1 mmol) is added and stirred for 4 h at room temperature. On completion of the reaction as indicated by Thin Layer Chromatography (ethyl acetate-hexane, 3:7, v/v; vanillin in perchloric acid 50%), the reaction mixture is quenched with 10% sodium thiosulphate solution (2x10 mL) and is extracted with CH_2Cl_2. The organic layer is washed with sodium bicarbonate and brine solution, dried over anhydrous sodium sulfate, and concentrated under a vacuum to give the crude product. This is purified by column chromatography (SiO_2, $CHCl_3$-EtOAc, 2:8 v/v) to afford the pure product.

3.2 Model reaction II: (25R)-Spirost-5α-hydroxy-3,6-dione

(87%), mp. 283°C (methanol). FTIR (KBr): ν max 3345 (ν OH), 2953 (ν CH), 1713 (ν C=O), 1456 and 1376 (δ CH), 1057 (ν C-O), 980 and 899 (spiroketal); ^{1}H-NMR (δ, ppm): 2.10/1.86 (m, 2H, H-1), 2.40 (m, 2H, H-2), 2.78/2.25 (m, 2H, H-4), 2.96/2.34 (m, 2H, H-7), 1.99 (m, 1H, H-8), 1.90 (m, 1H, H-9), 1.50 (m, 2H, H-11), 1.81/1.31 (m, 2H, H-12), 1.46 (m, 1H, H-14), 1.99/1.32 (m, 2H, H-15), 4.47 (m, 1H, H-16), 1.85 (m, 1H, H-17), 0.83 (s, 3H, CH3–18), 1.04 (s, 3H, CH3–19), 1.90 (m, 1H, H-20), 1.01 (d, 3H, CH3–21), 1.69 (m, 2H, H-23), 1.50 (m, 2H, H-24), 1.66 (m, 1H, H-25), 3.5/3.38 (m, 2H, H-26), 0.82 (d, 3H, CH3–27). ^{13}C-NMR (δ, ppm): 31.8 (C-1), 37.3 (C-2), 210.2 (C-3), 41.9 (C-4), 82.7 (C-5), 210.4 (C-6), 44.8 (C-7), 36.8 (C-8), 44.7 (C-9), 43.1 (C-10), 21.3 (C-11), 39.5 (C-12), 41.0 (C-13), 56.0 (C-14), 31.6 (C-15), 80.4 (C-16), 62.0 (C-17), 16.4 (C-18), 13.9 (C-19), 41.6 (C-20), 14.5 (C-21), 109.3 (C-22), 31.3 (C-23), 28.8 (C-24), 30.3 (C-25), 66.9 (C-26), 17.1 (C-27) (4, **Figure 2**).

4. Conclusions

It is necessary to highlight some important aspects:

- Simple procedure does not require special laboratory conditions such as extra-dry solvents, inert atmosphere, or additional treatment of the generated derivatives.

- The catalytic system is easily recovered from the reaction mixture by simple filtration.

- The yield of carbonyl derivatives exceeds 75%.

- The described process can be applied in undergraduate and graduate courses of implementation of catalytic processes in organic synthesis.

- This catalytic oxidation system, hypervalent iodine (III)/TEMPO-4-N-acetoxyamine, is applied for the first time to spirostane derivatives obtained from diosgenin for the generation of carbonyl compounds widely used as plant growth promoters and metamorphosis control in insects.

In conclusion, the present method, using hypervalent iodine (III)/Tempo-4-N-Acetoxyamine system, is a simple, efficient, convenient, and high yielding (>75%) process for the oxidative transformation of 3β-secondary steroidal alcohols without any over oxidation, isomerization of the double bond or oxidative cleavage of spirocetalic fragment in different substrates. Therefore, this method does not involve toxic heavy metals and is expected to have wide utility in the oxidation process of these compounds in eco-sustainable conditions.

Research programs related to the optimization of these catalytic oxidation processes under heterogeneous conditions using condensed pyroclastic flow, molecular oxygen, and non-conventional energy sources (microwave, infrared, and ultrasound technologies) are currently being developed.

Acknowledgements

The authors greatly acknowledge the technical, logistic, and administrative support from the Technical University of Esmeraldas "Luis Vargas Torres", Republic of Ecuador, during 2020-2021. One of the authors (JET) thanks Prof. J. Bobbitt (University of Connecticut, School of Liberal Arts, USA) for his generosity in donating oxoammonium salts and literature, and Indian Institute of Chemical Technology (IICT-Hyderabad, Dr. Yadav S. and Dr. Ahmed Kamal) for its kindness in welcoming him as a postdoctoral fellow.

Conflict of interest

The authors declare no conflict of interest.

Author details

Joseph Cruel Sigüenza[1], Carla Bernal Villavicencio[1], María Elizabeth Canchingre[1], Christie Durán García[2] and Juan E. Tacoronte[1*]

1 Faculty of Technology, Chemical Engineering Division, Campus Bellos Horizontes, Technical University of Esmeraldas "Luis Vargas Torres" Ecuador

2 Universidad Nacional Santiago Antúnez de Mayolo, Perú

*Address all correspondence to: juan.tacaronte.morales@utelvt.edu.ec

References

[1] Ivankovic A, Talic S. Review of 12 principles of green chemistry in practice. Int. J Suit. & Greenener. 2017;**6**(3):39-48. DOI: DOI.org/10.11648/j.ijrse.20170603.12

[2] Anastas P, Warner J. Green Chemistry, Theory and Practice. Oxford: UK: Oxford University Press; 1998

[3] Sheldon R. Fundamentals of green chemistry: Efficiency in reaction design. Chemical Society Reviews. 2012;**41**:1437-1451. DOI: 10.1039/C1CS15219J

[4] Sheldon R. Green chemistry and resource efficiency: Towards a green economy. Green Chemistry. 2016;**18**(11):3180-3183. DOI: 10.1039/c6gc90040b

[5] Mike L. Green Chemistry, an Introductory Text. 3rd ed. London UK: RSC; 2016. pp. 1-373

[6] Asif M. Green Synthesis, Green Chemistry, and Environmental Sustainability: An Overview on Recent and Future Perspectives of Green Chemistry in Pharmaceuticals. Green Chemistry & Technology Letters. 2021;**7**(1):18-27. DOI: 10.18510/gctl.2021.713

[7] Cerminara I, Chiummiento L, Funicello M, Lupattelli P, Scafato P, Scorza F, et al. Green chemistry, circular economy and sustainable development: An operational perspective to scale research results in SMEs practices. Computational Science and Its Applications–ICCSA. 2020;**12255**:206-213. DOI: 10.1007/978-3-030-58820-5_16

[8] Sharma S, Mudhoo A, Zhang W. Chemistry and Engineering. In: Green Chemistry for Environmental Sustainability. USA: CRC Press; 2011. DOI: 10.1201/EBK1439824733

[9] Patel M, Patel H, Mevada S, Patel O. Chemistry goes green: A review on current and future perspectives of pharmaceutical green chemistry. World J. Pharm. & Med. Res. 2020, 2020;**6**(7):125-131

[10] O'Brien K, Myers J, Warner J. Green chemistry: Terminology and principles. Environ Health Perspect. 2009;**117**(10):A434. DOI: 10.1289/ehp.0900835

[11] Kharissova Q, Kharisov B, González C, Méndez Y, López I. Greener synthesis of chemical compounds and materials. Royal Society Open Science. 2019;**6**(11):191378. DOI: 10.1098/rsos.191378

[12] Kim S, Hong S, Ahn K, Gong S. Priority survey between indicators and analytic hierarchy process analysis for green chemistry technology assessment. Environ. Health Toxicol. 2015;**30**(Suppl):s2015003. DOI: 10.5620/eht.s2015003

[13] Varma R, Iravani S. Greener synthesis of lignin nanoparticles and their applications. Green Chemistry. 2020;**3**:1-69. DOI: 10.1039/c9gc02835h

[14] Anastas P, Zimmerman J. Design Through the 12 Principles of Green Engineering. Environmental Science & Technology. 2003;**37**(5):94A-101A. DOI: 10.1021/es032373g

[15] Rodewald A, Mills D, Gebhart M, Jirikowski G. Steroidal pheromones and their potential target sites in the vomeronasal organ. Steroids. 2019;**2**(142):14-20. DOI: 10.1016/j.steroids.2017.09.010

[16] Bridgham J, Carroll S, Thornton J. Evolutionofhormone-receptorcomplexity by molecular exploitation. Science. 2006;**312**(5770):97-101. DOI: 10.1126/science.1123348

[17] Ning J, Stahl S, editors. Green oxidation in organic synthesis. John Wiley & Sons; 2019. p. 543. DOI: 10.1002/9781119304197

[18] Javaid R, Qazi U. Catalytic oxidation process for the degradation of synthetic dyes: an overview. Int J Environ Res Public Health. 2019;**16**(11):2066-2093. DOI: 10.3390/ijerph16112066

[19] Backvall J-E, editor. Modern Oxidation Methods. 2nd ed. Germany: Wiley-VCH Verlag & Co. KGaA; 2010 p. 483

[20] Salvador J, Silvestre S, Moreira V. Catalytic Oxidative Processes in Steroid Chemistry: Allylic Oxidation, β-Selective Epoxidation, Alcohol Oxidation and Remote Functionalization Reactions. Current Organic Chemistry. 2006;**10**(17):2227-2257. DOI: 10.2174/138527206778742641

[21] Tojo G, Fernandez M, editors. Oxidation of Alcohols to Aldehydes and Ketones, A Guide to Current Common Practice. Springer Science Business Media Inc.; 2006. p. 376. DOI: 10.1007/b135954

[22] Korde S, Udasi R, Trivedi G. Oxidation of Steroidal 5-En-3β-ol with Pyridinium Chlorochromate: Isolation of Key Intermediate, Steroidal 6β-Hydroxy-4-en-3-one. Synthetic Communications. 1997;**27**(19):3419-3430. DOI: 10.1080/00397919708005643

[23] Wendell S, Edward J. A Short Review of Methods for the Allylic Oxidation of Δ^5 Steroidal Compounds to Enones. J Steroids Horm Sci. 2016;**7**(2):171. DOI: 10.4172/2157-7536.1000171

[24] Silvestre S, Silva M, Jorge AR, Salvador J. Green Oxidative Chemical Processes Applied to Steroid Chemistry. Recent Highlights. In: Larramendy M, Soloneski S, editors. Green Nanotechnology - Overview and Further Prospects. IntechOpen; 2016. DOI: 10.5772/62381 Available from: https://www.intechopen.com/chapters/50132

[25] Zhdankin V. Hypervalent Iodine Chemistry. Preparation, Structure and Synthetic Applications of Polyvalent Iodine Compounds. India: John Wiley & Sons, Ltd; 2014. p. 480. DOI: org/10.1002/9781118341155

[26] Zhdankin V. Hypervalent iodine compounds: reagents of the future. ARKIVOC. 2020;**4**:1-11. DOI: 10.24820/ark.5550190.p011.145

[27] Bobbitt J, Endo T, Tacoronte J. Oxoammonium salt oxidations of alcohols in the presence of an on the surface of, an electrode, an ion-exchange resin and silica gel. In: Third International Symposium "Catalysis and Supported Reagents", Univ. Lecmirk, Ireland. Vol. 216. Issue 1. London: London Royal Chemical Society; 1998. pp. 200-206. The Royal Society of Chemistry. Special Publication

[28] Crich D, Zou Y. Catalytic allylic oxidation with a recyclable, fluorous seleninic acid. Organic Letters. 2004;**6**(5):775-777. DOI: 10.1021/ol036501h

[29] Moriarty R, Hu H, Gupta S. Direct α-hydroxylation of ketones using iodosobenzene. Tetrahedron Letters. 1981;**22**(14):1283-1286. DOI: 10.1016/S0040-4039(01)90297-7

[30] Moriarty R, John L, Du P. Hypervalent iodine in organic synthesis. A novel route to the dihydroxyacetone

side-chain in the pregnene series. Chemical Communications. 1981;**13**:641-642. DOI: 10.1039/C39810000641

[31] Hazi A, Beejapur A, Qi ZH, Kecheng Z, Li J, Wang J, et al. TEMPO in Chemical Transformations: From Homogeneous to Heterogeneous. ACS Catalysis. 2019;**9**(4):2777-2830. DOI: 10.1021/acscatal.8b05001

[32] Bobbitt J. Oxoammonium Salts. 6. 1 4-Acetylamino-2,2,6,6-tetramethylpiperidine-1-oxoammonium Perchlorate: A Stable and Convenient Reagent for the Oxidation of Alcohols. Silica Gel Catalysis. J. Org. Chem. 1998;**63**(25):9367-9374. DOI: 10.1021/jo981322c

[33] Ullman E, Osiecki J, Boocok D, Darey R. Stable free radicals. X. Nitronyl nitroxide monoradicals and biradicals as possible small molecule spin labels. J. Am. Chem. Soc. 1972;**94**(20):7049-7059. DOI: 10.1021/ja00775a031

[34] Tsubokawa N, Kimoto T, Endo T. Oxidation of alcohols with copper (II) salts mediated by nitroxyl radicals immobilized on ultrafine silica and ferrite surface. J. Mol. Cat. A. 1995;**101**(1):45-50. DOI: 10.1016/1381-1169(95)00060-7

[35] Semmelhack M, Chou C, Cortes D. Nitroxyl-mediated electrooxidation of alcohols to aldehydes and ketones. J. Am. Chem. Soc. 1983;**105**(13):4492-4494. DOI: 10.1021/ja00351a070

[36] Semmelhack F, Schmid C, Cortes D, Chou S. Oxidation of alcohols to aldehydes with oxygen and cupric ion, mediated by nitrosonium ion. J. Am. Chem. Soc. 1984;**106**(11):3374-3376. DOI: 10.1021/ja00323a064

[37] Zhang S, Miao C, Xia C, Sun W. 4-CH_3CONH-TEMPO/Peracetic Acid System for a Shortened Electron-Transfer-Cycle-Controlled Oxidation of Secondary Alcohols. Chem Cat Chem. 2015, 2015;**7**(12):1865-1870. DOI: 10.1002/cctc.201500214

[38] Steves J, Stahl S. Stable TEMPO and ABNO Catalyst Solutions for User-Friendly (bpy)Cu/Nitroxyl-Catalyzed Aerobic Alcohol Oxidation. Journal of Organic Chemistry. 2015;**80**(21):11184-11188. DOI: 10.1021/acs.joc.5b01950

[39] Kamernitsky AV. Ecdisteroids. Moskva-Khimija. 1989;**4**:380

[40] Suksamrarn A, Charoensuk S, Yongyongnarongkul B. Synthesis and biological activity of 3-deoxyecdysteroid analogues. Tetrahedron. 1996;**52**(32):10673-10684. DOI: 10.1016/0040-4020(96)00589-3

[41] Koyama T, Texada M, Halberg K, et al. Metabolism and growth adaptation to environmental conditions in Drosophila. Cell. Mol. Life Sci. 2020;**77**:4523-4551. DOI: 10.1007/s00018-020-03547-2

[42] Cluter HG, Yokota T, Adam G. (Eds), in "Brassinosteroids. Chemistry. Bio-activity and Applications". ACS Symposium. ACS, Washington DC: 1991. p. 474

[43] Deng S, Yu B, Lou Y, Hui Y. First Total Synthesis of an Exceptionally Potent Antitumor Saponin, OSW-1. Journal of Organic Chemistry. 1999;**64**(1):202-208. DOI: 10.1021/jo981685c

[44] Gupta P, Mahajan A. Sustainable approaches for steroid synthesis. Environmental Chemistry Letters. 2018:1-17. DOI: 10.1007/s10311-018-00845-x

[45] McMorris T. Recent developments in the field of plant steroid hormones.

Lipids. 1997;**32**(12):1303-1308. DOI: 10.1007/s11745-006-0168-z

[46] Anelli P, Biffi C, Montanari F, Quici S. Fast and selective oxidation of primary alcohols to aldehydes or to carboxylic acids and of secondary alcohols to ketones mediated by oxoammonium salts under two-phase conditions. Journal of Organic Chemistry. 1987;**52**(12):2559-2562. DOI: 10.1021/jo00388a038

[47] Anelli P, Banfi S, Montanari F, Quici S. Oxidation of diols with alkali hypochlorites catalyzed by oxammonium salts under two-phase conditions. Journal of Organic Chemistry. 1989;**54**(12):2970-2972. DOI: 10.1021/jo00273a038

[48] Miller S, Nandi J, Leadbeater N, Eddy N. Probing the effect of counterion on the oxoammonium salt oxidation of alcohols. European Journal of Organic Chemistry. 2019;**1**:108-112. DOI: 10.1002/ejoc.201901369

[49] Laeini A, Shaabani A. Transition-metal-free homogeneous TEMPO-based catalyst: Aerobic oxidation of alcohols in aqueous. Media Chemistry Select. 2017;**2**:9084-9087. DOI: 10.1002/slct.201701428

[50] Kashparova V, Kashparov I, Zhukova I, Astakhov A, Kagan E. Salt Effects in the Reaction of Alcohols Oxidation with a Nitroxyl Radical–Iodine Catalytic System. Russ. J. Gen. Chem. 2015;**85**(3):567-570. DOI: 10.1134/S1070427210040324

[51] Schulze J, Migenda J, Becker M, Sören M, Schuler M, Wende R, et al. TEMPO-functionalized mesoporous silica particles as heterogeneous oxidation catalysts in flow. Journal of Materials Chemistry A. 2020;**8**(7):4107-4117. DOI: 10.1039/C9TA12416K

Chapter 6

Potassium Persulfate as an Eco-Friendly Oxidant for Oxidative Transformations

Bilal Ahmad Mir and Suresh Rajamanickam

Abstract

The formation of carbon-carbon/carbon-heteroatom bonds by oxidative transformations is a hotly debated topic in chemistry. $K_2S_2O_8$ has emerged as a cost-effective inorganic oxidant for a wide range of oxidative reactions in this setting. This book chapter covers oxidative reactions facilitated by $K_2S_2O_8$ in the absence of a metal catalyst in detail. Organic chemists may find this book chapter valuable in formulating the mechanistic pathways involving the sulphate radical anion, as well as in the quick and environmentally friendly synthesis of novel chemical species.

Keywords: potassium persulfate, oxidant, eco-friendly, oxidative transformations, Minisci reaction, C-C and C-X bond formation

1. Introduction

Heterocyclic compounds are fascinating for several reasons, the most notable of which is that they have biological activities, and many drugs are heterocycles. Because of their biological properties, nitrogen heterocycles [1] and oxygen heterocycles (such as coumarins and analogues, [2–4] as well as chromone-based compounds [5]) have long aroused chemists' interest. As a result, organic chemists have been hard at work developing new and efficient synthetic transformations to make these heterocyclic compounds. Carbon carbon (C-C)/carbon-heteroatom (C-X) formations are generally used in the synthesis of these heterocycles. The formation of carbon carbon (C-C)/carbon heteroatom (C-X) bonds between two C-H/C-H bonds or C-H/X – H bonds *via* oxidative transformation has become a central focus in C-C/C-X bond forming reactions in this setting, obviating the use of prefunctionalized substrates and reducing salt waste generation, resulting in superior sustainability and environmental compatibility [6–8]. Over the last two decades, the reliance on these direct oxidative C-H functionalizations that offer higher atom economy and sustainability has steadily increased, indicating the importance of and growing interest in this synthetic methodology. Despite the inherent difficulty of generating high regioselectivity, effective realizations of numerous regioselective C – H functionalizations have been accomplished [9–10]. Transition-metal-catalyzed oxidative reactions are one of the most cutting-edge aspects in organic chemistry. In these reactions, an oxidant is always used to regenerate the catalyst. In several of these metal-catalyzed reactions, it has been discovered that selecting the right terminal oxidant is critical to achieve the desired catalytic result. Molecular oxygen, 2,3-Dichloro-5,6-dicyano-1,4-benzoquinone (DDQ), *p*-benzoquinone, *tert*-Butyl hydroperoxide (TBHP), $PhI(OAc)_2$, Iodine, metal oxidants, oxone, persulfates, and other oxidants have all been found to be useful in these processes. Among these inorganic and organic oxidants, potassium persulfate ($K_2S_2O_8$) has emerged as a good inorganic oxidant for a wide range of oxidative transformations, with applications spanning from laboratory studies to industrial processes. Since the discovery of the Minisci reaction, $K_2S_2O_8$ has demonstrated its special utility as an inexpensive, readily available oxidant [11–12]. Among various peroxygen families of compounds, such as H_2O_2, $KHSO_4$, and others, the peroxydisulfate ion ($S_2O_8^{2-}$) is the most effective oxidant. In aqueous solution, the standard redox potential is predicted to be 2.01 V [11]. Under mild circumstances, thermal, photolysis, radiolysis, or redox breakdown of $S_2O_8^{2-}$ produces the sulphate radical anion ($SO_4^{\bullet-}$) additionally, transition-metal ions can activate $K_2S_2O_8$ to generate $SO_4^{\bullet-}$. With a redox potential of 2.531 V, $SO_4^{\bullet-}$ is considered an extremely strong one-electron oxidant [13–14]. Because of their predilection for electron transfer processes, it has a longer lifetime (4 s) than hydroxyl radicals. Exergonic or endergonic electron transport processes exist. Because of the high activation energy, the endergonic process could be sluggish. Various metal salts and complexes, anions, nucleophilic radicals, and neutral organic molecules can all be oxidized by $K_2S_2O_8$ [11]. In recent years, a plethora of literature has emerged, highlighting the potential use of this oxidant in a variety of metal-catalyzed and metal-free organic reactions. Not only has it found widespread application in palladium catalysis as a result of Minisci's work, but it has also been used to carry out a variety of novel oxidative transformations without the aid of any metal catalyst. In comparison to $K_2S_2O_8$, other versions such as $Na_2S_2O_8$ and $(NH_4)_2S_2O_8$ are significantly less widely utilized. This is due to the potassium salt's higher solubility in organic solvents, allowing for more efficient transformation. The C – C to C – X (X = N, O, S, P, B, Si, F, Br, I) bond forming reaction is covered by the spectrum of transformations accomplished with $K_2S_2O_8$ in metal-catalyzed and metal-free methods across a wide variety of substrates. $K_2S_2O_8$ has also been shown to be quite effective at degrading organic pollutants, especially aromatic pollutants [15]. Despite the fact that selective organic transformations involving electron transfer to $SO_4^{\bullet-}$ have been studied in the literature [11, 16, 17], no comprehensive investigation on metal-free oxidative transformations involving persulfate $K_2S_2O_8$ has been published, including updates on recent progress in the field. The focus of the current book chapter is on the use of $K_2S_2O_8$ in oxidative organic transformations. Because there is so much information about this oxidant, just the most important examples illustrating a wide range of bond types are provided here. This review is separated into groups based on the types of bonds produced. We hope that chemists working on or planning to work on developing $K_2S_2O_8$-based approaches for metal-free oxidative processes will find this book chapter useful, permitting considerable scientific advancement in this area.

2. Metal-free oxidative transformations with $K_2S_2O_8$

$K_2S_2O_8$ is used as the major oxidant in the creation of a variety of C – C/C – N/C – S/C – O bonds, allowing access to a variety of cyclic and acyclic compounds. In the process of oxidation $S_2O_8^{2-}$ is either directly involved or the radical anion sulphate ($SO_4^{\bullet-}$), which, in turn, is produced by the breakdown of $K_2S_2O_8$.

2.1 C–C bond formation

Direct radical acylation, alkylation, and arylation reactions (through cross-dehydrogenative coupling reactions and decarboxylative processes), cascade radical addition cyclization processes, multifold bond-cleavage-bond-forming reactions, and photoredox reactions are among the various types of C-C bond-forming reactions reported with $K_2S_2O_8$ as the sole oxidant. $K_2S_2O_8$-mediated hydroxyalkylation of 2*H*-benzothiazoles with aliphatic alcohols in aqueous solution was reported by Weng and co-workers [18]. This mild and convenient approach produced a variety of hydroxyalkylated benzothiazoles in moderate to good yields. In addition, benzimidazole and ethers were compatible in this reaction, resulting in C-2 ether-substituted heteroarenes. $K_2S_2O_8$ not only works as an oxidant in this case, but it also aids in the formation of radicals. In the reaction, there would be cross-coupling between the radicals. In addition, additional radicals would target benzothiazole, so two feasible pathways are proposed in **Figure 1**. To begin, homolytic cleavage of $K_2S_2O_8$ produced sulphate radical anions ($SO_4^{\bullet-}$), which stripped hydrogens from benzothiazole and alcohol, yielding benzothiazole radical and hydroxyl radical, respectively. The two radicals then had a cross-coupling reaction, resulting in the desired product. The hydroxyl radical, on the other hand, would attack the 2-position of benzothiazole to create radical cation intermediates, which were then deprotonated by $SO_4^{\bullet-}$ to provide the products (**Figure 1**) [18].

$K_2S_2O_8$-mediated Minisci acylation on electron-rich pyrroles was used to establish regioselective monoacylation of (NH)-free pyrroles (**Figure 2**). Under initial heating circumstances, homolytic cleavage of $K_2S_2O_8$ might form a sulphate radical anion ($SO_4^{\bullet-}$), which could then be decarboxylated to produce acyl radical. Two mechanisms could lead to the synthesis of benzoylated product. From the acyl

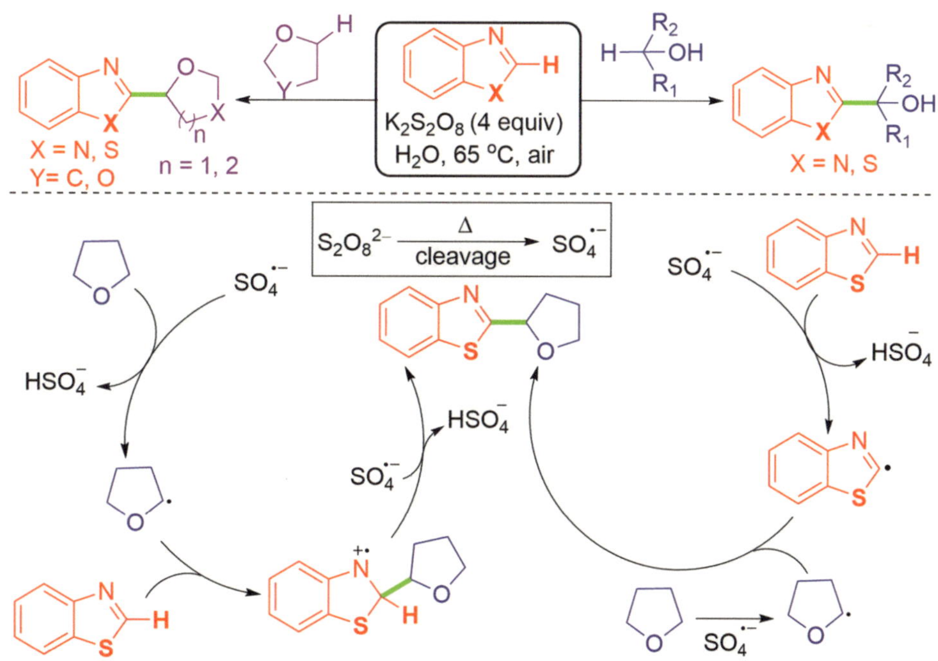

Figure 1.
$K_2S_2O_8$-mediated hydroxyalkylation of benzothiazoles with alcohols.

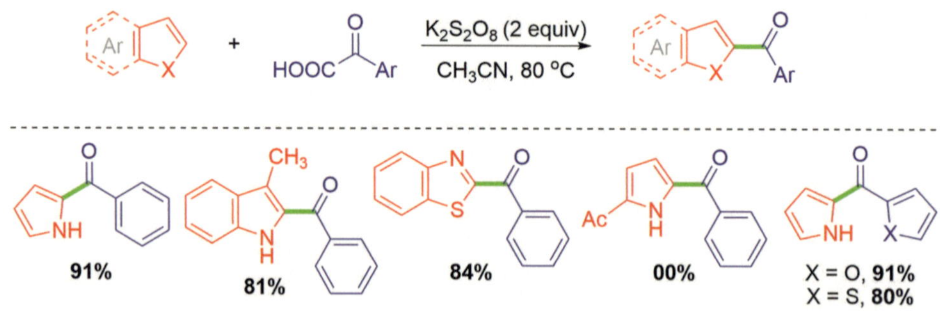

Figure 2.
$K_2S_2O_8$-mediated aroylation of electron-rich pyrroles.

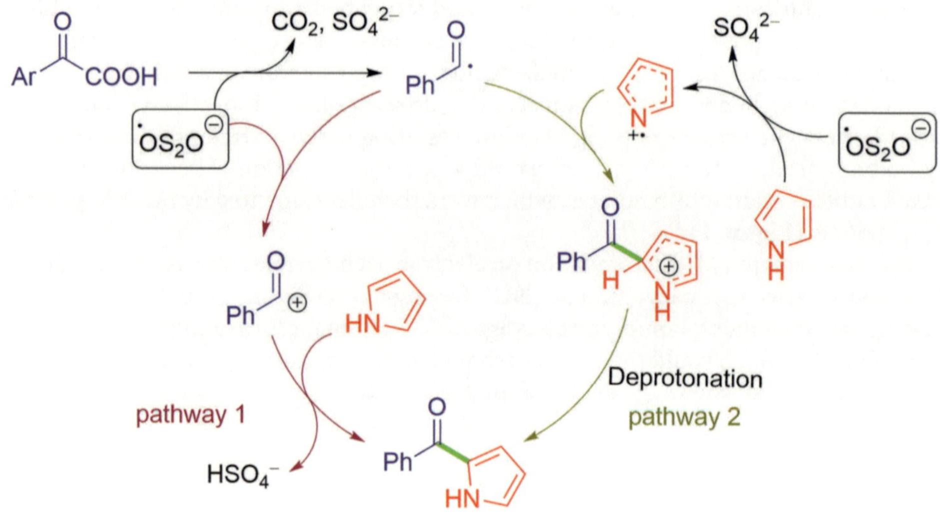

Figure 3.
A mechanism for $K_2S_2O_8$-mediated aroylation of electron-rich pyrroles.

radical, the acylium ion might arise, which could then be electrophilically substituted with pyrrole to create nitrogen radical cation. In other pathway the sulphate radical anion ($SO_4^{\bullet -}$) can capture one electron from pyrrole, resulting in pyrrole radical cation, which could produce adduct when reacting with nucleophilic acyl radical. Deprotonation of the intermediate could generate 2-benzoylpyrrole (**Figure 3**) [19].

Wei and colleagues demonstrated a simple, environmentally friendly, and effective method for undertaking radical cyclizations of enynes/dienes in water. This methodology was developed to employ mild reaction conditions with no catalyst, and it was easy to scale up. It was also designed to use $K_2S_2O_8$ as a green oxidant and water as the solvent, resulting in a process that is both clean and simple to operate, meeting the green chemistry criterion. This reaction undergoes a sequential radical addition, intramolecular cyclization and H-abstraction to give the final product (**Figure 4**) [20].

Zhang and Chen jointly reported $K_2S_2O_8$—/tetrabutylammonium hydrogen sulfate (TBAHS) promoted cascade oxidative aryl-alkylation of *N*-Aryl-alkylation of *N*-aryl acrylamides for functionalized oxindole synthesis (**Figure 5**) [21]. Under the heating condition, $K_2S_2O_8$ interacted with TBAHS to produce bis(tetrabutylammonium)

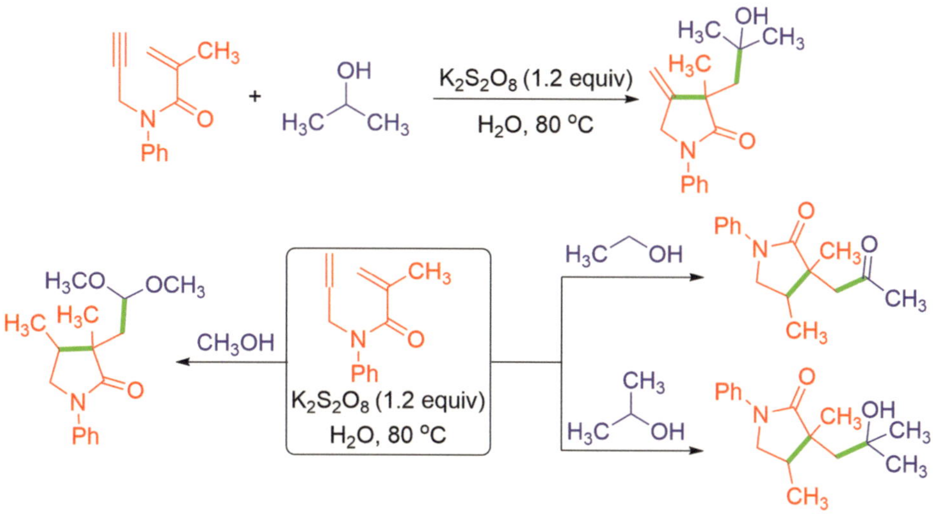

Figure 4.
$K_2S_2O_8$-mediated radical cyclizations of enynes/dienes with alcohols.

peroxydisulfate, which might undergo homolytic O-O bond cleavage to yield two molecules of tetrabutylammonium sulphate radical anions (n-$Bu_4N^+SO4^{\bullet-}$) [22, 23]. The radical anions then extract a hydrogen atom from a variety of C_{sp3}-H compounds, resulting in the C-centered radical. Alkene would then grab the resultant radical, which would then be cyclized to form an aryl radical. Finally, a carbocation is formed by single-electron oxidation of resultant aryl radical by radical anions, which is deprotonated by the produced sulphate dianion (n-$Bu_4N^+SO_4^{2-}$) to obtain the desired product. The active n-$Bu4N^+SO4^{\bullet-}$ produced in the first stage of the reaction can provide higher solubility and a suitable oxidation potential for the product.

Under metal-, photocatalyst-, and light-free circumstances, the Baishya group described two simple and successful C-3 arylation protocols of quinoxalin-2(1*H*)-ones with arylhydrazines and aryl boronic acids, respectively, *via* free radical cross-coupling reactions. Under two separate reaction conditions, $K_2S_2O_8$ has been employed as an effective oxidant to create aryl radicals from arylhydrazines and aryl boronic acids. The process starts when persulfate $S_2O_8^{2-}$ decomposes into the sulphate radical anion $SO_4^{\bullet-}$, which interacts with phenylhydrazine to create the phenylhydrazine radical. Another sulphate anion radical combines with $SO_4^{\bullet-}$ to produce phenyldiazene, which then reacts with still another sulphate anion radical to produce phenyldiazene radical. The phenyl radical is formed when N_2 gas is removed from phenyldiazene radical, and it conducts a Minisci-type radical addition reaction on the C-3 position of quinoxalin-2(1*H*)-one, yielding the desired product as shown in **Figure 6** [24].

In the presence of persulfate, Ryu's group found that a wide range of unactivated acyclic and alicyclic substrates cleanly undergo site-selective alkenylation of unactivated C(sp^3)-H bonds with 1,2-bis(phenylsulfonyl)ethene. The sulphate radical formed by thermally induced homolysis of the persulfate anion abstracts a hydrogen from the β-position of cyclopentanone to create alkyl radical. After that, this radical combines with the C-C double bond of 1,2-bis(phenylsulfonyl)ethene to generate another radical, which then undergoes β-scission to yield the desired product as shown in **Figure 7** [25].

Inorganic oxidants such as potassium persulfate ($K_2S_2O_8$) have been frequently employed in oxidative transformations because they are inexpensive and readily available. Tang and Chang's group published a method for selective intramolecular

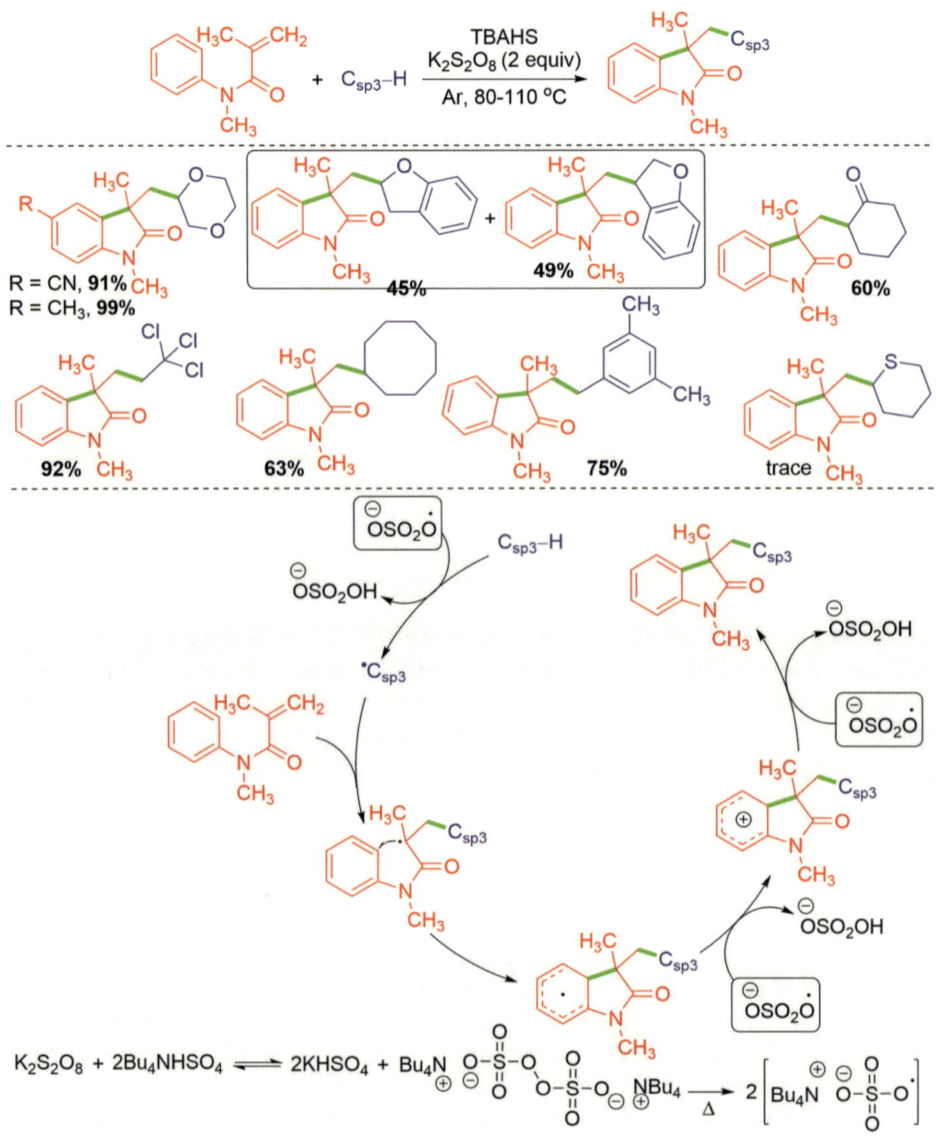

Figure 5.
$K_2S_2O_8$ and TBAHS promoted synthesis of oxindoles.

radical trifluoromethylacylation of alkenes with low-cost CF_3SO_2Na and $K_2S_2O_8$ to produce CF_3-functionalized chroman-4-ones (**Figure 8**) [26]. The rate-determining phase entailed the production of trifluoromethyl radical (CF_3) from CF_3SO_2Na *via* the oxidation of $K_2S_2O_8$. The reaction was started by CF_3 rather than the acyl radical from the aldehyde, according to control experiments and DFT calculations.

Under metal-free circumstances, the Xaio group disclosed a novel and simple approach for the synthesis of 3-(2-oxo-2- arylethyl)chroman-4-ones as shown in **Figure 5**. Using the radical method, aromatic or aliphatic aldehydes react with various

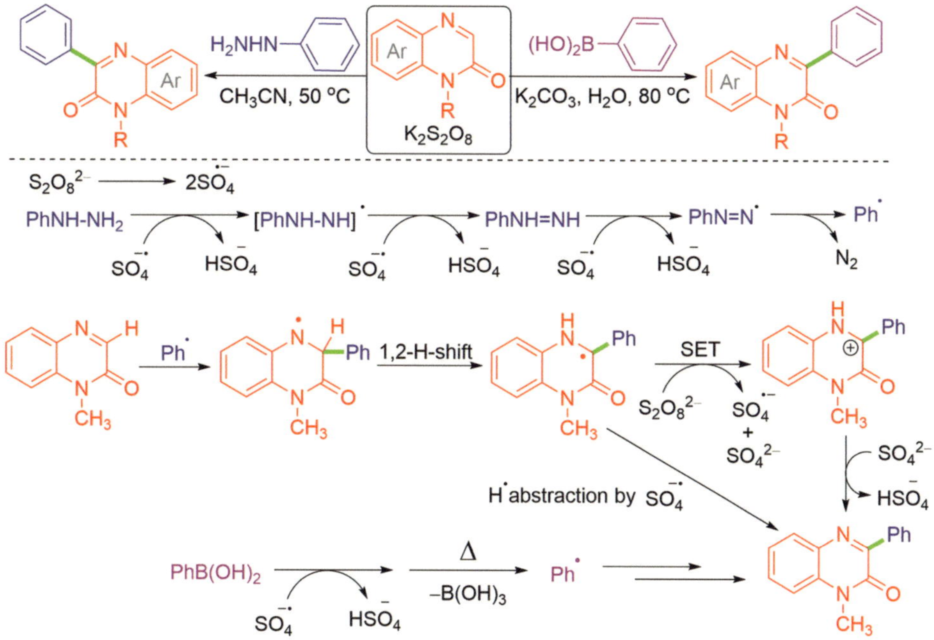

Figure 6.
$K_2S_2O_8$Mediated C-3 arylation of quinoxalin-2(1H)-ones.

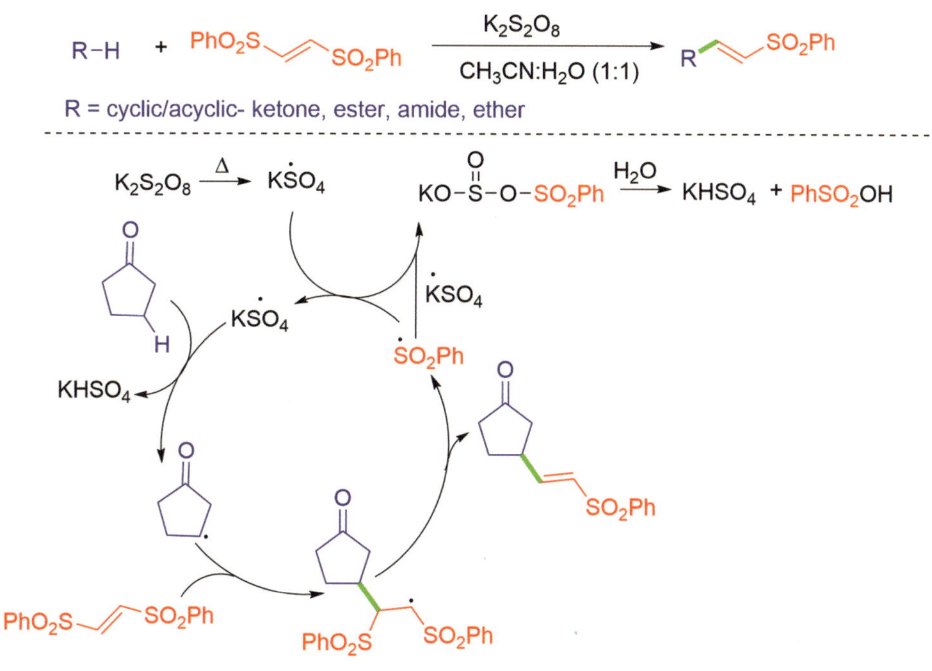

Figure 7.
Persulfate anion-induced $C_{(sp3)}$-H alkenylation by 1,2-bis(phenylsulfonyl)ethane.

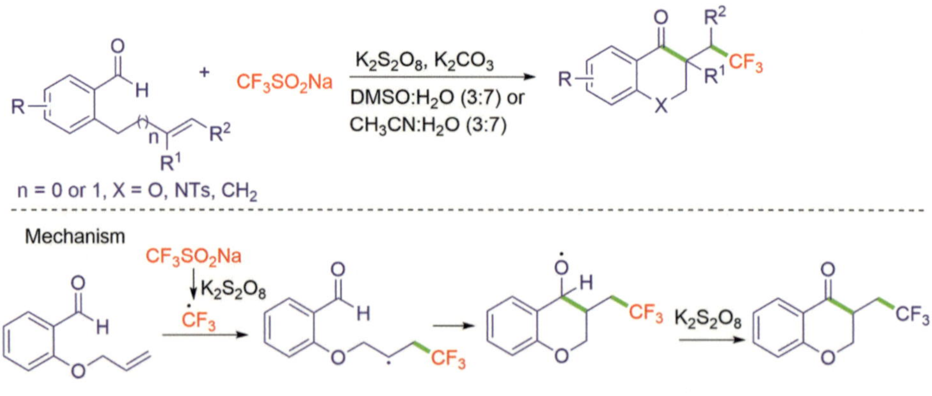

Figure 8.
Synthesis of CF_3-functionalized chroman-4-ones.

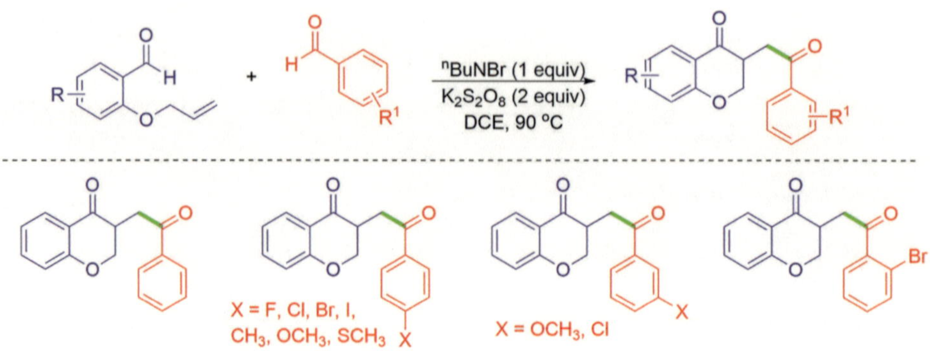

Figure 9.
$K_2S_2O_8$- mediated synthesis of chroman-4-one derivatives.

2-(allyloxy)arylaldehydes to make chroman-4-one derivatives in a moderate to good yields. The procedure is metal-free and has a step-by-step approach, as well as readily available starting materials, demonstrating its some physiologically active chemicals having practical synthetic use (**Figure 9**) [27].

2.2 C–N bond formation

Under metal-free circumstances, several nitration, azidation, and intramolecular C-N bond-forming reactions with $K_2S_2O_8$ have been described. Unlike the decarboxylation of alkyl radicals from carboxylic acids or the generation of sulfur-centered radicals from the corresponding metal salts, silver or other metal catalysts are not required for the generation of nitrogen dioxide or azide radicals (redox potentials of nitrogen dioxide and azide radicals are +1.04 V and + 1.33 V, respectively), [28, 29] and $K_2S_2O_8$ alone could suffice.

By employing TBN and various internal alkenes, Patel group developed a metal-free approach with $K_2S_2O_8$ and quinolone for the synthesis of 1,2,5-oxadiazole-N-oxides (furoxans) and nitrolefins from various internal alkenes as shown in **Figure 10** [30]. In this method, the TBN undergoes thermal heterolytic cleavage of *tert*-butyl nitrite, resulting in the formation of a NO radical and a *tert*-butoxy

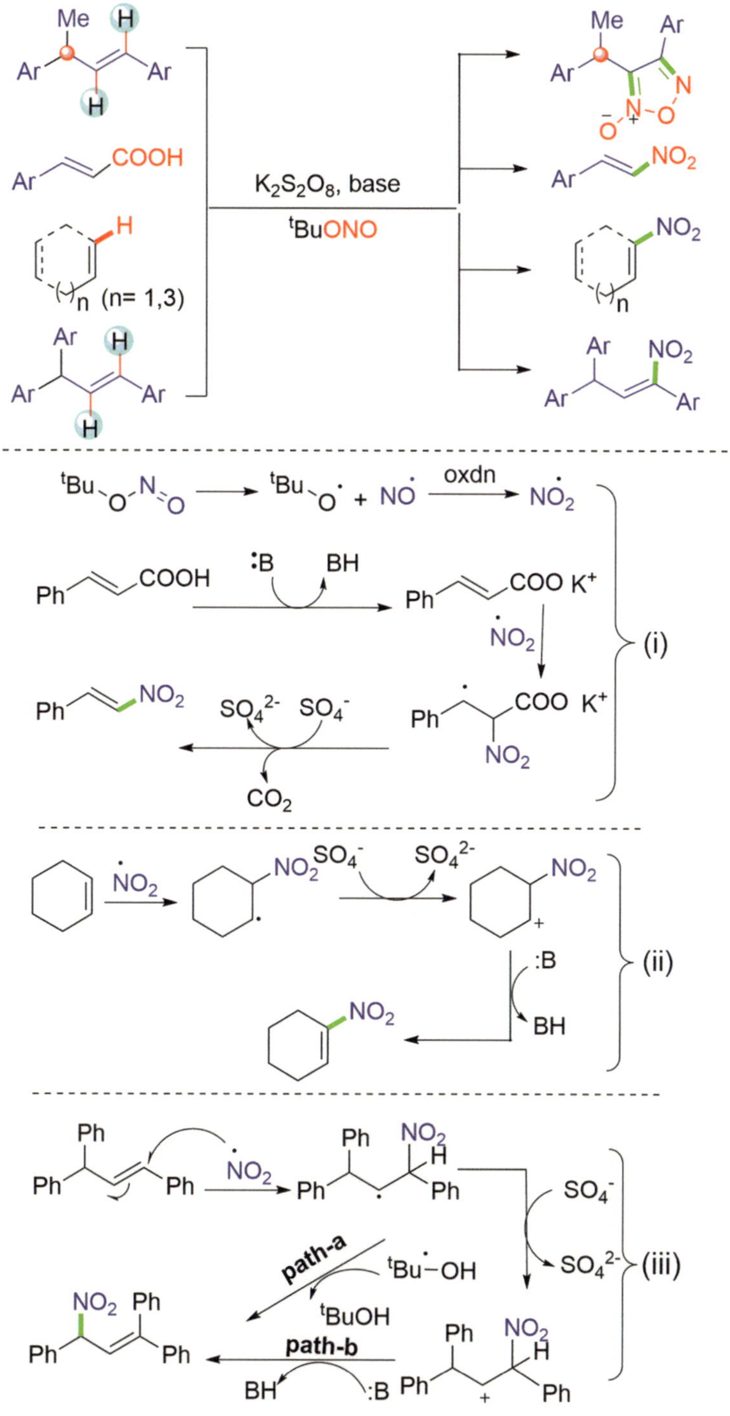

Figure 10.
Tert-butyl nitrite mediated differential functionalizations of internal alkenes.

radical. Under aerobic reaction conditions, the NO radical is transformed to a NO_2 radical. As shown in **Figure 10**, the NO_2 radical generated attacks the various alkenes and produces the desired products [30].

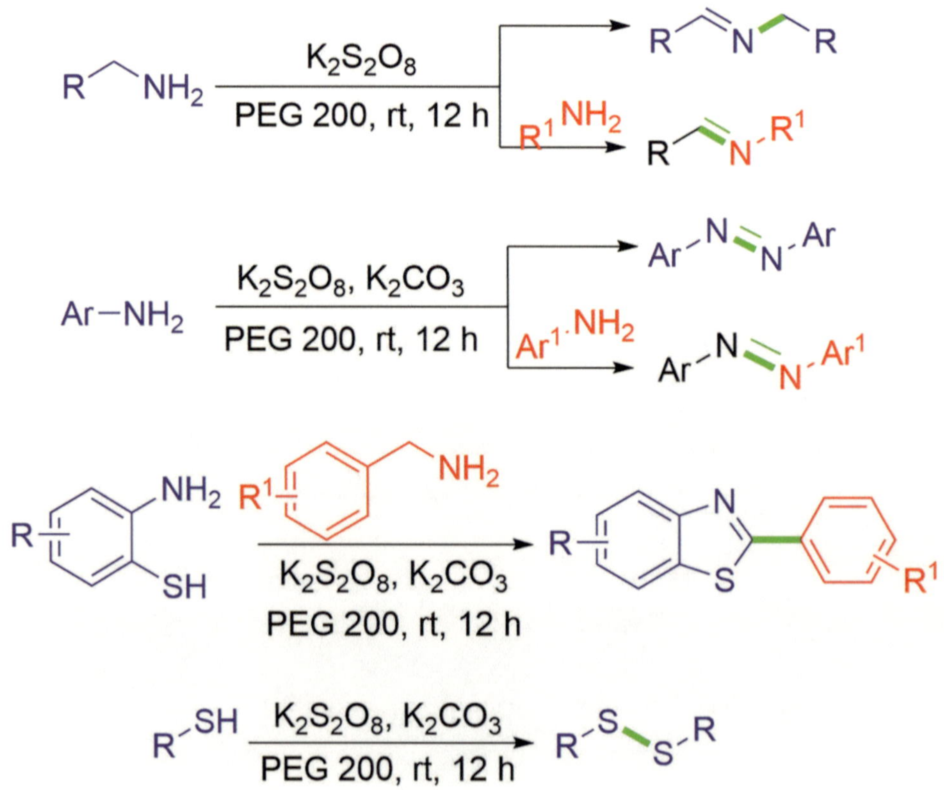

Figure 11.
$K_2S_2O_8$-mediated synthesis of imines, azobenzenes, benzothiazoles, disulfides.

The Sawant group published an excellent yielding transition metal-free technique for oxidative coupling of primary amines to imines and azobenzenes, thiols to disulfides, and 2-aminothiophenols to benzothiazoles. The use of biocompatible and green reaction conditions such as solvent, room temperature reactions, and a transition metal-free approach are among the advantages of the current ecologically friendly process. It also has a wider range of substrate scope (**Figure 11**) [31].

2.3 C–O/C–S/C–Se/C–halogen bond formation

The oxidative C-O, C-S, C-Se and C-halogen bond-forming processes utilizing $K_2S_2O_8$ as the only oxidant are described in some detail. For the synthesis of 1,2-diketones from internal alkynes, Chao and colleagues established a $K_2S_2O_8$-mediated, transition-metal-free approach [32]. For diaryl- and aryl-alkyl acetylenes, Chao's procedure is quite convenient. However, under this $K_2S_2O_8$-mediated reaction state, the aldehyde functionality connected to the aryl-alkyne is unwelcome, resulting in a very poor yield of the desired 1,2-diketone product. Transition-metal (Pd, Ru, Au, Ag, and Cu) catalyzed reactions are generally used to convert alkynes to 1,2-diketones [33, 34]. This $K_2S_2O_8$-mediated process is a good transition-metal catalyzed reaction alternative. The author concluded from ^{18}O-isotope labelling tests that oxygen incorporated into alkyne came from $K_2S_2O_8$ and molecular oxygen rather than water (H_2O^{18}) as shown in **Figure 12** [32].

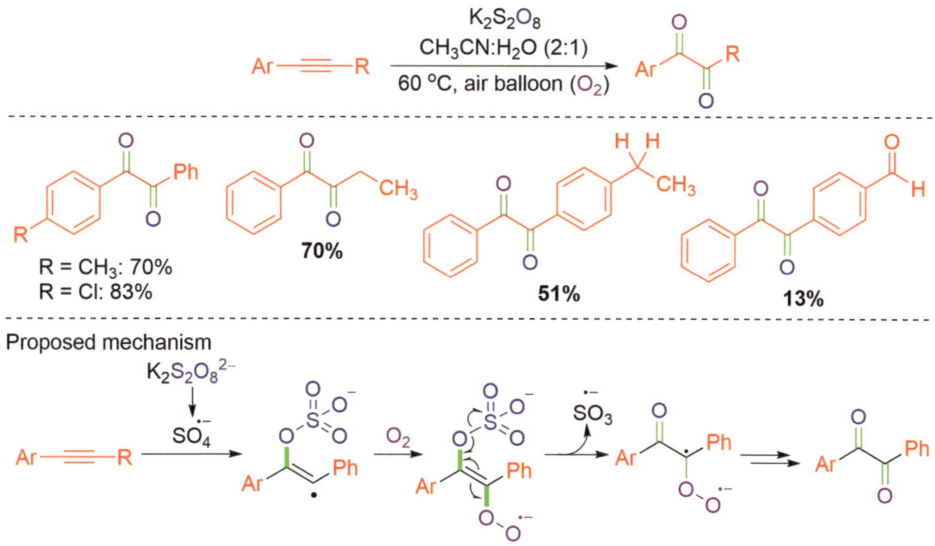

Figure 12.
$K_2S_2O_8$-mediated synthesis of 1,2-diketones.

Tetrahydro-carbolines were oxidized by persulfate, according to the Chen group. In moderate to good yields, this reaction promotes the synthesis of a range of 2-formyl N-substituted tryptamines and related derivatives as important intermediates. The approach can be used to perform direct last-stage oxidation of Cialis and evodiamine, two interesting medicines (**Figure 13**). Under thermolysis in the DMSO solvent, breakdown of $S_2O_8^{2-}$ results in the creation of the sulphate radical anion $SO_4^{\bullet-}$ single electron transfer (SET) from tetrahydro-carboline to $SO_4^{\bullet-}$ yields the carbon-centered radical, which is then further oxidized to provide the iminium intermediate. *N*-Boc-2-formyl-Trp-OH is produced *via* intermolecular nucleophilic addition to an iminium intermediate (**Figure 13**) [35].

In order to synthesis the related flavanones and chalcones in good to excellent yields, a novel $K_2S_2O_8$-mediated approach for the oxidative deoximation of flavanone and chalcone oximes was developed. Flavanone oximes, chalcone oximes, ketoximes, and aldoximes have all been effectively deoximated using this approach. This approach works for both inhibited and functionalized aldoximes as well as ketoximes (**Figure 14**) [36].

Bhat recently reported paraselective thiocyanation of phenol and aniline driven by $K_2S_2O_8$. For heterocycles like indoles, high regioselectivity was also seen (C3-thiocyanation) as shown in **Figure 15** [37]. To learn more about the mechanism, the radical scavenger TEMPO (2,2,6,6- tetramethyl-1-piperidinyloxyl) was treated with the reaction mixture. Even after an extended reaction period, the thiocyanation reaction did not progress, indicating that a free radical route was most likely engaged during the process. $K_2S_2O_8$ is well recognized for producing a powerful, short-lived oxidant-sulphate radical anion ($SO_4^{\bullet-}$). When the sulphate radical anion (E^o = 2.6 V) interacts with aromatics with low ionization potential, it produces a radical cation [38, 39].

Yu reported using thiocyanation and C-O cyclization in the presence of $K_2S_2O_8$ to obtain 3-thiocyanato-4*H*-chromen-4-ones from different 2-hydroxyaryl enaminones [40].

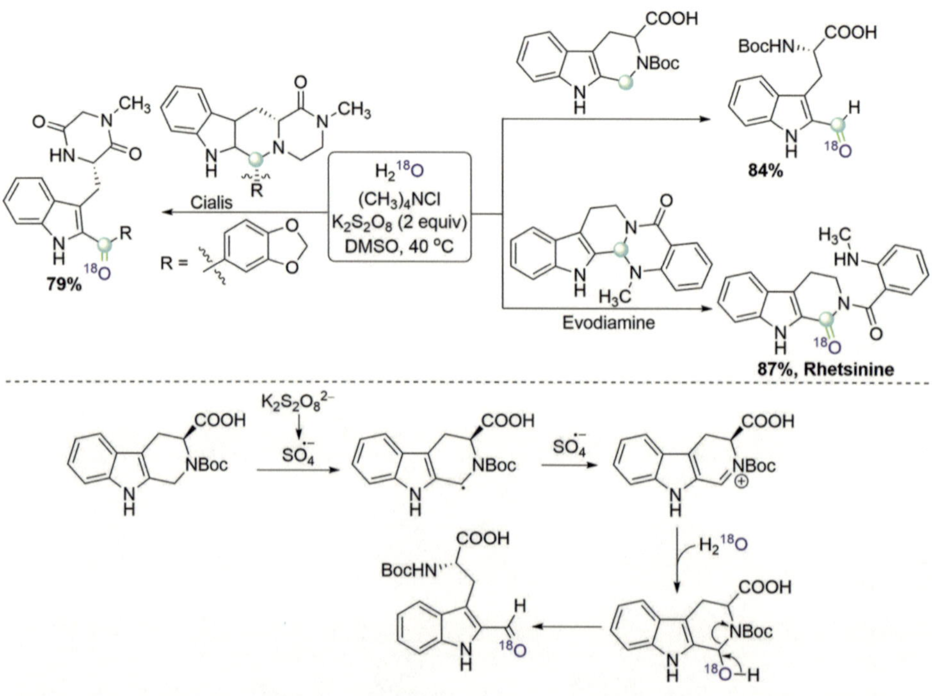

Figure 13.
$K_2S_2O_8$-mediated oxidation of tetrahydro-β-carbolines by persulfate.

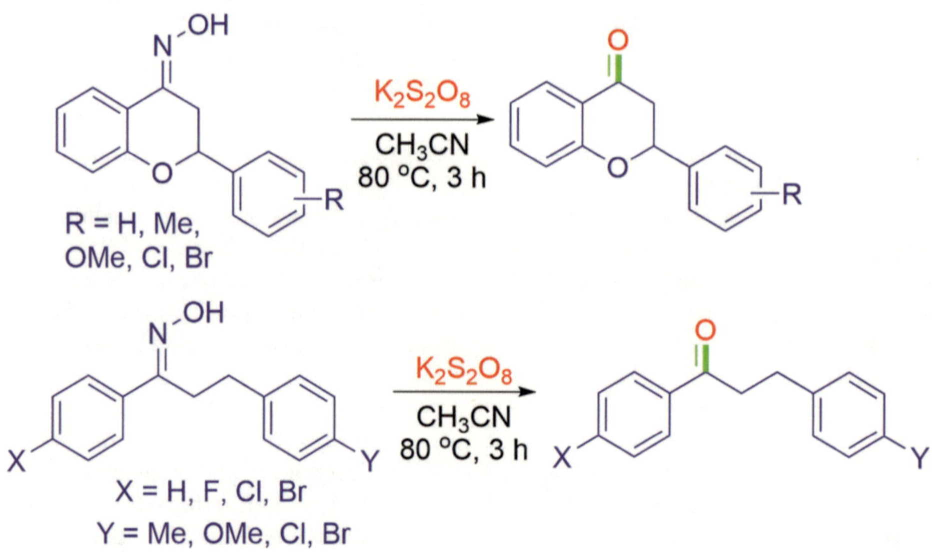

Figure 14.
$K_2S_2O_8$-mediated oxidative deoximation of oximes.

The production of a nucleophilic SCN radical was hypothesized as the next step in this process as shown in **Figure 16**.

Sun and co-workers described a $K_2S_2O_8$-mediated selenoamination of alkenes using diphenyl diselenide and several nitrogen containing compounds such as saccharin,

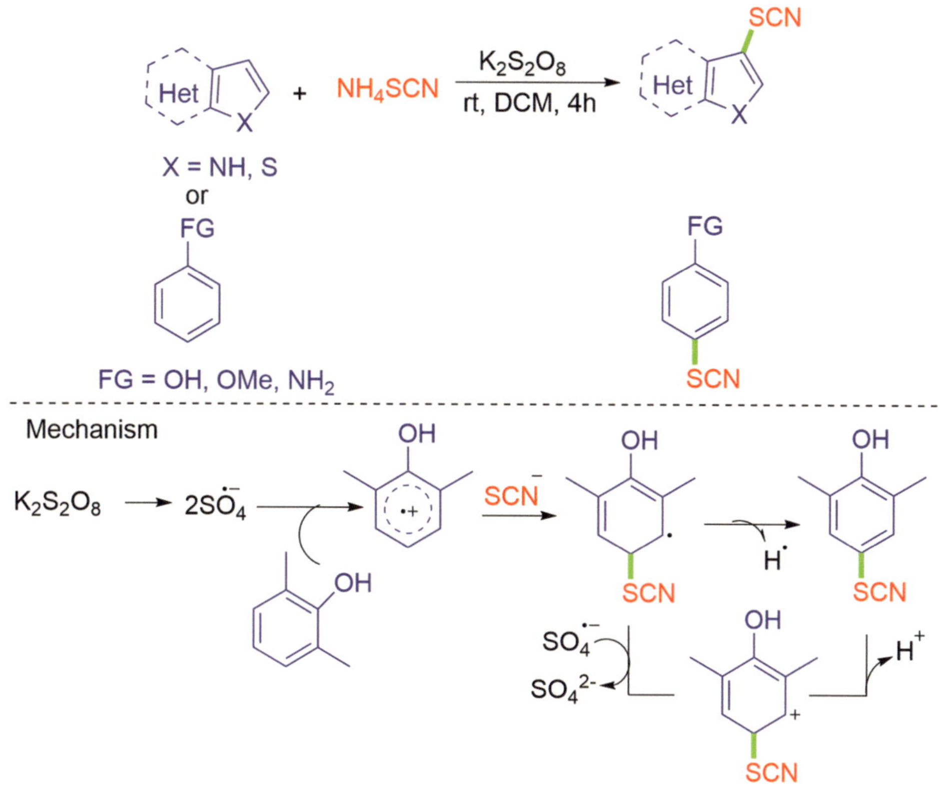

Figure 15.
$K_2S_2O_8$-mediated Thiocyanation of phenols, anilines and heterocycles.

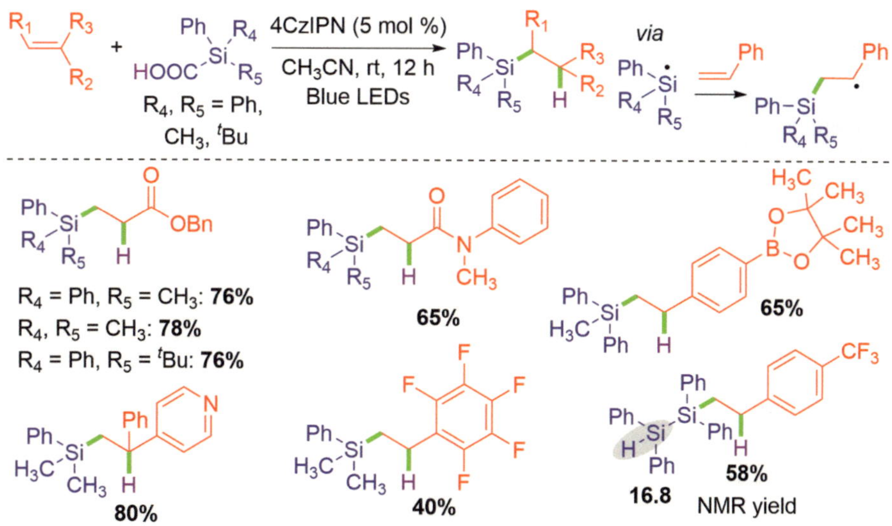

Figure 16.
$K_2S_2O_8$-mediated synthesis of 3-thiocyanato-4H-chromen-4-ones.

dibenzenesulfonimide, benzotriazole, pyrazole, 1,2,4-triazole, 6-chloropurine, and others as shown in **Figure 17** [41].

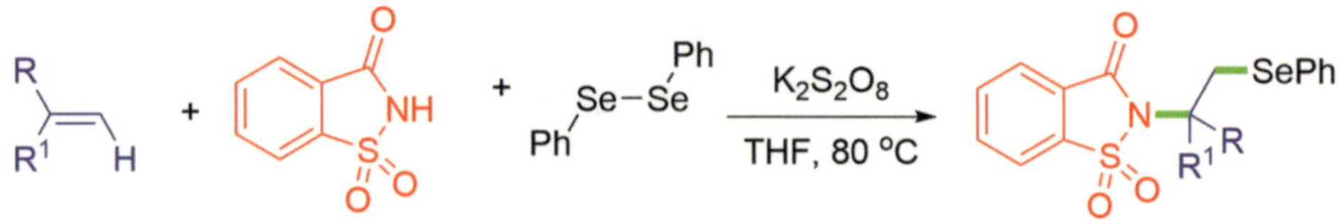

Figure 17.
$K_2S_2O_8$-promoted selenoamination of alkenes.

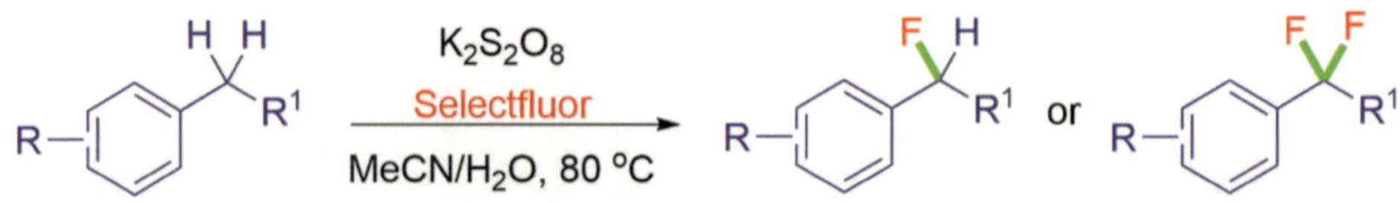

Figure 18.
$K_2S_2O_8$-promoted benzylic monofluorination and difluorination.

Yi and co-workers used Selectfluor and $K_2S_2O_8$ to develop a transition-metal-free technique for direct benzylic C-H fluorination [31]. Despite the existence of different techniques for this transformation, the use of $K_2S_2O_8$ as a cheap oxidant was shown to be the most effective. The most plausible scenario was the formation of a benzylic radical with the sulphate radical ion, which then interacted with a F atom from Selectfluor (**Figure 18**) [39].

3. Conclusions

Oxidative transformations that result in the formation of C-C/C-X bonds are an important class of reactions that has made significant progress in recent years. The oxidative reactions carried out under metal-free conditions using $K_2S_2O_8$ as the major oxidant are highlighted in this book chapter. Overall, this book chapter covers a wide range of greener metal-free transformations (C-C, C-N, C – O/C – S/C – Se/C – Halogen bond formations) with $K_2S_2O_8$ and the mechanisms that underpin them. Their applications in visible light and photoredox-catalyzed reactions have recently been discovered. Nonetheless, given the current state of knowledge about the use of this oxidant in diverse transformations, a comeback of new techniques is very likely in the near future, which can be hastened even further with a complete grasp of mechanistic pathways.

Acknowledgements

Bilal Ahmad Mir acknowledges the support of this chapter by University Grants Commission (CH/20-21/0228) for Fellowship. Suresh R acknowledges the support of this chapter by SERB for funding under the National Post-Doctoral Fellowship scheme SERB-NPDF (PDF/2021/002055) and Rajamalli P, MRC, IISc Bangalore for providing RA position.

Conflict of interest

The authors declare no conflict of interest.

Author details

Bilal Ahmad Mir[1*] and Suresh Rajamanickam[2]

1 Department of Chemistry, Pondicherry University, Pondicherry, India

2 Materials Research Centre, Indian Institute of Science, Bangalore, Karnataka, India

*Address all correspondence to: ahmadmir.bilal6@gmail.com

References

[1] Müller CE, Hansen FK, Gütschow M, Lindsley CW, Liotta D. New drug modalities in medicinal chemistry, pharmacology, and translational science: Joint virtual special issue by journal of medicinal chemistry, ACS medicinal chemistry letters, and ACS Pharmacology & Translational Science. Journal of Medicinal Chemistry. 2021;**64**:13935-13936. DOI: 10.1021/acs.jmedchem.1c01548

[2] Borges F, Roleira F, Milhazes N, Santana L, Uriarte E. Simple Coumarins and analogues in medicinal chemistry: Occurrence, synthesis and biological activity. Current Medicinal Chemistry. 2005;**12**:887-916. DOI: 10.2174/0929867053507315

[3] Keri RS, Budagumpi S, Pai RK. Chromones as a privileged scaffold in drug discovery: A review. European Journal of Medicinal Chemistry. 2014;**78**:340-374. DOI: 10.1016/j.ejmech.2014.03.047

[4] Gaspar A, Matos MJ, Garrido J. Chromone: A valid scaffold in medicinal chemistry. Chemical Reviews. 2014;**114**:4960-4992. DOI: 10.1021/cr400265z

[5] Emami S, Ghanbarimasir Z. Recent advances of chroman-4-one derivatives: Synthetic approaches and bioactivities. European Journal of Medicinal Chemistry. 2015;**93**:539-563. DOI: 10.1016/j.ejmech.2015.02.048

[6] Lyons TW, Sanford MS. Palladium-catalyzed ligand-directed C–H functionalization reactions. Chemical Reviews. 2010;**110**:1147-1169. DOI: 10.1021/cr900184e

[7] Ping L, Chung DS, Bouffard J, Lee S-G. Transition metal-catalyzed site-and Regio-divergent C–H bond functionalization. Chemical Society Reviews. 2017;**46**:4299-4328. DOI: 10.1039/C7CS00064B

[8] He J, Wasa M, Chan KSL, Shao Q, Yu J-Q. Palladium-catalyzed transformations of alkyl C–H bonds. Chemical Reviews. 2017;**117**:8754-8786. DOI: 10.1021/acs.chemrev.6b00622

[9] Iwai T, Sawamura M. Transition-metal-catalyzed site-selective C–H functionalization of Quinolines beyond C-2 selectivity. ACS Catalysis. 2015;**5**:5031-5040. DOI: 10.1021/acscatal.5b01143

[10] Murakami K, Yamada S, Kaneda TI, K. C–H functionalization of Azines. Chemical Reviews. 2017;**117**:9302-9332. DOI: 10.1021/acs.chemrev.7b00021

[11] Minisci F, Citterio A, Giordano C. Electron-transfer processes: Peroxy disulfate, a useful and versatile reagent in organic chemistry. Accounts of Chemical Research. 1983;**16**:27-32. DOI: 10.1021/ar00085a005

[12] Mandal S, Bera T, Dubey G, Saha J, Laha JK. Uses of $K_2S_2O_8$ in metal-catalyzed and metal-free oxidative transformations. ACS Catalysis. 2018;**8**:5085-5144. DOI: 10.1021/acscatal.8b00743

[13] Matzek LW, Carter KE. Activated persulfate for organic chemical degradation: A review. Chemosphere. 2016;**151**:178-188. DOI: 10.1016/j.chemosphere.2016.02.055

[14] Brienza M, Katsoyiannis IA. Sulfate radical technologies as tertiary treatment for the removal of emerging contaminants from wastewater. Sustainability. 2017;**9**:1604. DOI: 10.3390/su9091604

[15] Pari S, Wang IA, Liu H, Wong BM. Sulfate radical oxidation of aromatic contaminants: A detailed assessment of density functional theory and high-level quantum chemical methods. Environmental Science: Processes & Impacts. 2017;**19**:395-404. DOI: 10.1039/C7EM00009J

[16] Minisci F, Fontana F, Vismara E. Substitutions by nucleophilic free radicals: A new general reaction of Heteroaromatic bases. Journal of Heterocyclic Chemistry. 1990;**27**:79-96. DOI: 10.1002/jhet.5570270616

[17] Duncton MAJ. Minisci reactions: Versatile C-H Functionalizations for medicinal chemists. MedChemComm. 2011;**2**:1135-1161. DOI: 10.1039/C1MD00134E

[18] Xu W-X, Dai X-Q, Weng J-Q. ACS omega $K_2S_2O_8$-mediated Hydroxyalkylation of Benzothiazoles with alcohols in aqueous solution. ACS Omega. 2019;**4**:11285-11292. DOI: 10.1021/acsomega.9b01695

[19] Laha JK, Kaur M, Hunjan, Hegde S, Gupta A. Aroylation of electron-rich pyrroles under Minisci reaction conditions. Organic Letters. 2020;**22**:1442-1447. DOI: 10.1021/acs.orglett.0c00041

[20] Wang D-K, Fang Y-L, Zhang J, Guan Y-T, Huang X-J, Zhang J, et al. Radical cyclizations of enynes/dienes with alcohols in water using a green oxidant. Organic & Biomolecular Chemistry. 2020;**18**:8491-8495. DOI: 10.1039/D0OB01902J

[21] Zhang M-Z, Li W-T, Li Y-Y, Wang Q, Li C, Liu Y-H, et al. Discovery of an oxidative system for radical generation from Csp^3–H bonds. A synthesis of functionalized Oxindoles. The Journal of Organic Chemistry. 2021;**86**:15544-15557. DOI: 10.1021/acs.joc.1c02032

[22] Mi X, Wang C, Huang M, Wu Y, Wu Y. Preparation of 3-Acyl-4-arylcoumarins via metal-free tandem oxidative acylation/cyclization between Alkynoates with aldehydes. The Journal of Organic Chemistry. 2015;**80**:148-155. DOI: 10.1021/jo502220b

[23] Siddaraju Y, Lamani M, Prabhu KR. A transition metal-free Minisci reaction: Acylation of Isoquinolines, Quinolines, and Quinoxaline. The Journal of Organic Chemistry. 2014;**79**:3856-3865. DOI: 10.1021/jo500294z

[24] Dutta NB, Bhuyanab M, Baishya G. $K_2S_2O_8$ mediated C-3 arylation of quinoxalin-2(1*H*)-ones under metal-, photocatalyst- and light-free conditions. RSC Advances. 2020;**10**:3615-3624. DOI: 10.1039/D0RA00013B

[25] Ueda M, Kamikawa K, Fukuyama T, Wang Y-T, Wu Y-K, Ryu I. Site-selective Alkenylation of Unactivated $C(sp^3)$–H bonds mediated by compact sulfate radical. Angewandte Chemie, International Edition. 2021;**60**:3545-3550. DOI: 10.1002/anie.202011992

[26] Tang L, Yang Z, Chang X, Jiao J, Ma X, Rao W, et al. $K_2S_2O_8$-mediated selective Trifluoromethylacylation and Trifluoromethylarylation of alkenes under transition-metal-free conditions: Synthetic scope and mechanistic studies. Organic Letters. 2018;**20**:6520-6525. DOI: 10.1021/acs.orglett.8b02846

[27] Xiao Y-M, Liu Y, Mai W-P, Mao P, Yuan J-W, Yang L-R. A novel and facile synthesis of Chroman-4-one derivatives *via* Cascade radical cyclization under metal-free condition. ChemistrySelect. 2019;**4**:1939-1942. DOI: 10.1002/slct.201900147

[28] Armstrong DA, Huie RE, Lymar S, Koppenol WH, Merenyi G, Neta P, et al. Standard electrode potentials involving

radicals in aqueous solution: Inorganic radicals. Bioinorganic Reaction Mechanisms. 2013;**9**:59-61. DOI: 10.1515/irm-2013-0005

[29] DeFelippis MR, Faraggi M, Klapper MH. Redox potentials of the Azide and Dithiocyanate radicals. The Journal of Physical Chemistry. 1990;**94**:2420-2424. DOI: 10.1021/ja016600v

[30] Mir BA, Singh SJ, Kumar R, Patel BK. *Tert*-butyl nitrite mediated different Functionalizations of internal alkenes: Paths to Furoxans and Nitroalkenes. Advanced Synthesis and Catalysis. 2018;**360**:3801-3809. DOI: 10.1002/adsc.201800668

[31] Hudwekar AD, Verma PK, Kour J, Balgotra S, Sawant SD. Transition metal-free oxidative coupling of primary amines in polyethylene glycol at room temperature: Synthesis of imines, Azobenzenes, Benzothiazoles, and disulfides. European Journal of Organic Chemistry. 2019;**2019**:1242-1250. DOI: 10.1002/ejoc.201801610

[32] Shen D, Wang H, Zheng Y, Zhu X, Gong P, Wang B, et al. Catalyst-free and transition-metal-free approach to 1,2-Diketones *via* aerobic alkyne oxidation. The Journal of Organic Chemistry. 2021;**86**:5354-5361. DOI: 10.1021/acs.joc.0c03010

[33] Sawama Y, Asai S, Monguchi Y, Sajiki H. Versatile oxidation methods for organic and inorganic substrates catalyzed by platinum-group metals on carbons. Chemical Record. 2016;**16**:261-272. DOI: 10.1002/tcr.201500217

[34] Yuan L-Z, Hamze A, Alami M, Provot O. Synthesis of substituted Benzils from Diarylalkyne oxidation. Synthesis. 2017;**49**:504-525. DOI: 10.1055/s0036-1588608

[35] Chen H, Ye F, Luo J, Gao Y. Oxidation of Tetrahydro-β-carbolines by persulfate. Organic Letters. 2019;**21**:7475-7477. DOI: 10.1021/acs.orglett.9b02772

[36] Waheed M, Ahmed N, Alsharif MA, Alahmdi MI, Mukhtar S. $K_2S_2O_8$-mediated efficient oxidative Deoximation of flavonoid oximes under mild reaction conditions. ChemistrySelect. 2019;**4**:7572-7576. DOI: 10.1002/slct.201901554

[37] Mete TB, Khopade TM, Bhat RG. Transition-metal-free Regioselective Thiocyanation of phenols, anilines and heterocycles. Tetrahedron Letters. 2017;**58**:415-418. DOI: 10.1016%2Fj.tetlet.2016.12.043

[38] Hey DH, Jones GH, Perkins MJ, lnternuclear Cyclisation. Part XXIX. Oxidation of some N-Methylbiphenyl-2-carboxamides with Persulphate. Journal of Chemical Society Perkin I. 1972;**1772**:118-124. DOI: 10.1039/P19720000118

[39] Zhang XZ, Ge DL, Chen SY, Yu XQA. Catalyst-free approach to 3-Thiocyanato-4H-chromen-4-ones. RSC Advances. 2016;**6**:66320-66323. DOI: 10.1039/c6ra13303g

[40] Sun K, Wang X, Lv Y, Li G, Jiao H, Dai C, et al. Peroxodisulfate-mediated Selenoamination of alkenes yielding Amidoselenide-containing Sulfamides and azoles. Chemical Communications. 2016;**52**:8471-8474. DOI: 10.1039/C6CC04225B

[41] Ma JJ, Yi W, Lu GP, Cai C. Transition-metal-free C–H oxidative activation: Persulfate-promoted selective benzylic Monoan Difluorination. Organic & Biomolecular Chemistry. 2015;**13**:2890-2894. DOI: 10.1039/C4OB02418D

Chapter 7

Thermally Activated Delayed Fluorescence (TADF) Compounds as Photocatalyst in Organic Synthesis: A Metal-Free Greener Approach

Suresh Rajamanickam and Bhisma K. Patel

Abstract

Thermally activated delayed fluorescent (TADF) molecules undergo efficient intersystem crossing (ISC) and reverse intersystem crossing (RISC) processes, making them as third-generation emitters in organic light-emitting diodes (OLEDs), photodynamic therapy (PDT) and time-resolved luminescence imaging. Apart from these applications, recently, TADF molecules have been used extensively as photocatalysts in light-mediated synthesis. In general, highly expensive complexes of Rh, Ir, Ru and organic dyes (Eosin Y, Rose Bengal, 9-mesityl-10-methylacridinium perchlorate [Acr-Mes]$^+ClO_4^-$) are commonly used in the photocatalysis process. Organic-TADF based molecules help to avoid these costly metal catalysts and frequently used organic dyes, making the reaction economical and greener. This chapter will briefly summarize the photocatalytic properties of organic-TADF compounds in organic synthesis.

Keywords: thermally activated delayed fluorescence (TADF), photocatalysis, 4CzIPN, organo photoredox catalysis, radical chemistry, single electron transfer (SET), halogen atom transfer (XAT), cross dehydrogenative coupling (CDC), Minisci reaction, cyclopropanation, cyclization reaction, ring opening reaction, deuteration reaction, hydroformylation reaction

1. Introduction

The term 'photocatalysis' is derived from the concepts of photochemistry. Previously, ultraviolet (UV) irradiation was commonly applied in classical photochemical reactions. The use of high energy ultraviolet light has selectivity issues and requires a designer reaction setup. However, the recent photochemical reaction uses low energy and selective wavelength of visible light from Light-Emitting Diodes (LED). Due to low energy usage, modern photochemical reactions are highly selective. In general, visible light has low absorptivity, so it can not drive the organic reaction competently. A secondary substrate, usually a photocatalyst is introduced to enhance the light absorptivity. This photocatalyst absorbs visible light and provides stable and long photoexcited states, which induces the substrates or reagents to participate in the chemical reaction. The photoexcited catalyst either donates or removes a single electron from the reacting partners, which triggers further reaction *via* oxidative or reductive quenching. The current need of modern organic chemistry is the effective utilization of raw materials, energy resources. Further, elimination of waste, toxic, hazardous solvents, reagents and replacement of expensive and less efficient processes are the basic tenets of Green Chemistry. In this regard, visible-light assisted photo-catalyzed reactions provides a smooth pavement for sustainable chemical synthesis. In addition, the visible-light assisted photo-catalyzed reaction received much attention in the synthetic community due to the simplicity of reaction setup and broad applicability to various reactions. However, most visible light-mediated reactions use expensive metal-based Ru or Ir based photoredox complexes making the process economically unviable. Nevertheless, the use of bench-stable inexpensive Thermally Activated Delayed Fluorescence (TADF) organic material as photocatalysts obviate many of the problems.

Polish physicist Aleksander Jabłoński studied the molecular absorbance and emission of light. He developed the famous Jablonski diagram to explain the spectra and kinetics of fluorescence and phosphorescence. This diagram illustrates the excited states energy level of a molecule and their radiative and non-radiative transitions. A typical Jablonski diagram is shown in **Figure 1a**. Under appropriate light irradiation, the molecules excite to an excited singlet state (Sn), and then the excitons migrate to the lowest excited singlet state (S_1) *via* internal conversion (IC). The lowest excited singlet state (S_1) excitons subsequently migrate from the S_1 state to the ground state

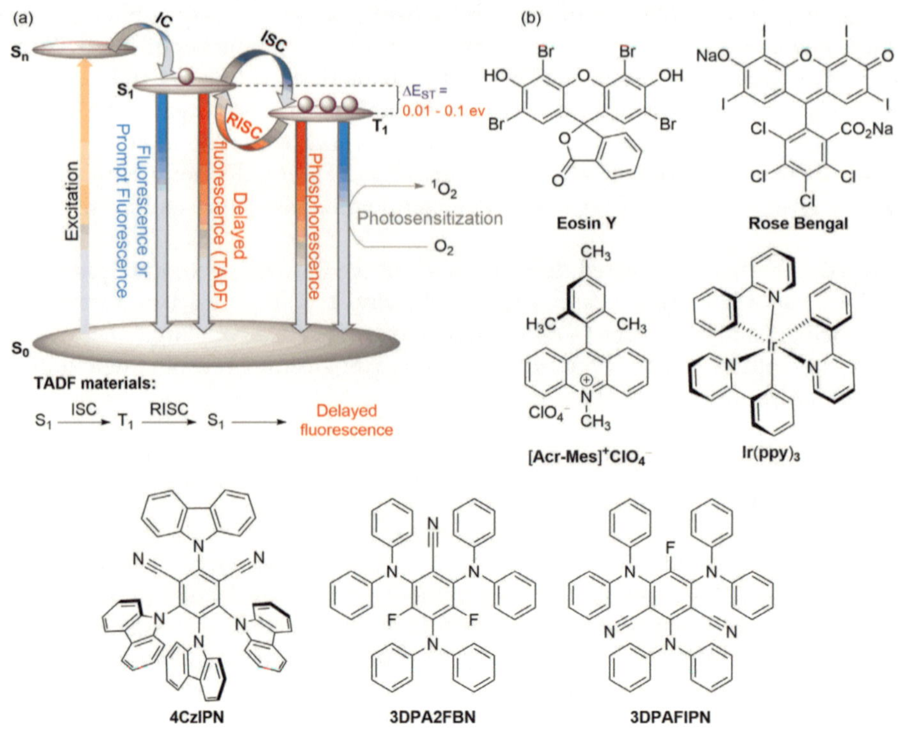

Figure 1.
Simplified Perrin-Jablonski diagram and commonly used TADF motif in photocatalysis.

(S_0) by releasing radiative light energy. This process is known as fluorescence or prompt fluorescence. Meanwhile, the singlet state excitons (S_1) can further migrate to the lowest triplet excited state (T_1) through intersystem crossing (ISC), which can undergo a radiative decay transition process from the lowest triplet excited state (T_1) to the ground state (S_0), ($T_1 \rightarrow S_0$ transition) which is termed as phosphorescence.

Followed by Jabłoński, French physicists Jean Baptist Perrin, the winner of 1926 Nobel Prize in Physics, and his son Francis Perrin rationalized a third type of radiative transition known as delayed fluorescence. This occurs when a molecule in the lowest triplet excited state (T_1) transitions to the lowest excited singlet state (S_1) *via* reverse intersystem crossing (RISC) followed by a radiative transition to the ground state (S_0). The reverse intersystem crossing (RISC) from T_1 to S_1 is only possible when the molecule has a very small singlet-triplet energy gap ΔE_{ST} (upto <0.6 ev) between the lowest-lying singlet S_1 and triplet excited states T_1. When the ΔE_{ST} is very small, the environment heat helps to achieve the reverse intersystem crossing ($T_1 \rightarrow S_1$), resulting in delayed fluorescence. The overall process is termed as Thermally Activated Delayed Fluorescent (TADF). These TADF molecules have huge applications in the third-generation Organic Light-Emitting Diode (OLED) display technology. According to Transparency Market Research, the global OLED displays market is expected to reach $100 billion by 2030 from $4.9 billion in 2012. Recently, the TADF process is also incorporated in the Jablonski diagram and the diagram is termed as Perrin-Jablonski diagram.

The TADF proceeds *via* the result of multiple cycles between intersystem crossing (ISC) from S_1 to T_1 and the reverse intersystem crossing (RISC) from T_1 to S_1. So a delayed fluorescence is observed after the short-lived fluorescence. TADF emission is identical in wavelength to prompt fluorescence, but it occurs on a longer timescale. The emission lifetime of fluorescence or prompt fluorescence is shorter (nanosecond scale). On the other hand, the delayed fluorescence has a longer emission lifetime (microsecond or even millisecond scale).

Eosin Y is the first organic compound identified to show Thermally Activated Delayed Fluorescence (TADF) property. This inexpensive organic dye is widely used in photocatalysis (PC) due to its moderate redox potentials (in 1:1 ratio of acetonitrile and water ratio the ground oxidation and reduction potentials of Eosin Y are E_{ox} = 0.78 V and E_{red} = −1.06 V respectively. The excited state reduction and oxidation potentials are $E_{ox}{}^{*}$ = −1.11 V and $Er_{ed}{}^{*}$ = 0.83 V) and visible region absorption (λ_{abs} = 520 nm in MeOH). Followed by Eosin Y, other organic TADF molecules such as Rose Bengal, 9-mesityl-10-methylacridinium perchlorate $[Acr\text{-}Mes]^{+}ClO_4^{-}$ and polypyridyl metal complexes $Ir(ppy)_3$, $(Ir[dF(CF_3)ppy]_2(dtbpy))PF_6$ and $[Ru(bpy)_3](PF_6)_2$) were extensively studied in photocatalysis process (**Figure 1**).

In 2012, Adachi group prepared a conformationally twisted electron-donor and acceptor TADF material, 2,4,5,6-tetra(carbazol-9-yl)benzene-1,3-dicarbonitrile (4CzIPN) by single-step reaction between 2,4,5,6-tetrafluoroisophthalonitrile and carbazole *via* nucleophilic aromatic substitution (S_NAr) reaction (**Figure 2**) [1]. This molecule has a very small energy gap (ΔE_{ST} = 0.08 eV in toluene) between S_1 and T_1 levels. This small energy difference allows reverse intersystem crossing (RISC) from T_1 to S_1 to qualify it as a TADF molecule [1]. 4CzIPN is a poor single-electron oxidant or reductant in the ground states. On the other hand, it is a potent single electron transfer reagent in the excited states under visible-light irradiation. This TADF molecule has high photoluminescence quantum yield (94.6%) and a long lifetime at an excited state (5.1 μs). One of the primary advantages of 4CzIPN is its low synthetic cost [2].

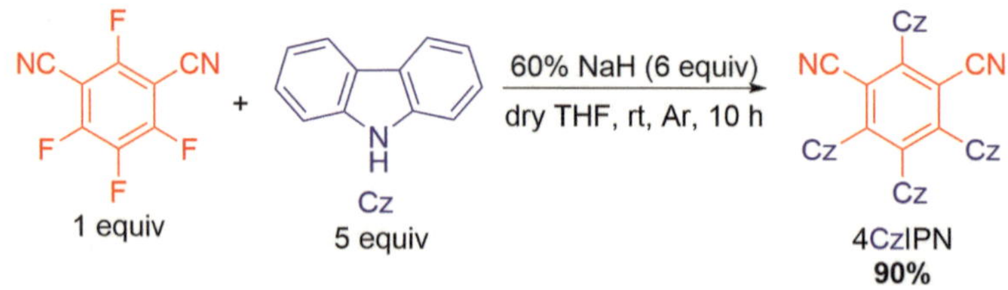

Figure 2.
Synthesis of 2,4,5,6-tetra(carbazol-9-yl)benzene-1,3-dicarbonitrile (4CzIPN).

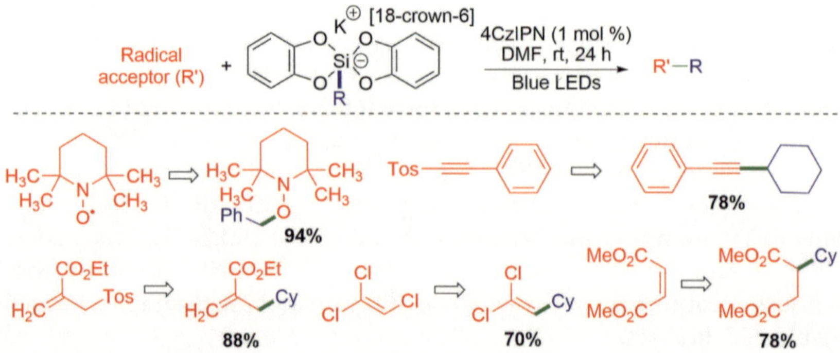

Figure 3.
4CzIPN-catalyzed radical functionalization of silicates with various radical acceptors.

In 2016, Ollivier and Fensterbank group used 4CzIPN as a photocatalyst for an organic transformation [3]. Under blue LEDs irradiation, 4CzIPN undergo photoexcitation, the photoexcited 4CzIPN* generated benzyl radical from benzyl *bis* (caecholato)silicates, the generated radical is trapped by (2,2,6,6-tetramethylpiperidin-1-yl)oxyl (TEMPO) (**Figure 3**). The author further extended the scope of this methodology for the preparation of alkynylation, vinylation and Giese-type products by treating alkylsilicates with a variety of radical acceptors (**Figure 3**). This work open-up a new window for many organic transformations using 4CzIPN as an economically cheaper, greener organo photoredox catalyst.

Followed by Ollivier and Fensterbank radical-mediated synthesis [3], various organic transformations were documented using 4CzIPN as a photocatalyst under the irradiation of visible light or blue LEDs. Besides this, many reactions were reported using 4CzIPN in combination with a transition metal. This chapter excludes transition metal assisted (synergic catalysis) synthesis and mainly focuses on 4CzIPN as an independent photocatalyst without any transition metals.

2. Cyclopropanation reactions

In 2018, a group of Gutierrez and Molander demonstrated a redox-neutral photocatalytic cyclopropanation of olefins with triethylammonium *bis*(catecholato) iodomethylsilicate using a combination of 4CzIPN and blue LEDs light. Triethylammonium *bis*(catecholato)iodomethylsilicate serve as an iodomethyl radical precursor (**Figure 4**) [4].

From mechanistic aspects, the photocatalytically generated halomethyl radical is trapped by the alkene and generate a stable tertiary radical. This radical accepts a

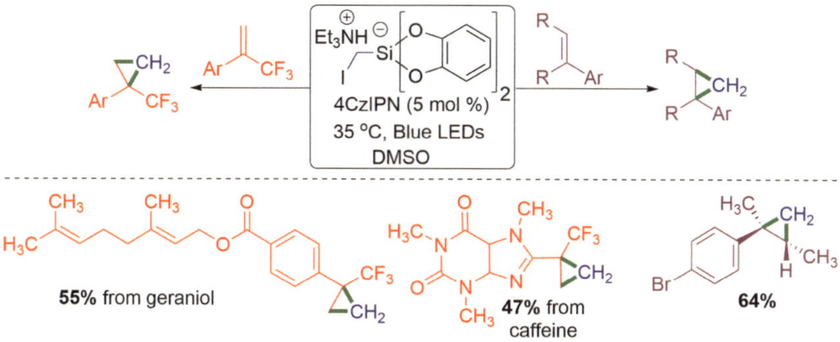

Figure 4.
4CzIPN-catalyzed cyclopropanation of alkene using iodomethylsilicate.

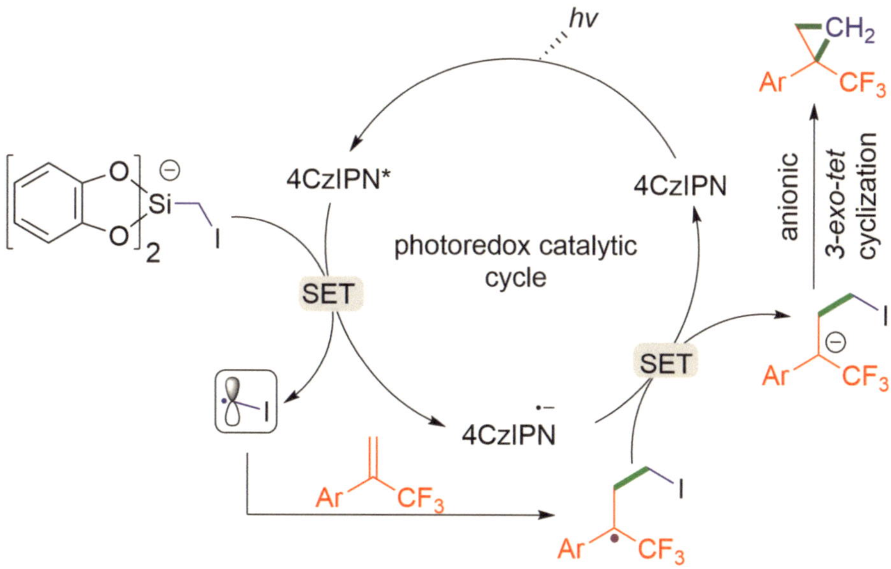

Figure 5.
4CzIPN-catalyzed cyclopropanation of alkene using iodomethylsilicate.

single electron from the 4CzIPN$^{\cdot -}$, to form an anion and regenerate the catalyst. It underwent an anionic *3-exo-tet* ring closure that leads to a cycopropylation product. With the support of both computational and control experiments performed, the authors concluded that the reaction proceeds *via* an anionic *3-exo-tet* ring closure (**Figure 5**) [4].

In the same year, the Molander group further extended their aforementioned cyclopropanation methodology [4] to the homoallylic tosylates system (**Figure 6**) [5]. In their previous report, the leaving group (iodo) is attached to the radical precursor motif itself [4]. In their follow-up work, the Molander group incorporated the leaving group into the alkene core. They treated three different alkyl radical precursors such as *bis*(catecholato)alkylsilicates, alkyltrifluoroborates and 4-alkyl dihydropyridines, with an array of linear homoallylic tosylates systems [5]. All leads to 1,1-disubstituted cyclopropanes derivatives in moderate to good yields *via in-situ* generated anionic cyclization (**Figure 6**). The classical electrophilic carbenoid Simmons-Smith

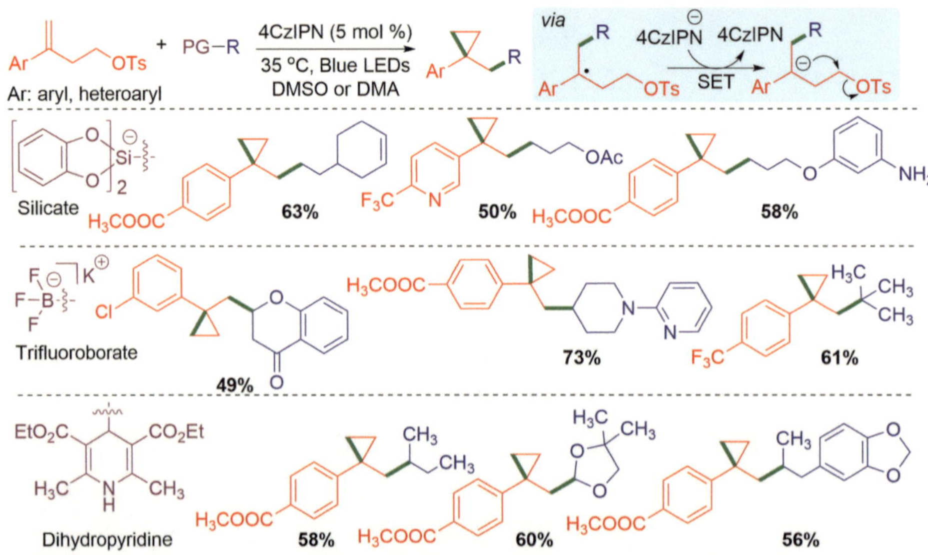

Figure 6.
4CzIPN-catalyzed cyclopropanation of linear homoallylic tosylates using alkyl radical precursors.

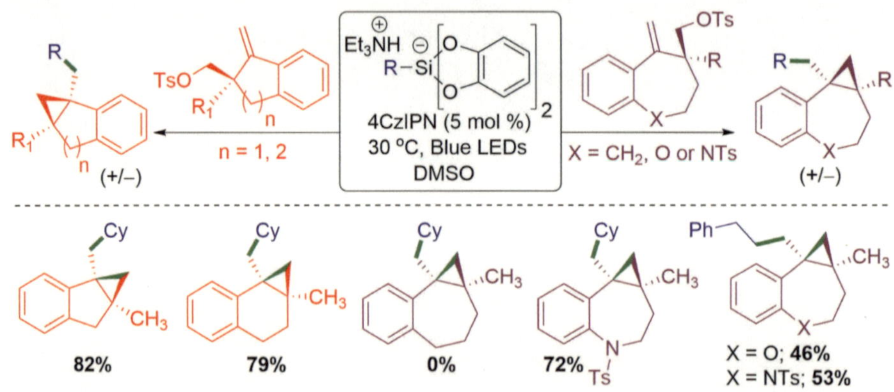

Figure 7.
4CzIPN-catalyzed cyclopropanation of exocyclic homoallylic tosylates using bis(catecholato)alkylsilicates.

cyclopropanation reaction is incompatible for Lewis basic heterocycle and free amine functional groups. However, this radical-polar annulation reaction (RPARs) protocol is highly comfortable with those substrates (Lewis basic heterocycles and free amines substrates). α-Heteroatom, secondary, tertiary and benzylic radical generated from trifluoroborate reagents all smoothly delivered cyclopropanation derivatives [5].

In successive work, the reaction between *bis*(catecholato)alkylsilicates and exocyclic homoallylic tosylates in the presence of photocatalyst 4CzIPN under blue LEDs irradiation afforded polycyclic cyclopropanes (**Figure** 7) [6]. This reaction proceeds smoothly with five (indanone) and six-membered (tetralone) carbocylic tosylate alkenes. But, seven-membered *i.e.* cycloheptanone-based tosylate failed to provide a cyclopropanation product. However, the introduction of heteroatoms (nitrogen and oxygen) in the benzylic position of seven-membered ring *i.e.* benzoazepines and benzoxepines-derivatives, smoothly afforded Radical-Polar Crossover (RAC) annulation products (**Figure 7**) [6].

Around the same time, Noble and Aggarwal's group jointly documented 4CzIPN-catalyzed cyclopropane reaction by treating aliphatic carboxylic acids with electron-deficient internal and external chloro alkenes (**Figure 8**) [7]. The reaction proceeded *via* decarboxylative radical addition followed by a polar cyclization cascade pathway (**Figure 8**). This methodology shows broad substrate scope for both acids (including cyclic, acyclic carboxylic acids, amino acids and dipeptides) and electron-withdrawing groups (carboxylate esters, nitriles, primary amides, sulfones and phosphonate esters) attached chloro alkenes.

Homoallyl chlorides provided good yields of 1,1-disubstituted cyclopropanes. On the other hand, allyl chlorides lead to vicinal substituted cyclopropanes with moderate yields. The slightly lowered yield obtained is due to the formation of an allylic ester by-product *via* S_N2 reaction between allylic chloride and carboxylate (**Figure 8**) [7]. It is noteworthy to mention that, this mild organic photocatalyst protocol is applied for the late-stage cyclopropylation of a variety of acid-containing bioactive natural products namely, dehydroabietic acid (terpenes), biotin (vitamin B7), trolox (vitamin E analogue), cholic acid (bile acid) and gemfibrozil (fibrate drug) (**Figure 8**) [7].

From a mechanistic perspective, the excited photocatalyst 4CzIPN* underwent SET with the carboxylate to form a carbon center radical by reduction of the excited photocatalyst to radical anion ($4CzIPN^{\cdot -}$). The carbon center radical underwent Giese-type addition into the homoallyl chloride to generate the stabilized alkyl radical. This stabilized alkyl radical accept a single electron from $4CzIPN^{\cdot -}$ leading to a stabilized carbanion. Polar 3-exotet cyclization of stabilized carbanion afforded cyclopropane product (**Figure 9**) [7].

The above cyclopropanation reactions have considerable advantages over other reagents such as diazomethane (respiratory irritant) and highly pyrophoric diethylzinc $(C_2H_5)_2Zn$, used in the Simmons-Smith reaction.

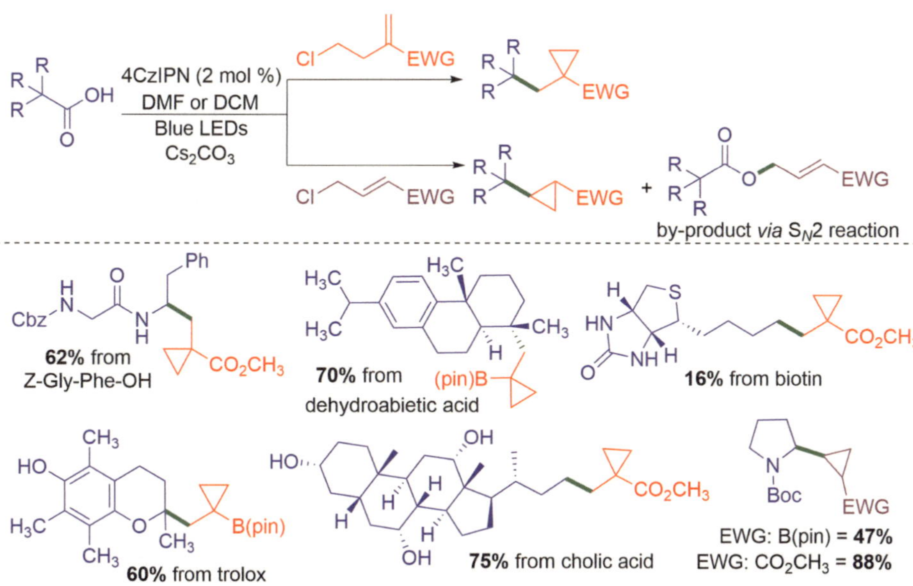

Figure 8.
4CzIPN-catalyzed cyclopropanation of allyl and homo allyl chlorides with carboxylic acid.

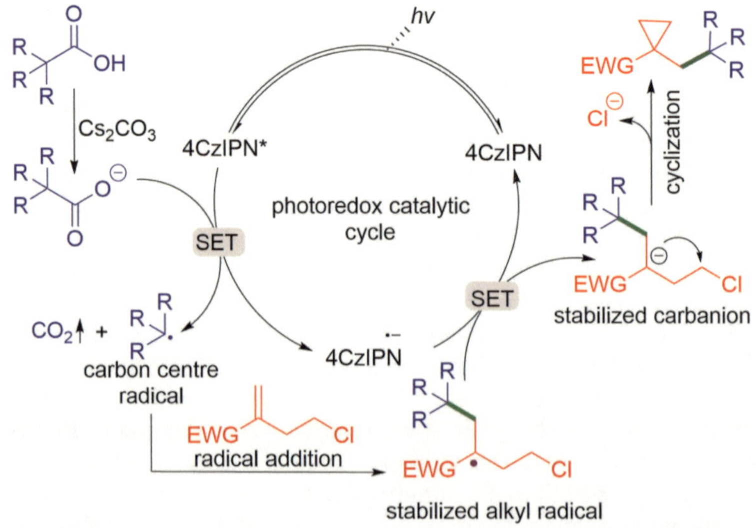

Figure 9.
A mechanism for 4CzIPN-catalyzed decarboxylative cascade radical addition–polar cyclization reaction towards substituted cyclopropanes synthesis.

Subsequently, various research groups generated 4CzIPN photo catalyzed decarboxylative carbon-centered radicals from a carboxylic acid and added it into alkenes (*via* Giese-type addition) [8–11] or aromatic heterocyclic (Minisci reaction) [12, 13].

For example, Wang group generated a carbon-centered radical (**10B**) *in-situ via* a decarboxylative path from the anionic form of diethoxyacetic acid (**10A**) by photoirradiation of 4CzIPN. The generated radical (**10B**) adds regioselectively into aryl olefins (styrene derivatives), at the less substituted site which leads to a stable benzylic radical (**10C**). This benzyl radical then accepts a single electron from the radical anion of photocatalyst $4CzIPN^{\cdot-}$ to form a benzyl anion (**10D**). Protonation of benzyl anion followed by acid hydrolysis of the acetal provided hydroformylation product (**10E**) (**Figure 10**) [8]. In sum, Wang group developed a 4CzIPN catalyzed hydroformylation of aryl olefins with diethoxyacetic (**Figure 10**). In this reaction, diethoxyacetic acid acts as the formylation reagent (formyl radical equivalent). It is noteworthy to mention that this is the first radical-based hydroformylation strategy. The author avoided competitive radical polymerization reaction by performing the reaction in a continuous flow system for electron-deficient olefins, and batch method applied for electron neutral and rich olefinic systems. This radical formylation protocol was successfully extended towards late-stage formylation of biologically relevant complex olefins (**Figure 10**) [8].

The authors studied the competitive reaction between aryl and alkyl olefin in both inter and intra-molecular manners. The reaction showed higher chemoselective at the aryl olefin site. Alkyl olefin site remains intact in both continuous flow and batch method (**Figure 11**) [8].

Schubert group employed 4CzIPN-mediated decarboxylative radical conjugate addition to C=C bonds of dehydroalanine (Dha) and its derivatives peptides (**Figure 12**) [9]. This protocol opens up new avenues to a diastereoselective synthesis of unnatural amino acids and the late-stage derivatization of a tripeptide (**Figure 12**) [9].

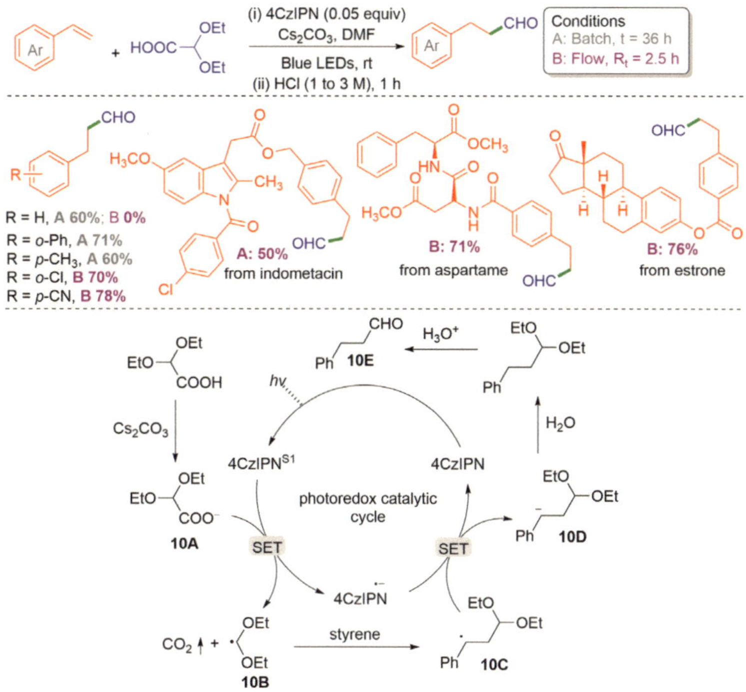

Figure 10.
4CzIPN-photoredox catalyzed hydroformylation of olefins.

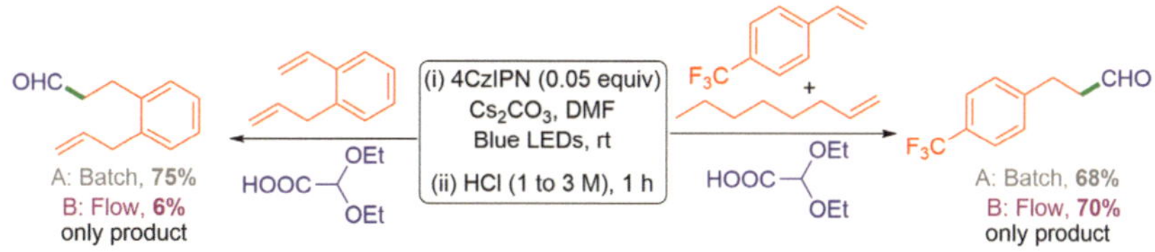

Figure 11.
4CzIPN-catalyzed intra- and intermolecular chemoselectivity hydroformylation reactions.

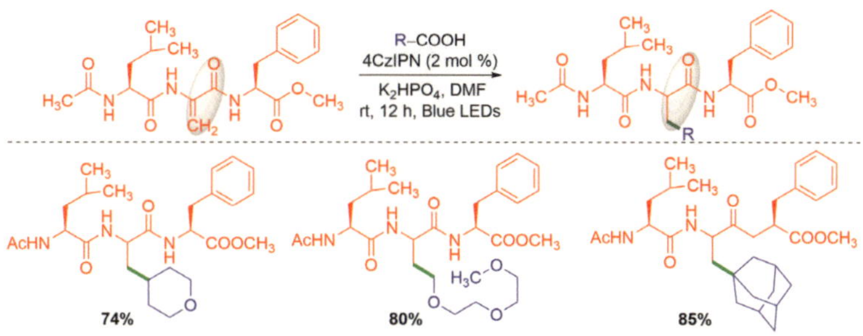

Figure 12.
4CzIPN-catalyzed radical conjugate addition to dehydroalanine containing tripeptides.

3. Three-component C—C and C—N bond formation reaction

So far, we discussed 4CzIPN catalyzed two compound reactions [3–9]. For the first time in the year 2019, the Studer group developed a 4CzIPN photocatalyzed three-component reaction for 1,2-amidoalkynylation of unactivated alkenes (**Figure 13**) [10]. Photoexcited TADF (4CzIPN*) generated an amidyl radical (**14B**) *in-situ* from the anionic form of Troc-protected α-amido-oxy acid (**14A**) *via* a single electron transfer followed by decarboxylation and extrusion of acetone (**Figure 14**). An amidyl radical (**14B**) is added into alkene, which generated an alkyl radical (**14C**), which then couples with an alkyne radical produced from ethynyl benziodoxolones (EBX) to form 1,2-amidoalkynylation product (**14D**) [10].

This three-component reaction showed broad substrate scope for mono, di and tri substituted terminal alkene and substituted benziodoxolones. In addition to these, vinyl ethers, esters and enamides are also compatible with these reaction conditions (**Figure 13**). The reaction provided a high level of chemo-selective product. The polar effect plays a major role in chemo-selective product formation. An amidyl radical is attached at the less substituted site of alkene and the alkyne radical is attached at the more substituted site of alkene (**Figure 13**) [10]. For a particular note, this is the first

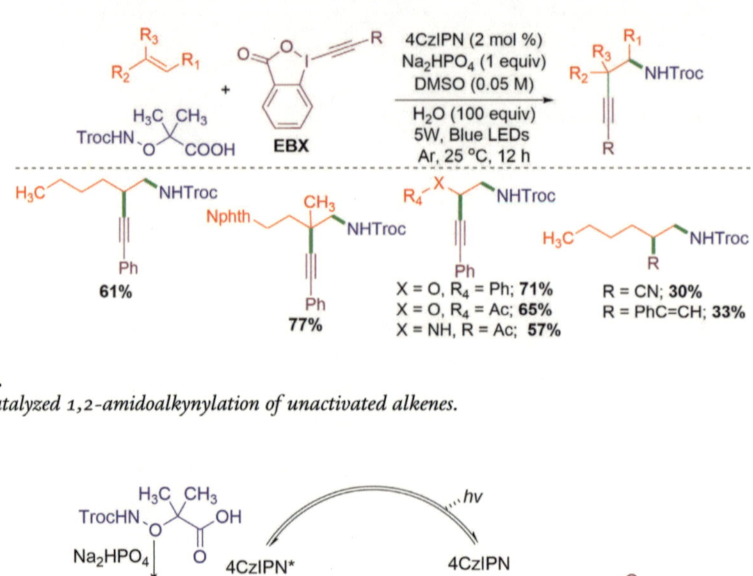

Figure 13.
4CzIPN-catalyzed 1,2-amidoalkynylation of unactivated alkenes.

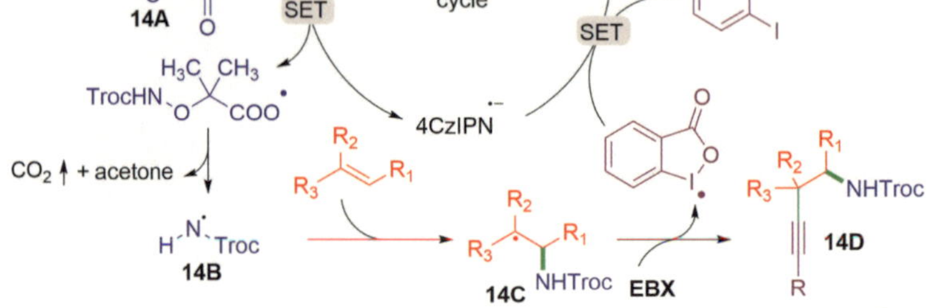

Figure 14.
Proposed mechanism of 4CzIPN-catalyzed 1,2-amidoalkynylation of unactivated alkenes.

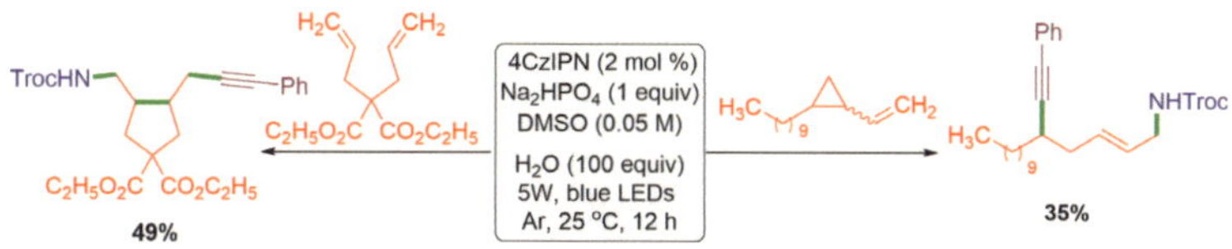

Figure 15.
Radical clock experiments in the synthesis of 1,2-amidoalkynylation of unactivated alkenes.

transition metal-free alkene aminoalkynylation. Before this report, alkene aminoalkynylation reactions are restricted to disubstituted alkenes.

The author concluded that the reaction proceeded through a radical pathway by performing two different radical clock experiments using 1,6-diene and vinylcyclopropane (**Figure 15**) [10].

4. Photoinduced C—Si bond formation via decarboxylation of silacarboxylic acids

All the aforementioned examples deal with photoinduced C—C bond formation *via* decarboxylation of carboxylic acids. In 2020, Uchiyama group generated a silyl radical by photoirradiation of silacarboxylic acids in the presence of 4CzIPN. This silyl radical is successfully trapped by an alkene, leading to silyl alkane *via* C—Si bond formation (**Figure 16**) [11]. Substrate scopes of the reaction were demonstrated with three silacarboxylic acids ($Ph_2MeSiCOOH$, $Ph_2{}^tBuSiCOOH$ and $PhMe_2SiCOOH$) and a broad range of electron-withdrawing substituted alkenes (**Figure 16**). Interestingly, 1,1,2,2-tetraphenyldisilane-1-carboxylic acid, (having Si—H and Si—COOH) also provided decarboxylative silyl radical coupled product, without affecting the Si—H group. (**Figure 16**, product no. 16.8) [11].

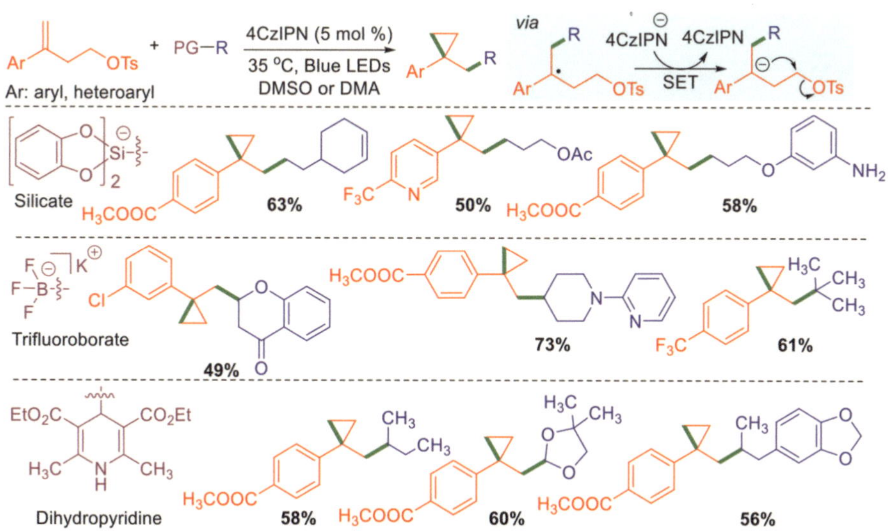

Figure 16.
4CzIPN-catalyzed decarboxylative hydrosilylation of alkenes from silacarboxylic acids.

5. TADF-photocatalyzed Minisci reactions

The addition of nucleophilic radical to electron-deficient nitrogen-containing heteroarenes bases followed by a formal hydrogen atom loss is known as **Minisci reaction** [14, 15]. The major problem associated with the Minisci reaction is regio-selectivity. For example, classical Minisci reaction on quinoline often provides a mixture of C2 and C4 addition products. In 2020, Phipps group developed a TADF-photocatalyzed protocol for regioselective C2 and C4 functionalization of quinoline with phenylalanine-derived redox-active esters (**Figure 17**) [12]. The authors systematically studied the effect of acid, solvents and photocatalyst in the regioselective decarboxylative functionalization (alkylation) of quinolone. The Brønsted acids, photocatalyst and solvent all play major roles in selectivity. The TADF-photocatalyst 4CzIPN with *p*-toluenesulfonic acid (PTSA) in a more polar solvent (DMA) favor C4-functionalized product (5.4:1 ratio of C4/C2). On the other hand, another TADF-photocatalyst 3DPAFIPN with structurally bulky 2,4,6-triisopropylbenzenesulfonic acid (TIPBSA) in a less polar solvent (dioxane) leads to the highest C2 selectivity (C4/C2 = 1:7.3) (**Figure 17**) [12].

Sherwood and co-workers employed a 4CzIPN-photocatalyzed Minisci reaction between a variety of electron-deficient *N*-containing heterocycles and the intermediate of an *in-situ* generated *N*-(acyloxy)phthalimides (NAP) from aliphatic carboxylic acid (**Figure 18**) [13]. Sherwood used various substituted heteroarenes namely

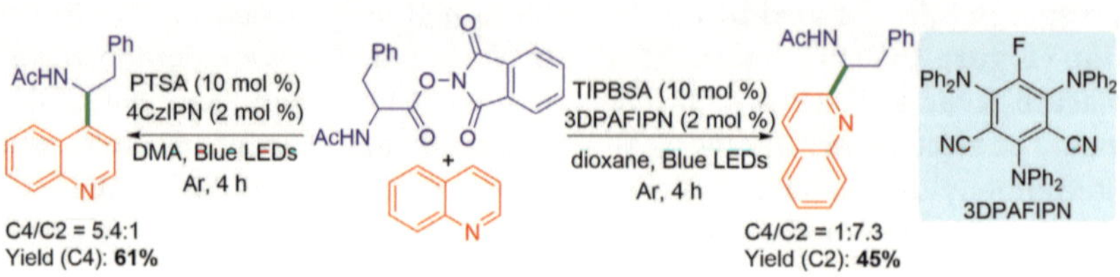

Figure 17.
4CzIPN and 3DPAFIPN-photocatalyzed regioselective Minisci reaction of quinoline with redox-active esters.

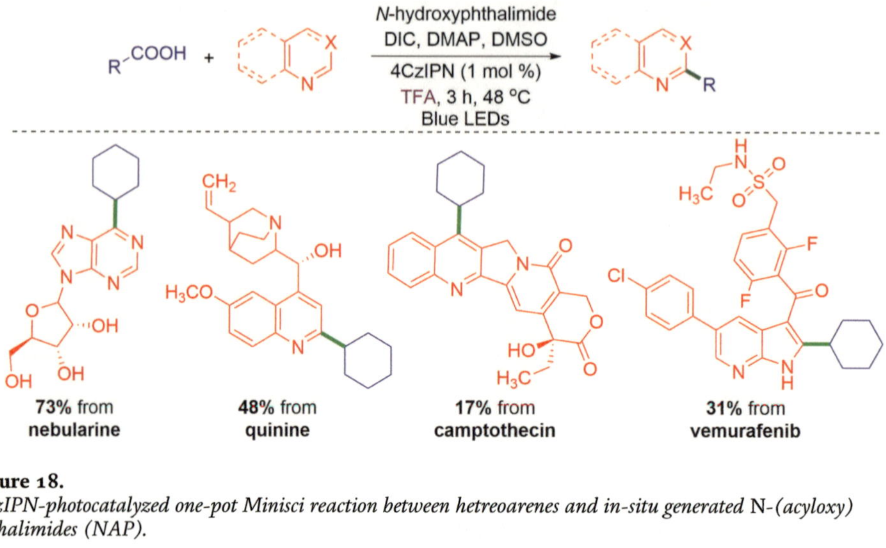

Figure 18.
4CzIPN-photocatalyzed one-pot Minisci reaction between hetreoarenes and in-situ generated N-*(acyloxy) phthalimides (NAP).*

pyridine, quinoline, isoquinoline, quinoxaline, quinaldine, quinazolinone, phthalazine, purine, azaindole, benzimidazole, benzothiazole, benzoxazole, indazole and caffeine to test the efficiency of the reaction protocol. Among these, azaindole, benzoxazole and indazole provided the least amount of product (5%), all other substrates afforded moderate to excellent yields of Minisci functionalized products. In addition, this one-pot reaction protocol showed a high degree of functional group tolerance towards late-stage functionalization (LSF) of nucleosides (nebularine and peracetylated nebularine, adenosine), alkaloids (quinine, camptothecin) and anticancer marketed drug scaffolds namely vemurafenib, imatinib (**Figure 18**) [13].

From mechanistic aspects, when blue LEDs light is exposed to organophotocatalyst 4CzIPN and *in-situ* synthesized *N*-(acyloxy)phthalimides (NAP) (**19A**), the latter underwent reductive fragmentation to generate an alkyl radical (**19B**) and phthalimide (by extrusion of carbon dioxide). This is followed by SET from excited 4CzIPN* to a phthaloyl radical. The alkyl radical (**19B**) was added to the electron-deficient *N*-containing protonated heteroarenes (**19C**) to form adducts (**19D, E**). The adduct (**19E**) further underwent a second SET with the oxidized form of the organophotocatalyst 4CzIPN$^{\bullet+}$ to give a Minisci product (**19F**) and regenerated the catalyst (**Figure 19**) [13].

In the afore-mentioned Minisci protocols (**Figures 17** and **18**), acid additives were used in the reaction medium [12, 13]. In 2019, Graham and Noonan demonstrated an acid additive-free, large-scale (67 g) photoredox catalyzed Minisci reaction towards the synthesis of 2,4-dichloro-6-[1-(methylsulfanyl)cyclopropyl]pyrimidine (**Figure 20**) [16]. This protocol reduces four reaction steps in the classical production

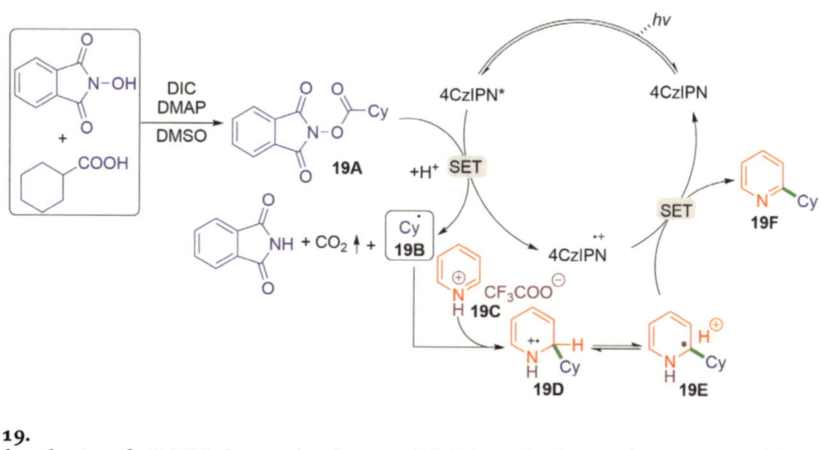

Figure 19.
Proposed mechanism of 4CzIPN-photocatalyzed one-pot Minisci reaction between hetreoarenes and in-situ generated N-*(acyloxy)phthalimides (NAP).*

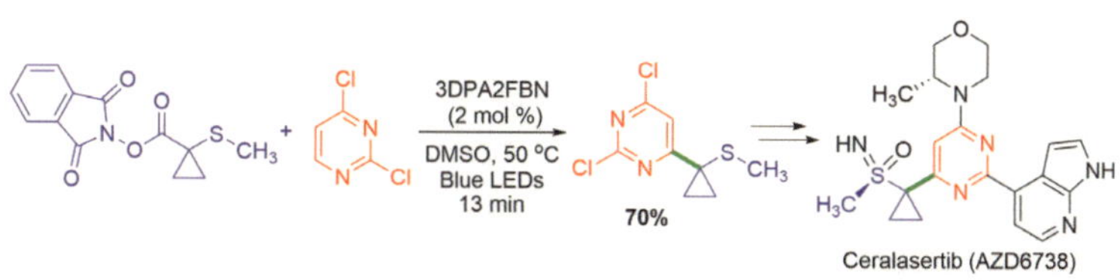

Figure 20.
3DPA2FBN-photocatalyzed one-pot Minisci reaction towards ceralasrtib synthesis.

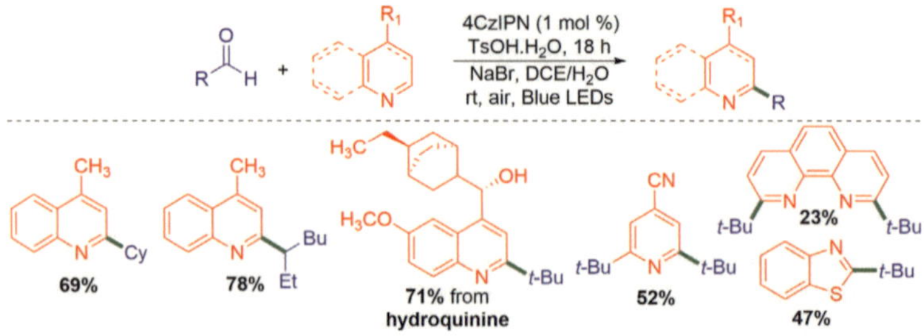

Figure 21.
4CzIPN-photocatalyzed decarbonylative Minisci reaction.

of cancer's phase II clinical trials molecule ceralasertib [16]. Compared to 4CzIPN (50% yield), 3-DPA2FBN (70%) is a more effective photocatalyst for the above-mentioned transformation.

In general most of the Minisci reactions proceeds through the decarboxylation (—CO_2) pathway [12, 13, 15, 16]. Large amounts of oxidants are generally required when aldehydes are used as the radical precursors. In 2019, Huang and Zhao groups disclosed a visible-light-induced photoredox decarbonylative (—CO) Minisci type C—C bond formation (alkylation) between aldehydes and *N*-heteroarenes (**Figure 21**). The reaction proceeded at room temperature and air (O_2) was used as the sole oxidant (**Figure 21**) [17].

This reaction is highly compatible with secondary and tertiary aldehydes. However, primary alkyl aldehydes and aromatic aldehydes failed to deliver decarbonylative Minisci-type alkylated products. The substrate scope of this aerobic photoredox decarbonylative alkylation reaction is decorated by various mono *N*-heteroarenes (quinolines, isonicotinonitrile, isoquinoline, phenanthridine, quinoxyfen, hydroquinine, and benzothiazole) and multi-nitrogen containing heteroarenes such as 1,10-phenanthroline, phthalazine, quinoxaline and imidazo[1,2-a]pyridine. Mono-nitrogen heteroarenes provided a good yield of Minisci-type alkylation products. On the other hand, multi-nitrogen-containing heteroarenes delivered a lower yield of products (**Figure 21**) [17].

The author proposed a plausible reaction mechanism, as shown in **Figure** 22. Visible light-induced photoexcited catalyst 4CzIPN* underwent SET with O_2 to form a superoxide radical anion ($O_2^{\cdot -}$). This superoxide radical anion abstracts a hydrogen atom from aldehyde to produce an acyl radical (**22A**). Decarbonylation of an acyl radical (**22A**) generates an alkyl radical (**22B**). The alkyl radical (**22B**) addition to the protonated *N*-heteroarene (**22C**) leads to *N*-heteroarene radical cation (**22D**). Further, sequential deprotonation of radical cation (**22D**) and SET between alpha alkyl radical and $[4CzIPN]^+$ afforded final alkylated quinolone product (**22E**) and regeneration of photocatalyst 4CzIPN (**Figure** 22) [17].

6. Cross-dehydrogenative Minisci type reactions

Cross dehydrogenative coupling (CDC) reaction is step and atom economical reaction. It plays a vital role in the construction of a diverse array of C—C and C—heteroatom bonds, by functionalizing C—H bonds of all types sp, sp^2, sp^3 [18–23].

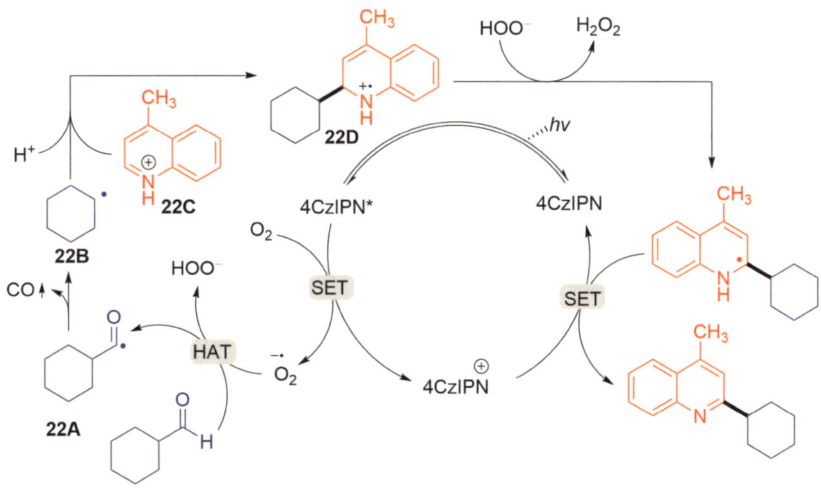

Figure 22.
A plausible mechanism for 4CzIPN-photocatalyzed decarbonylative Minisci reaction.

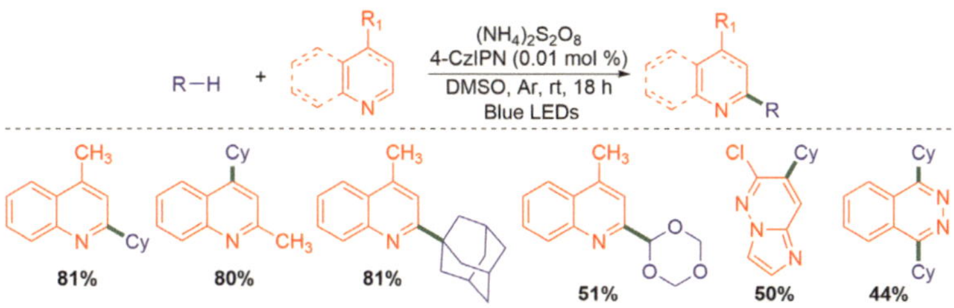

Figure 23.
A plausible mechanism for 4CzIPN-photocatalyzed decarbonylative Minisci reaction.

In 2020 Li and An group demonstrated an acid-free, 4CzIPN photocatalyzed, Minisci reaction between diverse Csp^3—H sources and *N*-heteroarenes (**Figure 23**) [24].

Sun group developed 4CzIPN and quinuclidine-catalyzed direct C—H silylation of quinoxalinones or electron-deficient heteroarenes *via* Cross Dehydrogenative Coupling (CDC) between quinoxalinones Csp^2—H and silane Si—H bond (**Figure 24**). The reaction proceeded with the combination of photoredox (4CzIPN) and hydrogen atom transfer (HAT) catalyst (quinuclidine) (**Figure 24**) [25]. From a mechanistic perspective, a single electron transfer (SET) between photoexcited catalyst 4CzIPN* and quinuclidine (**24.A**) generates 4CzIPN$^{\cdot -}$ radical anion and quinuclidinium radical cation (**24.B**). The hydrogen atom transfer between silane and quinuclidinium radical cation (**24.B**) produced silyl radical (**24.C**) and protonated quinuclidine (**24.D**). Pyridine base accepts hydrogen from protonated quinuclidine (**24.D**) to regenerate hydrogen atom transfer (HAT) catalyst quinuclidine (**24.A**). Meanwhile, the *in-situ* generated silyl radical couple with the quinoxalinones (**24.E**) to form a radical adduct (**24.F**). Parallelly, superoxide radical anion ($O_2^{\cdot -}$) is formed by another single electron transfer (SET) between 4CzIPN$^{\cdot -}$ and $^1O_2$. Finally, the intermediate **24.F** underwent direct hydrogen atom transfer (HAT) with superoxide radical anion ($O_2^{\cdot -}$) to give the desired CDC silylated product **24.G** (**Figure 24**) [25].

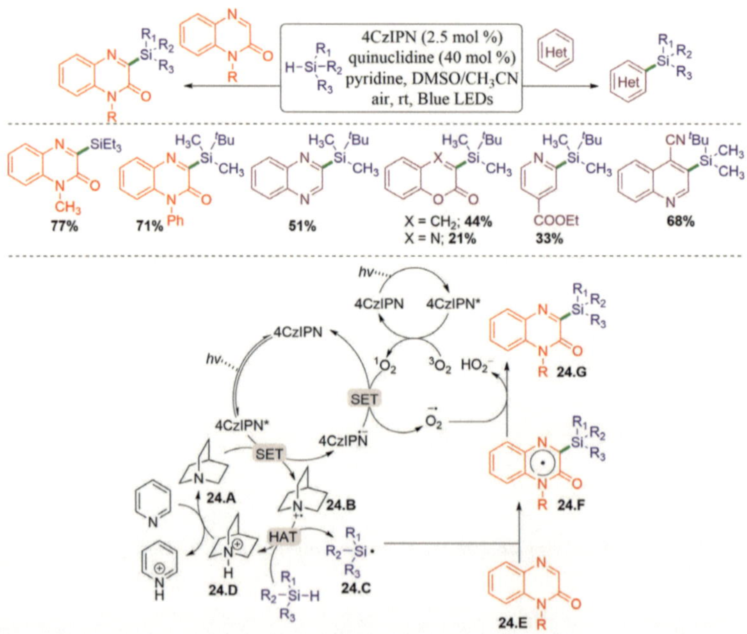

Figure 24.
4CzIPN and quinuclidine catalyzed silylation of quinoxalinones and electron-deficient heteroarenes with alkyl silane via CDC approach.

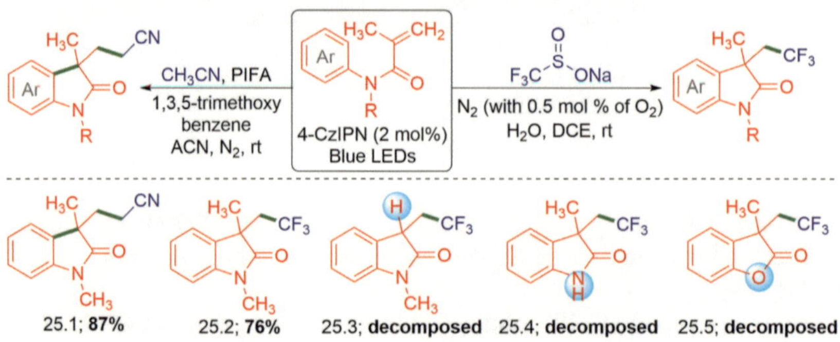

Figure 25.
4CzIPN-photocatalyzed cascade oxidative aryl-trifluoromethylations and aryl-methylcyanation of N*-aryl acrylamides.*

7. Cyclization reactions

Cai group established a 4CzIPN-photocatalyzed intramolecular cascade oxidative aryl-trifluoromethylations [26] and aryl-methylcyanation [27] of *N*-aryl acrylamides for the synthesis of functionalized oxindole (**Figure 25**). Cai used Langlois reagent (sodium triflinate, CF_3SO_2Na) and acetonitrile as trifluoromethyl and methylcyanation sources, respectively. The aryl-trifluoromethylations reaction proceeded without a strong oxidant. Oxygen present in industrial-grade nitrogen (with 0.5 mol% of oxygen) acts as the oxidant. The use of high purity argon (>99.999%) completely failed to provide any trifluoromethylations product (**Figure 25**). On the other hand, aryl-methylcyanation reaction proceeds in the presence of strong oxidant, phenyliodine *bis*(trifluoroacetate) (PIFA) and an additive 1,3,5-trimethoxybenzene (**Figure 25**). Both aryl-trifluoromethylations [26] and aryl-methylcyanation [27] reactions show broad substrate scope with *N*-aryl, alkyl acrylamides. However, substrates **25.3, 25.4** and **25.5** readily decomposed and failed to provide their desired product (**Figure 25**).

From a mechanistic perspective, the photoexcited 4CzIPN* catalyst decomposes sodium triflinate (CF_3SO_2Na) into CF_3 radical and SO_2. This CF_3 radical is added into alkene of *N*-aryl acrylamide to generate a tertiary carbon center radical (**26.A**) or its resonance oxygen radical (**26.B**). Radical cyclization of **26.A** or **26.B** followed by sequential oxidation and deprotonation steps leads to trifluoromethylated oxindole product **26.C** (**Figure 26**) [26].

The author proposed a plausible reaction mechanism of aryl-methylcyanation of *N*-aryl acrylamides, as shown in **Figure 27**. The additive 1,3,5-trimethoxy benzene reacts with phenyliodine *bis*(trifluoroacetate) PIFA to delivered the diaryliodonium salt **27.A**. Which underwent single electron transfer (SET) with excited photocatalyst 4CzIPN*, generated active iodanyl radical **27.B** species. Which abstract acetic hydrogen from acetonitrile or acetone or dimethylsulfoxide to provide the corresponding

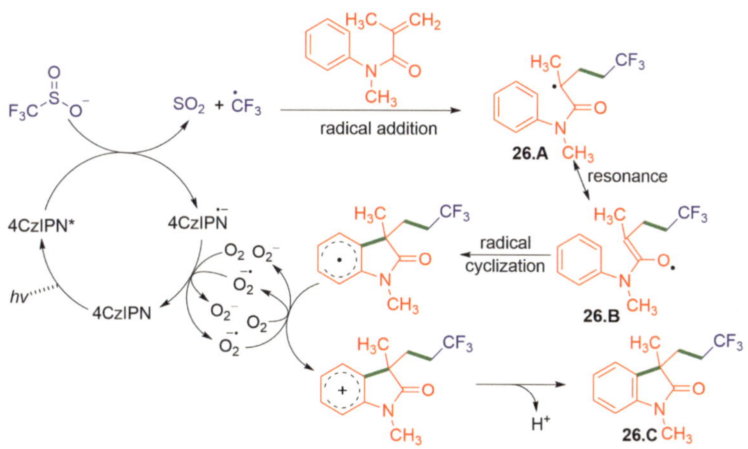

Figure 26.
A plausible mechanism for 4CzIPN-photocatalyzed cascade oxidative aryl-trifluoromethylations and aryl-methylcyanation of N*-aryl acrylamides.*

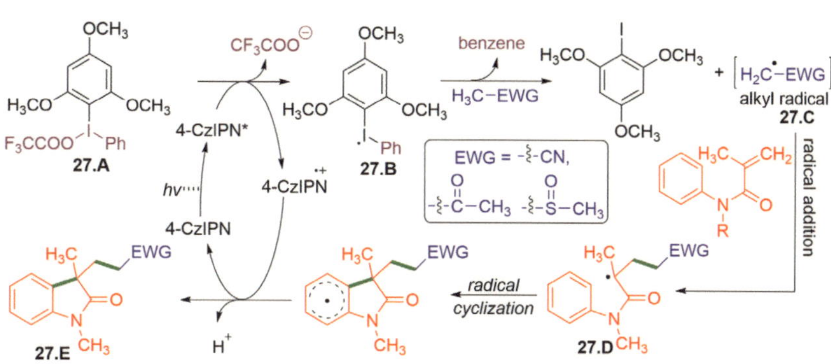

Figure 27.
A plausible mechanism for 4CzIPN-photocatalyzed cascade oxidative aryl-methylcyanation of N*-aryl acrylamides.*

alkyl radical **27.C**. This alkyl radical underwent radical addition into *N*-aryl acrylamide alkene, which leads to stable tertiary carbon center radical **27.D**. A sequential radical cyclization, followed by deprotonation, afforded alkyl functionalized oxindole product (**27.E**) (**Figure 27**) [27].

Cai group further extended the reaction protocol to *N*-benzoyl acrylamides to synthesize cyano, acetone, dimethylsulfoxide and trifluoromethyl substituted isoquinolinediones (**Figure 28**) [26, 27].

In 2021, Yu group demonstrated a 4CzIPN catalyzed cascade cyclization of *N*-arylpropiolamides to 3-phosphorylated, trifluoromethylated or thiocyanated azaspiro [4.5]trienones (spiro-γ-lactam derivatives). In this radical-initiated cascade annulation reaction, diphenylphosphine oxide or diethyl phosphite, 1-trifluoromethyl-1,2-benziodoxol-3(*1H*)-one (Togni's reagent II) or NH_4SCN have been used as phosphoryl, CF_3 and SCN sources respectively (**Figure 29**) [28]. The reaction showed broad substrate scope for *N*-(4-methoxyphenyl)propiolamides bearing different *N*-substituents (R^1) and Ar substituents (**Figure 29**). Interestingly *N*-arylpropiolamide

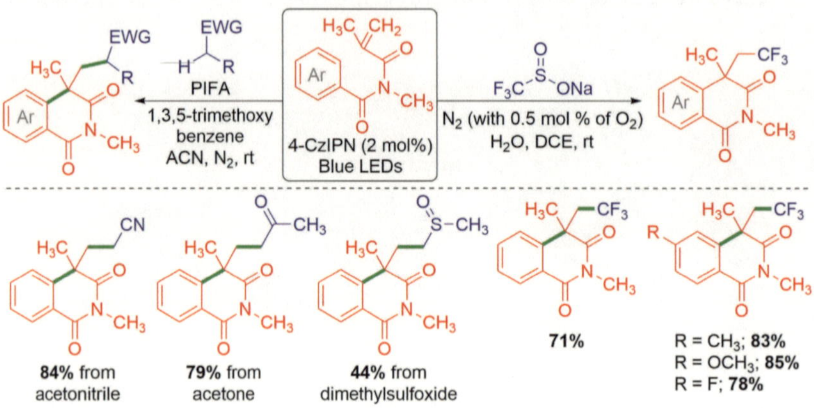

Figure 28.
4CzIPN-photocatalyzed cascade oxidative functionalization of N-*benzoyl acrylamides.*

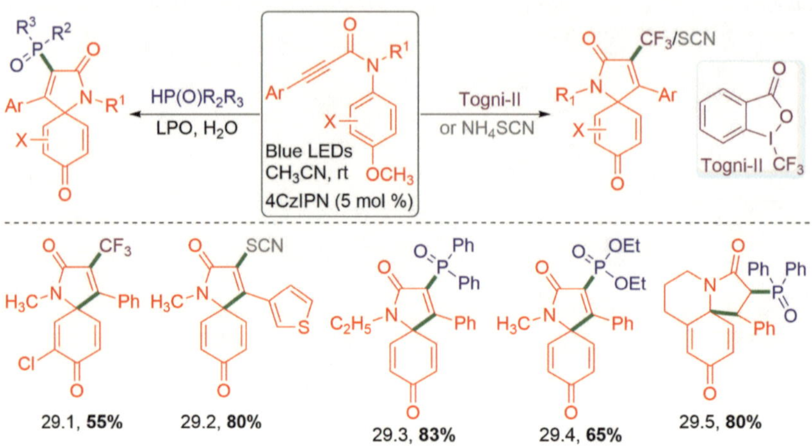

Figure 29.
4CzIPN-photocatalyzed cascade oxidative functionalization of N-*arylpropiolamides.*

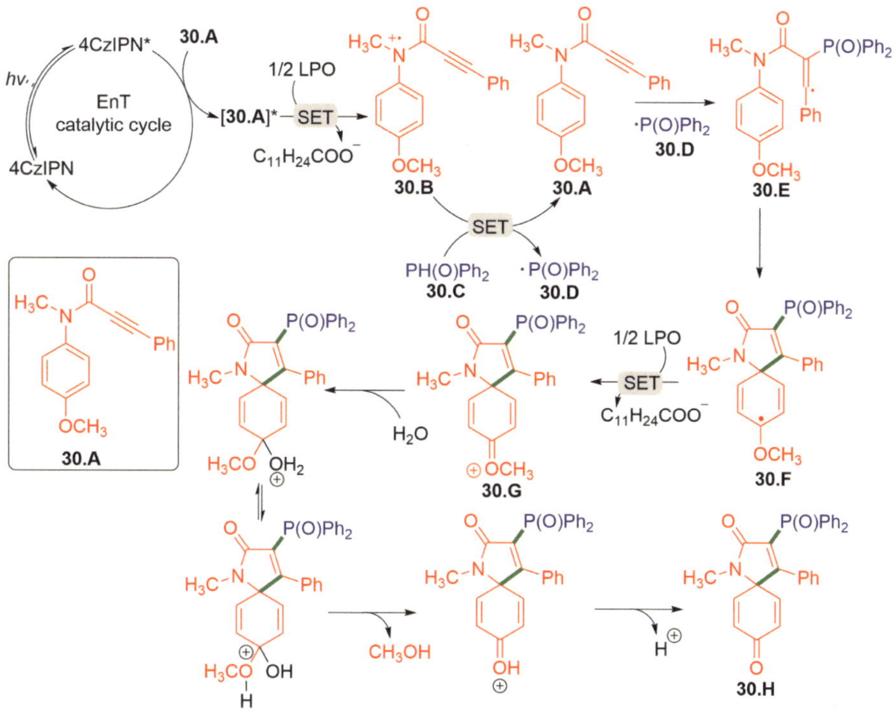

Figure 30.
Proposed mechanism for 4CzIPN-photocatalyzed cascade oxidative functionalization of N*-arylpropiolamides.*

containing a piperidine ring reacts with diphenylphosphine oxide, providing polyfused heterocycle product **29.5** in very good yield (80%) [28].

This cyclization reaction proceeds *via* the combination of energy transfer (EnT) and single electron transfer (SET) mechanism. The photo-excited catalyst 4CzIPN* transfer its energy to *N*-arylpropiolamide **30.A**, to form a high-energy-level **30.A***. Then a SET between lauroyl peroxide (LPO) and **30.A*** generated *N*-arylpropiolamide radical cation (**30.B**) and dodecanoate anion. Afterward, the second SET between diphenylphosphine oxide (**30.C**) and **30.B** provided *N*-arylpropiolamide **30.A** and diphenylphosphoryl radical (**30.D**). This phosphoryl radical (**30.D**) regioselectively added into alkyne of **30.A**, generated alkenyl radical intermediate (**30.E**). Subsequently it underwent intramolecular radical cyclization to form an azaspiro radical (**30.F**). The third single electron transfer reaction between **30.F** and lauroyl peroxide (LPO) leads to an azaspiro cation (**30.G**) and dodecanoate anion. Sequential addition of H_2O, followed by elimination of methanol and deprotonation, provided the desired phosphorylated azaspiro[4.5]trienones product (**30.H**) (**Figure 30**) [28]. A similar reaction mechanism was adopted for triflurомethylated and thiocyanated azaspiro [4.5]trienones synthesis (**Figure 30**) [28].

8. Ring opening reaction

He and co-workers demonstrated a 4CzIPZ catalyzed aerobic oxidative cleavage of unstrained Csp^3—Csp^3 bonds of morpholine derivatives using visible light as the energy source and O_2 as an oxidant (**Figure 31**) [29]. The author proposed that the

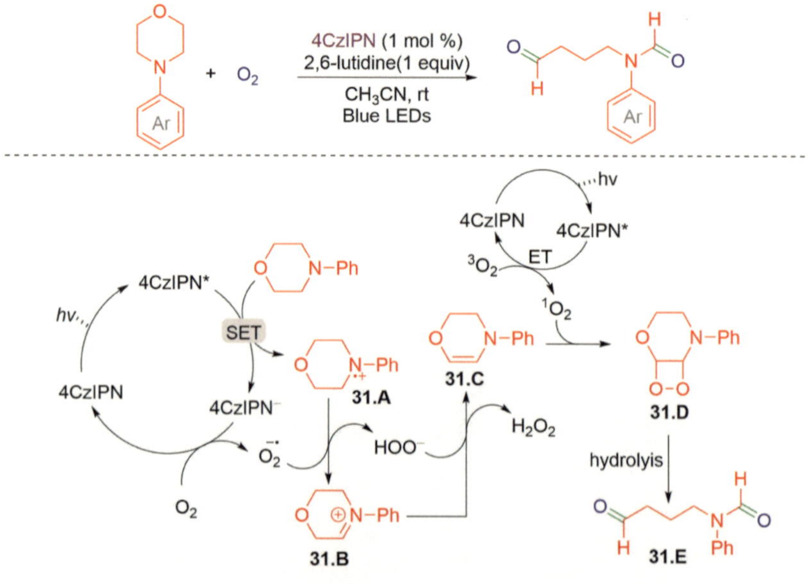

Figure 31.
4CzIPN-photocatalyzed oxidative cleavage of morpholine derivatives.

photoexcited 4CzIPN* was reduced by *N*-aryl morpholine *via* a single electron transfer to generate 4CzIPN$^-$ and aminium radical cation (**31.A**). Simultaneously, oxidation of 4CzIPN$^-$ by O_2 regenerate photocatalyst 4CzIPN and superoxide anion ($O_2^{\cdot -}$). This superoxide anion ($O_2^{\cdot -}$) abstracts a hydrogen atom from the aminium radical cation (**31.A**) to form an iminium intermediate (**31.B**) and HOO$^-$. Afterwards, the HOO$^-$ abstracts a proton from the iminium intermediate (**31.B**) and form an enamine species (**31.C**). Concurrently, singlet oxygen ($^1O_2$) is produced by sensitization with 4CzIPN* *via* energy transfer (ET). This singlet oxygen ($^1O_2$) is reacted with enamine (**31.C**) to form a dioxetane intermediate (**31.D**), which readily decompose and provide the desired oxidative cleaved product (**31.E**) [29].

9. Deuteration reaction

In 2020, Leonori and co-workers incorporated deuterium in unactivated 1°, 2° and 3° alkyl iodide using a combination of synergistic photoredox 4CzIPN catalyst and Bu_3N as the halogen atom transfer (XAT)-agent precursor and methyl thioglycolate—D_2O as the D-atom donor (**Figure 32**) [30].

From mechanistic aspects, the excited photocatalyst 4CzIPN* oxidize Bu_3N followed by deprotonation leads to α-aminoalkyl radical (**33.A**) and 4CzIPN$^{\cdot -}$. The α-aminoalkyl radical (**33.A**) abstract a halogen-atom from the alkyl halide to generate an alkyl radical (**33.B**) and α-iodoamine (**33.C**) *via* XAT. This α-iodoamine (**33.C**) species dissociate into the iminium iodide (**33.D**). Meanwhile, HAT, between deuteriated methyl thioglycolate (**33.E**) and alkyl radical (**33.B**) provided desired deuteriated product (**33.F**) and thiol radical (**33.G**). Finally, a single electron transfer between 4CzIPN$^{\cdot -}$ and thiol radical (**33.G**) regenerate photocatalyst 4CzIPN and methyl thioglycolate (**Figure 33**) [30].

In addition to deuteration reaction, Leonori and co-workers further utilized the *in-situ* generated alkyl radical (**34.B**) towards cross-electrophile coupling between

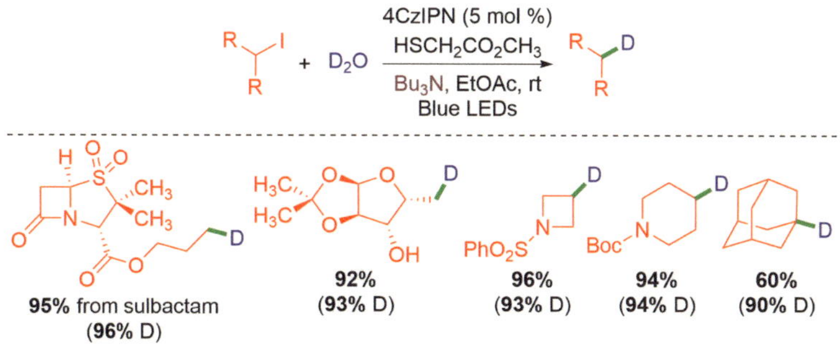

Figure 32.
4CzIPN-photocatalyzed deuteration of alkyl halides.

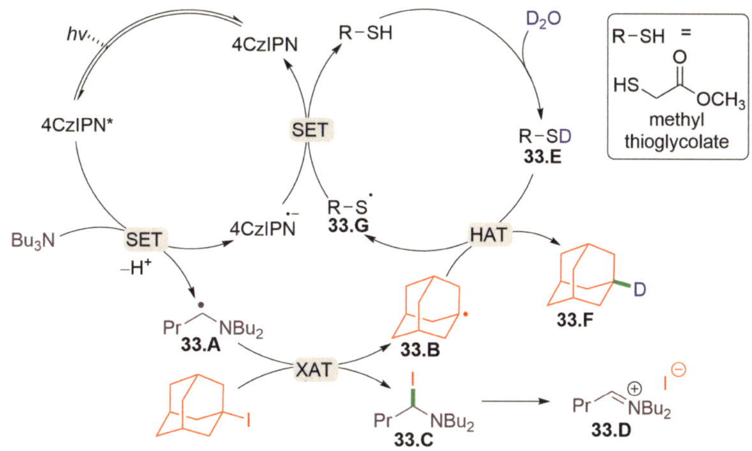

Figure 33.
Proposed mechanism for 4CzIPN-photocatalyzed deuteration of alkyl halides.

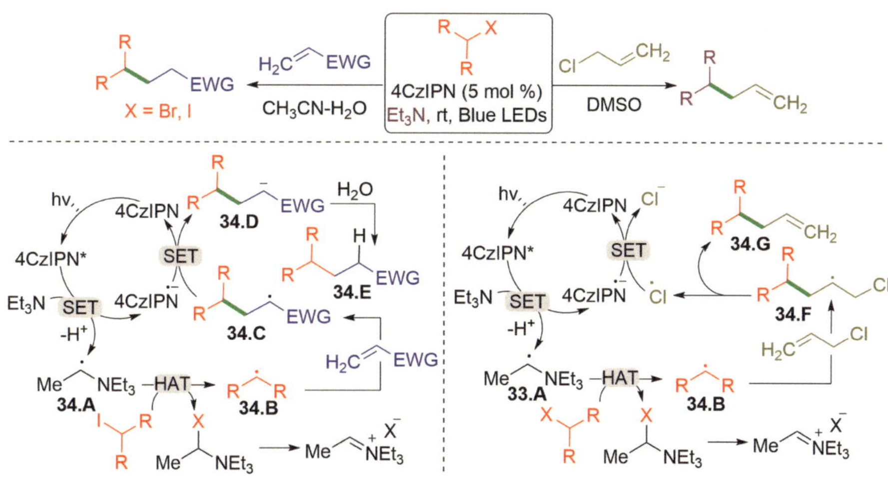

Figure 34.
4CzIPN-photocatalyzed deuteration of alkyl halides. Hydroalkylation and allylation.

electron-deficient olefins or allyl chlorides following Giese-type hydroalkylation mechanism (**Figures 33** and **34**) [30].

10. Conclusions

It is well-known that photocatalytic reactions are powerful tools for a wide range of organic transformations. In this regard, visible-light-induced metal complexes have gained huge attention in the last two decades. Recently TADF materials have been used as an alternative for metal photocatalyst. In this chapter, we summarized a few TADF materials, particularly 4CzIPN as photocatalyst for various radical-based organic transformation reactions. This inexpensive TADF photocatalyst is less toxic and greener. A large number of TADF materials are prepared and used in OLEDs applications. However, only very few TADF molecules are explored in visible light promoted organic transformations. This TADF catalyzed organic transformation reactions are still in its infancy. Many new organo photocatalysts should be discovered for milder organic transformation.

Acknowledgements

B.K. Patel. acknowledges the support of this chapter by SERB (EMR/2016/007042) and CSIR 02(0365)/19-EMR-II. R. Suresh acknowledges the support of this chapter by SERB for funding under the National Post-Doctoral Fellowship scheme SERB-NPDF (PDF/2021/002055) and MRC, IISc Bangalore.

Conflict of interest

The authors declare no conflict of interest.

Author details

Suresh Rajamanickam[1,2] and Bhisma K. Patel[1]*

1 Department of Chemistry, Indian Institute of Technology Guwahati, Guwahati, India

2 Materials Research Centre, Indian Institute of Science, Bangalore, Karnataka, India

*Address all correspondence to: patel@iitg.ac.in

References

[1] Uoyama H, Goushi K, Shizu K, Nomura H, Adachi C. Highly efficient organic light-emitting diodes from delayed fluorescence. Nature. 2012;**492**: 234-238. DOI: 10.1038/nature11687

[2] Wang Y, Carder HM, Wendlandt AE. Synthesis of rare sugar isomers through site-selective epimerization. Nature. 2020;**578**:403-408. DOI: 10.1030/s41586-020-1937-1

[3] Lévêque C, Chenneberg L, Corcé V, Ollivier C, Fensterbank L. Organic photoredox catalysis for the oxidation of silicates: Applications in radical synthesis and dual catalysis. Chemical Communications. 2016;**52**:9877-9880. DOI: 10.1039/C6CC04636C

[4] Phelan JP, Lang SB, Compton JS, Kelly CB, Dykstra R, Gutierrez O, et al. Redox-neutral photocatalytic cyclopropanation via radical/polar crossover. Journal of the American Chemical Society. 2018;**140**:8037-8047. DOI: 10.1021/jacs.8b05243

[5] Milligan JA, Phelan JP, Polites VC, Kelly CB, Molander GA. Radical/polar annulation reactions (RPARs) enable the modular construction of cyclopropanes. Organic Letters. 2018;**20**:6840-6844. DOI: 10.1021/acs.orglett.8b02968

[6] Milligan JA, Burns KL, Le AV, Polites VC, Wang Z-J, Molander GA, et al. Radical-polar crossover annulation: A platform for accessing polycyclic cyclopropanes. Advanced Synthesis and Catalysis. 2020;**362**:242-247. DOI: 10.1002/adsc.201901051

[7] Shu C, Mega RS, Andreassen BJ, Noble Z, Aggarwal VK. Synthesis of functionalized cyclopropanes from carboxylic acids by a radical addition-polar cyclization cascade. Angewandte Chemie, International Edition. 2018;**57**: 15430-15434. DOI: 10.1002/anie.201808598

[8] Huang H, Yu C, Zhang Y, Zhang Y, Mariano PS, Wang W. Chemo- and regioselective organo-photoredox catalyzed hydroformylation of styrenes via a radical pathway. Journal of the American Chemical Society. 2017;**139**: 9799-9802. DOI: 10.1021/jacs.7b05082

[9] Zhang O, Schubert JW. Derivatization of amino acids and peptides via photoredox-mediated conjugate addition. The Journal of Organic Chemistry. 2020;**85**:6225-6232. DOI: 10.1021/acs.joc.0c00635

[10] Jiang H, Studer A. Transition-metal-free three-component radical 1,2-amidoalkynylation of unactivated alkenes. Chemistry-A European Journal. 2019;**25**:516-520. DOI: 10.1002/chem.201805490

[11] Xu N-X, Li B-X, Wang C, Uchiyama M. Sila- and germacarboxylic acids: Precursors for the corresponding silyl and germyl radicals. Angewandte Chemie, International Edition. 2020;**59**: 10639-10644. DOI: 10.1002/anie.202003070

[12] Hadrys BW, Phipps RJ. Acid and solvent effects on the regioselectivity of Minisci-type addition to quinolines using amino acid derived redox active esters. Synlett. 2021;**32**:179-184. DOI: 10.1055/s-0040-1707888

[13] Sherwood TC, Li N, Yazdani AN, Dhar TGM. Organocatalyzed, visible-light photoredox-mediated, one-pot Minisci reaction using carboxylic acids via *N*-(acyloxy)phthalimides. The Journal of Organic Chemistry. 2018;**83**:

3000-3012. DOI: 10.1021/acs.joc.8b00205

[14] Minisci F, Bernardi R, Bertini F, Galli R, Perchinummo M. Nucleophilic character of alkyl radicals-VI: A new convenient selective alkylation of heteroaromatic bases. Tetrahedron. 1971;**27**:3575-3579. DOI: 10.1016/S0040-4020(01)97768-3

[15] Proctor RSJ, Phipps RJ. Recent advances in Minisci-type reactions. Angewandte Chemie, International Edition. 2019;**58**:13666-13699. DOI: 10.1002/anie.201900977

[16] Graham MA, Noonan G, Cherryman JH, Douglas JJ, Gonzalez M, Jackson LV, et al. Development and proof of concept for a large-scale photoredox additive-free Minisci reaction. Organic Process Research and Development. 2021;**25**:57-67. DOI: 10.1021/acs.oprd.0c00483

[17] Wang Z, Ji X, Zhao J, Huang H. Visible-light-mediated photoredox decarbonylative Minisci-type alkylation with aldehydes under ambient air conditions. Green Chemistry. 2019;**21**:5512-5516. DOI: 10.1039/c9gc03008e

[18] Rajamanickam S, Saraswat M, Venkataramani S, Patel BK. Intermolecular CDC amination of remote and proximal unactivated Csp^3–H bonds through intrinsic substrate reactivity—expanding towards a traceless directing group. Chemical Science. 2021;**12**:15318-15328. DOI: 10.1039/D1SC04365J

[19] Rajamanickam S, Majji G, Santra SK, Patel BK. Bu_4NI catalyzed C–N bond formation via cross-dehydrogenative coupling of aryl ethers (Csp^3–H) and tetrazoles (N–H). Organic Letters. 2015;**22**:5586-5589. DOI: 10.1021/acs.orglett.5b02749

[20] Mir BA, Banerjee A, Santra SK, Rajamanickam S, Patel BK. Iron(III)-catalyzed peroxide-mediated C-3 functionalization of flavones. Advanced Synthesis and Catalysis. 2016;**358**:3471-3476. DOI: 10.1002/adsc.201600565

[21] Majji G, Rout SK, Rajamanickam S, Guin S, Patel BK. Synthesis of esters via sp^3 C-H functionalization. Organic and Biomolecular Chemistry. 2016;**14**:8178-8211. DOI: 10.1039/C6OB01250G

[22] Majji G, Rajamanickam S, Khatun N, Santra SK, Patel BK. Generation of *bis*-acyl ketals from esters and benzyl amines under oxidative conditions. The Journal of Organic Chemistry. 2015;**80**:3440-3446. DOI: 10.1021/jo502903d

[23] Rajamanickam S, Sah C, Mir BA, Ghosh S, Sethi G, Yadav V, et al. Bu_4NI-catalyzed, radical-induced regioselective *N*-alkylations and arylations of tetrazoles using organic peroxides/peresters. The Journal of Organic Chemistry. 2020;**85**:2118-2141. DOI: 10.1021/acs.joc.9b02875

[24] Tian H, Yang H, Tian C, An G, Li G. Cross-dehydrogenative coupling of strong $C(sp^3)$-H with *N*-heteroarenes through visible-light-induced energy transfer. Organic Letters. 2020;**22**:7709-7715. DOI: 10.1021/acs.orglett.0c02912

[25] Dai C, Zhan Y, Liu P, Sun P. Organic photoredox catalyzed C-H silylation of quinoxalinones or electron-deficient heteroarenes under ambient air conditions. Green Chemistry. 2021;**23**:314-319. DOI: 10.1039/D0GC03697H

[26] Lu M, Liu Z, Zhang J, Tian Y, Qin H, Huang M, et al. Synthesis of oxindoles through trifluoromethylation of *N*-aryl acrylamides by photoredox catalysis. Organic and Biomolecular Chemistry.

2018;**16**:6564-6568. DOI: 10.1039/c8ob01922c

[27] Lu M, Zhang T, Tan D, Chen C, Zhang Y, Huang M, et al. Visible-light-promoted oxidative alkylarylation of *N*-aryl/benzoyl acrylamides through direct C-H bond functionalization. Advanced Synthesis and Catalysis. 2019;**361**: 4237-4242. DOI: 10.1002/adsc.201900712

[28] Zeng F-L, Chen X-L, Sun K, Zhu H-L, Yuan X-Y, Liu Y, et al. Visible-light-induced metal-free cascade cyclization of *N*-arylpropiolamides to 3-phosphorylated, trifluoromethylated and thiocyanated azaspiro[4.5]trienones. Organic Chemistry Frontiers. 2021;**8**: 760-766. DOI: 10.1039/D0QO01410A

[29] Donga C-L, Huanga L-Q, Guana Z, Huangb C-S, He Y-H. Visible-light-mediated aerobic oxidative C(sp^3)–C(sp^3) bond cleavage of morpholine derivatives using 4CzIPN as a photocatalyst. Advanced Synthesis and Catalysis. 2021;**363**:3803-3811. DOI: 10.1002/adsc.202100455

[30] Constantin T, Zanini M, Regni A, Sheikh NS, Julia F, Leonori D. Aminoalkyl radicals as halogen-atom transfer agents for activation of alkyl and aryl halides. Science. 2020;**367**: 1021-1026. DOI: 10.1126/science.aba2419

Chapter 8

Green Chemistry Applied to Transition Metal Chalcogenides through Synthesis, Design of Experiments, Life Cycle Assessment, and Machine Learning

Alexandre H. Pinto, Dylan R. Cho, Anton O. Oliynyk and Julian R. Silverman

Abstract

Transition metal chalcogenides (TMC) is a broad class of materials comprising binary, ternary, quaternary, and multinary oxides, sulfides, selenides, and tellurides. These materials have application in different areas such as solar cells, photocatalysis, sensors, photoinduced therapy, and fluorescent labeling. Due to the technological importance of this class of material, it is necessary to find synthetic methods to produce them through procedures aligned with the Green Chemistry. In this sense, this chapter presents opportunities to make the solution chemistry synthesis of TMC greener. In addition to synthesis, the chapter presents different techniques of experimental planning and analysis, such as design of experiments, life cycle assessment, and machine learning. Then, it explains how Green Chemistry can benefit from each one of these techniques, and how they are related to the Green Chemistry Principles. Focus is placed on binary chalcogenides (sulfides, selenides, and tellurides), and the quaternary sulfide Cu_2ZnSnS_4 (CZTS), due to its application in many fields like solar energy, photocatalysis, and water splitting. The Green Chemistry synthesis, characterization, and application of these materials may represent sustainable and effective ways to save energy and resources without compromising the quality of the produced material.

Keywords: transition metal chalcogenides, green chemistry, synthesis, design of experiments, life cycle assessment, machine learning

1. Introduction

The term Green Chemistry refers to the strategies for the production and use of safer chemical products as replacements for hazardous substances. In this sense, hazardous substances can be defined in a broad way as any substance representing any physical, such as injury as a result of short or long term exposure; environmental, such as water or air pollution; or toxicological risks, such as mutations or cancer [1].

Although the search for safer reagents and solvents has been an ongoing process in modern chemistry, the term Green Chemistry was coined at the beginning of the 1990's decade. Soon after the establishment of the Pollution Prevention Act of 1990 [2]. Among different proposals, this Act included source reduction as desirable in opposition to waste management in addition to pollution control and more cost-effective production and operation procedures to reduce or prevent pollution generated by industries.

In 1998, Anastas and Warner published the book Green Chemistry: Theory and Practice. This book presented for the first time the *12 Principles of Green Chemistry* [3]. These Principles serve as guidelines for good practices regarding minimization of chemical waste production, mitigation of harmful or hard to treat byproducts, atom economy, development of materials with reasonable degradation period after the end of their lifecycles, and search for safer chemical sources, renewable feedstocks, and energy-efficient processes.

The 12 Principles of Green Chemistry function like an instruction manual for those professionals willing to develop products and processes more aligned with the Green Chemistry concept. These Principles also have contributed to the popularization of Green Chemistry, since they work as a concise and accessible consulting resource. As the 12 Principles of Green Chemistry will be continuously recalled throughout this chapter, they are presented here to provide a quick reference to the readers [3].

Principle 1—Prevent the Waste.
Principle 2—Atom Economy.
Principle 3—Less Hazardous Chemical Synthesis.
Principle 4—Designing Safer Chemicals.
Principle 5—Safer Solvents and Auxiliaries.
Principle 6—Design for Energy Efficiency.
Principle 7—Use of Renewable Feedstocks.
Principle 8—Reduce Derivatives.
Principle 9—Catalyst reagents are preferred over stoichiometric ones.
Principle 10—Design for Degradation.
Principle 11—Real-time Analysis for Pollution Prevention.
Principle 12—Inherently Safer Chemicals for Accident Prevention.

Since the proposal of the Pollution Prevention Act, the Green Chemistry field has grown substantially in the scientific literature.

Many of the terms and parameters related to Green Chemistry were defined considering a molecular structure as a model to assess the safety and toxicological properties. While dealing with nanomaterials, besides molecular structure, other factors such as being crystalline or amorphous, crystal structure, surface area, particle size, porosity, and so on can play substantial roles regarding how a nanomaterial should be evaluated in relation to its production, life cycle, toxicity, and disposal. Hutchison *et al.* [4] published a paper presenting a comparison between the materials context and the molecular context, which were adapted and presented in **Table 1**.

Based on this context, this chapter focus on the strategies, perspectives, and advancements of the greener preparation of transition metal chalcogenides (TMCs). TMCs are part of a broad class of materials comprising binary, ternary, quaternary, and multinary sulfides, selenides, and tellurides. These materials have application in different areas such as solar cells, photocatalysis, sensors, photoinduced therapy, and

Concept	Molecular Context	Materials Context
Composition	Defined by molecular formula	Core and surface composition difficult to define; may vary according to sample shape and size
Size/shape	Defined molecular structure and shape	Often a mixture of sizes and shapes, dependent on synthetic method
Dispersity	Single and continuous composition and structure	Characterized by distributions of composition and structural features
Purity	Purification procedure is intimately related to molecular structure (i.e. chromatography)	Small molecule impurities coming from surface coating or unreacted precursors significantly influence properties
Toxicity	Possible to assess based on molecular structure	It may be inherent to the composition but also can be related to particle size, shape, or surface coating

Table 1.
Comparison between molecular chemistry and materials chemistry context for different concepts.

fluorescent labeling [5–7]. Due to the technological importance of this class of material, it is necessary to find synthetic methods and sophisticated tools to help produce the TMCs nanomaterials through procedures aligned with Green Chemistry.

In this sense, we review the recent literature for the recent advances not only in the chemical synthesis of the TMCs, but also in emerging planning and analysis techniques, such as the design of experiments, life cycle assessment, and machine learning. These emerging techniques can contribute to the further advancement of Green Chemistry.

2. Strategies to make a nanoparticle synthetic process greener

Any change in the synthetic process that eliminates or replaces a hazardous reagent or solvent [8], or is consistent with one of the 12 Principles of Green Chemistry will likely result in an overall process that is more environmentally friendly and less hazardous. Also, it is important to point out that the changes necessary to make the process greener must not compromise the quality of the final product. Green Chemistry, when successfully implemented, results in the green production of high-performance products. If performance is compromised, then the process does not yield a functional product.

There are many opportunities to make the synthesis of transition metal chalcogenides greener. In this section we outline strategies to *green up* the planning, preparation, and analysis of transition metal chalcogenides alongside the Green Chemistry Principles addressed by making the change [4, 9].

2.1 Strategy 1: safer reagents and solvents

The selection of safer reagents and solvents that are unsuitable for producing high-quality materials represents a waste of time and resources. Thus, the best course of action is to first examine results from related work in order to reasonably predict whether a reaction or procedure will be successful using the greener precursors and solvents. Also, working at small scales in the initial stages can represent an economy of time and resources. If the procedure did not work well on a small scale, then one would not proceed to a larger scale procedure. Finally, careful examination of the safety materials associated with each chemical is crucial for preventing problems arising from the combination of incompatible materials or the production of toxic byproducts. These strategies specifically address Green Chemistry Principles 1, 5, and 12.

2.2 Strategy 2: use more efficient energy input sources

The wet-chemical synthesis of transition metal chalcogenides requires some source of energy input (to promote the diffusion process), which is often provided by heating the solution containing the starting materials to temperatures above 200°C. Often, this heating procedure is carried out using a reflux apparatus, which requires the consumption of many liters of water to cool the reflux column. Alternatives to refluxing include reactions assisted by microwaves or ultrasound [10]. Furthermore, procedures that enable the synthesis at lower temperature or even at room temperature represent a greener process. The use of more efficient energy input sources addresses Green Chemistry Principles number 1 and 6.

2.3 Strategy 3: eliminate or minimize byproducts

The reduction or elimination of byproducts can mean little to no post-synthesis purification is required. Indeed, the separation of the desired product from the reaction medium as well as from the undesired byproducts often represents the most waste-generating step. The reduction or elimination of byproducts addresses Green Chemistry Principles 1, 2, 6, and 8.

2.4 Strategy 4: avoid using unnecessary additives and steps

In the synthesis of transition metal chalcogenides, it is common to use capping agents, which are often surfactants, to obtain a certain size and anisotropic shape for the nanoparticles. In many cases, surfactants are necessary to obtain a particular anisotropic shape. However, in some cases, the growth can be controlled by the solvent, by varying the amount of a certain starting material, or adjusting other parameters like temperature, pH, or ionic strength. Avoiding unnecessary additives means less post-synthesis purification is required, and fewer reagents are required overall.

Nanoparticle synthesis methods commonly produce nanoparticles in some non-polar solvent. To use these nanoparticles for some applications often requires dispersion in a polar solvent. When this happens, it is necessary to replace the capping agent that makes the particle dispersible in the non-polar solvent with another capping agent that makes particle dispersible in a polar solvent. This ligand exchange procedure consumes time and additional solvent and reagents. In many cases, ligand exchange can be avoided by simply choosing a synthetic route that yields the nanoparticles with surface chemistry that is suitable for the final application. Avoiding unneeded additives and unnecessary ligand exchange steps directly address Principles number 2, 5, and 6.

2.5 Strategy 5: greener purification procedures

Commonly employed purification procedures include washing nanoparticles with a solvent that can solubilize only the byproducts. Other purification processes are based on the difference in size of the products and byproducts, for instance, the size-exclusion chromatography and dialysis. All these procedures require the use of additional reagents, particularly solvents, which makes the purification procedure one of the most difficult steps to *green up*.

An ideal synthetic procedure will produce the desired product in both high yield and high purity. Indeed, even trace impurities can drastically compromise the performance of devices. In order to reduce the total solvent required, in dialysis, for example, sequential dialysis against smaller volumes of pure solvent will generate less waste and a product with higher purity.

Judicious selection of solvent may also mean that the post-dialysis solvent could be recycled by passing through a purification column, for example. Alternative purification procedures should be investigated to select the method that will be the greenest possible without compromising the purity of the final product. The use of greener purification procedures addresses Green Chemistry Principles 3, 4, 7, and 12.

2.6 Strategy 6: the use of design of experiments

Design of Experiment (DoE) approach helps minimize the number of experiments. The experiment minimization agrees with the Green Chemistry Principles 1, 2, 6, 8, and 11. One way to efficiently decrease the number of experiments needed to fully analyze the data is to apply the concept of Design of Experiments (DoE). The DoE consists of a set of statistical techniques where the experiments are planned and performed according to a multivariate approach. The multivariate approach can be understood as an experimental plan where all the possible factors are varied simultaneously [11].

The multivariate approach contrasts with the univariate approach, which is generally known by the acronym OFAT, meaning one factor at a time. The OFAT approach is usually the standard approach in the chemical literature [12]. For instance, suppose that a research group is interested in studying the effect of temperature, pH, and concentration. And the goal of the research is to maximize the yield of the reaction.

According to the OFAT approach, the group would choose, for instance, five temperature levels, 4 pH values levels, and four concentration levels. And then, they would set a temperature and pH, and find an optimal concentration. Next, they will fix this optimal concentration and vary the temperature and pH in all levels, and find an optimal pH value. Then, finally, they will select the optimal concentration and pH, and vary the temperature in all five levels until finding the optimal temperature.

The OFAT approach, although widely used in the literature, has some drawbacks. The first one is the usual large number of experiments to be performed. The second one is that by fixing one level for all the variables, except the one that will be varied in all levels, can lead to a situation where not all possible experimental conditions were explored. Consequently, there is the chance that the most optimal condition determined is not the actual optimal condition. Another consequence of not varying all the variables at the same time is that all factors are not studied in a connected way [13]. Consequently, it is impossible to analyze the effect of the interaction among two or more factors. Ultimately, it hinders the obtaining of a mathematical model that would allow estimating the yield for conditions initially untested.

In contrast, in the multivariate approach for the same situation, only two levels would be selected for each factor (temperature, concentration, or pH). Then, all three factors are varied simultaneously, leading to a total of eight independent experiments. The results obtained for these eight experiments are analyzed according to a set of algorithms. The output of these calculations would allow estimating not only the effect of each factor independently. But also, the effect of all possible combinations of factors two by two, and the combination of the three factors. Then, after determining which factors and interactions have statistically significant effects, it is possible to refine the calculations and obtain an empirical model that would allow estimating the results for an experimental condition initially untested [14, 15].

This DoE explained in this example is called 2^k full factorial design, where the number two relates to the number of levels, in this case, two for each factor. And the exponent k relates to the number of factors, in this example $k = 3$, due to the factors temperature, concentration, or pH. Therefore, this example explains a 2^3 full factorial design, and the result of the calculation that names the factorial design is equal to the number of independent experiments necessary to complete the factorial design [16]. This is the reason why eight independent experiments were required to complete the factorial design from this example.

2.7 Strategy 7: The use of life cycle assessment (LCA)

The use of life cycle assessmentFollowing a system thinking approach Life Cycle Assessments (LCA) are designed to evaluate and assess the potential environmental, economic, and societal impacts related to the sourcing of reagents, processing, distribution, use, and disposal of materials [17–19]. By listing, mapping, and evaluating the safety and suitability of the material and energy inputs, products, and byproducts, it is possible to compare distinctive synthetic methods or different possible products by determining relevant metrics focusing on process intensity (similar to atom economy), toxicity, and cost [20–23].

LCAs focused on metal nanoparticles have linked high energy consumption to upstream metal refining and been used to screen reducing agents indicating how these methods serve to analyze the literature and help tailor synthetic protocols [24, 25]. LCAs of manufactured photovoltaic cells with chalcogenides have been used to determine whether other components in these systems such as steel or glass contribute to downstream impacts helping to place research in a wider context [18].

While LCA methods are specific and tailored to a given system, examining analyses of metal chalcogenides and related green nanoparticle systems through metrics-based assessments can inform the design of transition metal chalcogenide nanoparticles and help mitigate unwanted impacts [22, 25, 26]. Researchers may look to blend their own experimental data with literature data to support claims of innovation or sustainability with quantitative analyses using a life cycle approach towards making and using nanoparticulate matter [27–29]. Life Cycle Assessment touches on many of the principles of green chemistry and specifically principles 1, 4, 6, and 10.

2.8 Strategy 8: the role of machine learning in predicting materials properties

Data-driven approaches are found helpful in numerous fields of material science, especially when they are paired with computational methods [30, 31], where the data can be generated in a high-throughput fashion, with consistent quality. Data-driven methods are also beneficial to a traditional synthesis-oriented areas, especially due to digitalization of information (for example, lab notebook) processing the wealth of experimental notes becomes possible [32, 33]. The machine-learning applications in chemistry currently focus on property prediction (ranging from mechanical properties to electronic structure) [34–36] rather than structure prediction and exploratory synthesis guidance [37–41]. The areas of machine-learning application in materials science, include solar cells, perovskites, and non-centrosymmetric structures, which echoes with chalcogenides' typical industrial applications.

Being one of the most rapidly emerging fields nowadays, machine learning quickly went through typical stages of method development and crystallized in the list of best practices for applying machine learning in materials domain [42]. For example, sharing entire code of the model, along with the input data and pre-processing methods, gave research publications transparency and promoted sharing the ideas to the next level. Democratization of data, approaches, and informatics allows domain experts to be part of the machine learning community, even with limited knowledge in computer science.

The main benefits of machine learning methods for materials are: (i) analysis of complex correlations between parameters and output, *e.g.*, synthesis conditions and crystal structures or composition and property; (ii) optimization of synthesis conditions; (ii) prediction of candidates with desired properties. In short, machine learning allows fast and detailed analysis of the available data to provide a list of potential candidates, which we can synthesize with fewer experiments.

Machine-learning approaches can guide us towards the direction of narrowing materials candidate pool, which eventually results in less waste (Principle 1). Targeted

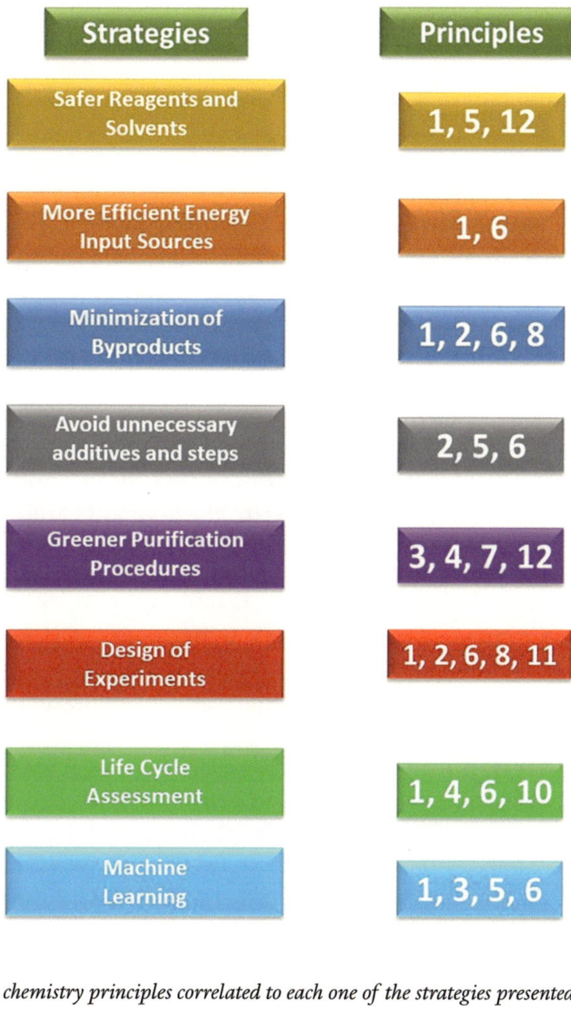

Figure 1.
Summary of the green chemistry principles correlated to each one of the strategies presented.

material selection minimizes the risk of exploring undesired composition, e.g. minimizing hazardous chemical content (Principle 3 and 5). Machine learning combined with DoE helps to optimize newly discovered material to improve their performance, which seconds the efficiency principle (Principle 6).

The correlation between each one of the strategies presented and the Green Chemistry Principles is shown in **Figure 1**.

3. Applications of green chemistry principles in the synthesis of transition metal chalcogenides

Transition metal chalcogenides (TMC) constitute an important class of materials and include different types of oxides, sulfides, selenides, and tellurides. Increasing interest has been devoted to TMC due to their technological applications in different fields such as photocatalysis, sensors, solar cells, supercapacitors, electrocatalysis, heterogeneous catalysis, and many other applications [43]. In view of the growing need and interest for this class of materials, it is necessary to find ways to produce them through solution chemistry synthetic routes that minimize the environmental impacts and health-related risks. In order to demonstrate that it is possible to produce TMC by greener routes, throughout this section, we will show some successful examples where changes in the synthetic process yielded substantial improvements by decreasing environmental impacts and biological risks.

3.1 Solution-based synthesis of transition metal chalcogenides nanoparticles

Quantum Dot is a class of oxides, sulfides, selenides, and tellurides with a particle size smaller than the Bohr radius for that certain material. Consequently, quantum dots are subjected to a phenomenon called quantum confinement, where their absorption and emission on UV–visible range happens at higher energy as the particle size decreases [44, 45].

Initially, the synthesis of quantum dots was established by Bawendi and co-workers, in 1993 [46]. They were able to prepare monodisperse quantum dots of CdS, CdSe, and CdTe, with controlled crystallite size between 1.2 and 11.5 nm. Two slightly different methods were used to produce these cadmium chalcogenides quantum dots. Both methods were based on the hot-injection of the organometallic Cd, and S, Se, or Te sources in a hot trioctylphosphine oxide (TOPO).

In the first method, the source of Cd was dimethyl cadmium (Me_2Cd), and elemental Se and Te. Me_2Cd was dissolved in trioctylphosphine (TOP), Se, and Te were also mixed with TOP to form the organometallic compounds trioctylphosphine selenide (TOPSe) and trioctylphosphine telluride (TOPTe). The source of Cd, plus TOPSe or TOPTe were injected at TOPO around 200°C, and growth proceeded for temperatures between 230 and 260°C.

In the second method, the solvent (TOPO) and source of Cd (Me_2Cd and TOP) were kept, but the sources of S, Se, and Te were replaced, respectively, by bis(trimethylsilyl) sulfide ($(TMS)_2S$), Bis(trimethylsilyl)selenium ($(TMS)_2Se$), and Bis(tert-butyldimethylsilyl)tellurium ($(BDMS)_2Te$). The growth temperature was between 290 and 320°C for larger particles, and around 100°C, for particles having size around 1.2 nm.

Since the pioneer paper by Bawendi et al. [46], this hot-injection method using TOP/TOPO as solvent has been used for the synthesis of many binary chalcogenides

(ZnS [47], ZnS doped by Mn [48], ZnSe [49], CdTe [50]). However, there are many opportunities to make this hot-injection synthesis more environmentally friendly and less harmful. One point for improvement is that the solvents TOP and TOPO are both considered to cause severe skin burns and eye damage [51, 52]. The Me_2Cd is a flammable and pyrophoric compound that can ignite spontaneously if exposed to air, release flammable gases when in contact with water, and cause severe skin burns and eye damage, and respiratory irritation [53]. Additionally, Peng *et al.* [54] reported that Me_2Cd supposedly decomposes generating some insoluble metallic cadmium when in contact with TOPO. To overcome all these problems related to the use of Me_2Cd, Peng's group developed a greener version of the hot-injection route in TOP/TOPO, where they replaced the Me_2Cd by cadmium oxide (CdO) [55]. Going more in depth, the CdO was solubilized by forming a complex with hexylphosphonic acid (HPA) or tetradecylphosphonic acid (TDPA) at temperatures around 300°C after injection in TOPO. This route enabled the production of monodisperse quantum dots of CdSe, CdS, and CdTe [55]. It also allowed a remarkable shape control, since it was able to produce CdSe quantum dots and rods, just by varying the precursors concentration and growth time and temperature [54].

Looking at Green Chemistry perspective, this CdO route also represents an accomplishment of the principles number 3, 4, 5, 8, and 12. Also, the fact that it was necessary to add, at least one more reagent (HPA or TDPA) does not represent a disagreement with principle 2, since neither HPA nor TDPA presents any known health hazard or toxicity issue [56, 57].

The Greenest solvent possible would probably be water, and theoretically, any pair of ions that can generate an insoluble product would be enough to produce a binary TMC. For instance, to produce CdS, it would be necessary just to find out some sources of Cd^{2+} and S^{2-}, and mix them in water in concentrations that would exceed the CdS solubility product for a certain temperature and pH condition. Although this strategy may work for certain cases, it is very likely to generate products having a broad size distribution, heterogeneous composition, and without any shape control. To avoid these drawbacks, it is necessary to use some capping agent. A completely Green synthesis of CdS was described by Unni *et al.*, [58] where they used $CdSO_4$ and $Na_2S_2O_3.5H_2O$, as cadmium and sulfur sources, and thioglycerol as capping agent. The reagents were mixed in water at 30°C and stirred at the same temperature for 1 hour, producing monodisperse quantum-confined CdS, having crystallite size around 3–5 nm. The photoluminescence (PL) of these CdS quantum dots could be controlled by successfully doping them with Cu^{2+} (PL red-shifted) or Zn^{2+} (PL blue-shifted) [58]. The use of unharmful solvents and room temperature synthesis complies with Green Chemistry Principles 3, 5, 6, and 12.

The aqueous based or hydrothermal syntheses of selenides and tellurides are more difficult to be carried out in comparison to the sulfides one, since it is hard to find non-toxic and stable sources of selenium and tellurium. For instance, both Na_2Se and Na_2Te are toxic and very reactive to be controlled during the reaction [5, 59, 60]. Reduction of elemental Se or Te by sodium borohydride ($NaBH_4$) in an oxygen-free atmosphere has been often used as a way to deliver Se^{2-} and Te^{2-} in aqueous synthesis [60, 61]. However, if this reduction does not happen completely, in case of selenides, amorphous elemental selenium can be generated, and it can crystallize to trigonal selenium, even at room temperature, in a range of few months [62].

Sodium selenite (Na_2SeO_3) or tellurite (Na_2TeO_3) have good solubility in water and can be easily purchased. However, as the oxidation state of Se and Te is +4, respectively, on Na_2SeO_3 and Na_2TeO_3, it is necessary to use an effective reducing agent because Se or Te will need to be first reduced to oxidation state 0, then further reduced to oxidation state –2. The most common reducing agent used in this case has been hydrazine (N_2H_4). The hydrothermal synthesis of selenides and tellurides from Na_2SeO_3 or Na_2TeO_3, using N_2H_4 has been successful preparing many compounds like: CdSe and CdTe nanorods [63, 64], ZnSe nanoflowers [65], ZnTe nanowires [66], NiSe nanoparticles [67], NiTe nanorods [68], among many other transition metal selenides and tellurides [68–70]. Although hydrazine is a versatile solvent enabling the dissolution of many different cations at room temperature [71], its high toxicity, flammability, pyrophoricity, and carcinogenicity make it a solvent hard to handle safely and therefore unsuitable for large scale applications [72].

Aiming to find new ways to deliver Se^{2-} and Te^{2-} to make the aqueous synthesis of selenides and tellurides, Xia *et al.* [73] prepared CdTe/CdSe core-shell quantum dots in water using oxygen-free NaHTe and Na_2SeSO_3 as sulfur sources, and mercaptopropionic acid (MPA) as capping agent. NaHTe can easily be prepared by heating elemental selenium in ethanol with excess $NaBH_4$, in oxygen-free atmosphere [74] also no toxicity information was found for NaHTe. The aqueous route using NaHTe to prepare CdTe proved to be very versatile regarding the capacity to produce nanomaterials from many different shapes, sizes, and PL emission, just by varying pH, counterion on Cd^{2+} source, the thiol capping agent, and the mole ratio between cadmium and tellurium source [75]. The route is adaptable regarding the variety of binary TMC able to be prepared, such as ZnTe [76], Ag_2Te [77], and PbTe [78].

Binary chalcogenides often contain cadmium as the metal cation, which is a toxic and carcinogenic element for humans [79, 80] and an environmental concern regarding its segregation and disposal [81]. Additionally, binary chalcogenides usually have band-gaps higher than 2.5 eV, which is suitable for the absorption of the UV radiation of the electromagnetic spectra, whereas ternary and quaternary chalcogenides typically have band-gaps ranging from 0.5 to 2.0 eV, which is suitable for the absorption of the visible light. Among different types of ternary and quaternary chalcogenides, we will focus our attention to a class of quaternary chalcogenide represented by the formula Cu_2ZnSnS_4, which will be hereafter referred to as CZTS. CZTS is comprised of non-toxic and earth-abundant elements, features that align with Green Chemistry Principles. In addition, its direct band-gap around 1.5 eV and high-absorption coefficient (> 10^4 cm^{-1}) make CZTS a suitable material for many applications, including use as the absorber layer in thin film solar cells [82], counter-electrode material in dye sensitized solar cells [83], visible-light photocatalyst for water pollutant degradation [84], as well as use in heterostructures with Pt or Au for H_2 evolution in water splitting systems [85].

Generally speaking, the successful hydrothermal synthesis of CZTS requires relatively long reaction times (ca. 12–24 h). To speed up the reaction and reduce the overall energy required, microwave heating represents a promising alternative to conventional heating. An important consideration is the capacity of a given solvent to absorb microwaves, which is measured by a parameter called dielectric loss factor ($\tan \delta$). In general, $\tan \delta$ can be classified as high ($\tan \delta > 0.5$), medium ($0.1 < \tan \delta < 0.5$), and low ($\tan \delta < 0.1$) [86]. Water has a $\tan \delta$ of 0.123, which is on the very low side of the medium range. Ethylene glycol, which does not have any toxicity if not ingested [87], has a high $\tan \delta$ of 1.350 [86], which has led to its use for the solvothermal synthesis of CZTS using microwave heating.

Pinto *et al.* developed a microwave-assisted method using ethylene glycol as the solvent to prepare CZTS nanoparticles by controlling the percentage of the kesterite and wurtzite crystal phases by varying the amount of thiourea, and the initial oxidation state of the tin acetate (between Sn^{+2} and Sn^{+4}) used as the tin source [88]. Later, the same group prepared CZTS doped with Co^{+2} ions ($Cu_2(Zn_{1-x}Co_x)SnS_4$) using a similar synthetic microwave-assisted route, having ethylene glycol as solvent. These powders were dispersed in water and deposited in molybdenum-coated substrates, followed by annealing at 600°C producing dense $Cu_2(Zn_{1-x}Co_x)SnS_4$ thin films [89].

In both papers by Pinto *et al.*, the microwave heating procedure was carried out at 160°C for 20 minutes, which is usually much shorter than the time used in the regular solvothermal heating procedures [88, 89].

Table 2 summarizes the main papers presenting some advance towards the Green Chemistry application to the solution-based synthesis of transition metal chalcogenides nanoparticles.

Material	Size and morphology	Reactants	Method and Temperature	Reference
CdS	Quantum confined nanoparticles and nanorods	CdO, HPA, TDPA, TOPO, and S	Hot-injection at 300°C	[55]
CdSe	Quantum confined nanoparticles and nanorods	CdO, HPA, TDPA, TOPO, and Se	Hot-injection at 300°C	[55]
CdTe	Quantum confined nanoparticles and nanorods	CdO, HPA, TDPA, TOPO, and Te	Hot-injection at 300°C	[55]
CdS	Quantum confined nanoparticles between 3 and 5 nm	$CdSO_4$, $Na_2S_2O_3$, water, thioglycerol	Co-precipitation at 30°C	[58]
$CdS{:}Zn^{+2}$	Quantum confined nanoparticles between 3 and 5 nm	$CdSO_4$, $Na_2S_2O_3$, water, $ZnSO_4$, thioglycerol	Co-precipitation at 30°C	[58]
$CdS{:}Cu^{+2}$	Quantum confined nanoparticles between 3 and 5 nm	$CdSO_4$, $Na_2S_2O_3$, water, $CuSO_4$, thioglycerol	Co-precipitation at 30°C	[58]
CdTe/CdSe	Core-shell quantum dots	$CdCl_2$, MPA, NaHTe, Na2SeSO3, water,	Reflux at 100°C to prepare the core CdTe, and around 78°C to prepare the shell CdSe	[73]
Cu_2ZnSnS_4 (CZTS)	Agglomerated particles around 20 nm	Copper (I) acetate, Copper (II) acetate, zinc acetate, tin (II) acetate, tin (IV) acetate, Ethylene glycol, thiourea, L-cysteine, thioglycolic acid, MPA	Microwave heating at 160°C	[88]
Cobalt-doped Cu_2ZnSnS_4	Agglomerated particles around 20 nm	Copper (II) acetate, zinc acetate, tin (II) acetate, cobalt (II), ethylene glycol, thiourea, sodium thioglycolate	Microwave heating at 160°C	[89]

Table 2.
Papers presenting some advance towards the green chemistry application to the solution-based synthesis of transition metal chalcogenides nanoparticles.

3.2 Biological approaches for the synthesis of transition metal chalcogenides

Generally speaking, the chemical synthesis of nanoparticles may involve the use of dangerous and non-biocompatible chemicals. This fact has created a demand for a greener approach to nanoparticles through biological synthesis.

Although the biological approach to quantum dots (QD) synthesis shows extreme promise with attributes such as control over the size and shape of the nanoparticles and biocompatibility, opening a market for the medical use of nanoparticles and TMCs. The promise of biologically synthesized nanoparticles fits right with the theme of *Green Chemistry* principles 1, 2, 3, 4, 8, 10, 11, and 12.

The biological approach typically uses cells within plants or fungi by surrounding the cells with a metal-ion solution that triggers a cell's defense mechanism. These specific cells can be engineered biologically to produce different shaped QDs and TMCs.

Bacterial, microbial, and viral methods offer a promising future for QDs and TMC-nanoparticles in terms of 'going greener'. The sheer variety of microorganisms provides a wide range of biological attributes from the natural selection from millions of years. As a result, these microorganisms are incredibly efficient at excreting the required enzymes to synthesize QDs and TMC-NPs both intra and extracellularly [90].

This synthesis method also generally occurs in relatively tame environments, bypassing the high heat and pressure usually associated with chemical synthesis. CdSe QDs, for example, typically require temperatures in the range of 240–300°C when chemically synthesized [91]. When performed biologically, the synthesis will occur close to room temperature, requiring less input work and falling right in line with Green Chemistry principles 6 and 12.

CdS QDs are other examples of a TMC that has high temperature and hazardous chemicals associated with it. A massive step forward in improving this process was the discovery of a genetically engineered strain of *Escherichia coli* was shown to synthesize CdS QDs *intracellularly* [92]. Intracellular synthesis is generally considered a shortcoming on the bacterial and microbial front of biosynthesis. When QDs are synthesized intracellularly, the steps required to harvest said QDs often proves to negate the benefits of biosynthesis in the first place. To separate the QDs from the *E. coli* cells, the cells were lysed with a hyper acoustic cell grinder and centrifuged. The re-suspended cells were then freeze-thawed at –70° C. The QDs then had to be purified using anion exchange resin columns [92]. This energy and time-consuming process is one of many associated with intracellular synthesis and thus may diminish the purpose of a greener synthesis.

Despite the additional steps necessary to obtain the CdS quantum dots, the intracellular preparation has proven to be capable of generating functional CdS quantum dots. For example, Yan and coworkers produced intracellularly from *E. coli* cells CdS QDs with a fluorescent emission at 470 nm, when excited by UV radiation. Additionally, antimicrobial susceptibility studies showed that the resistance of *E. coli* cells to eight different antibiotics is minimally changed when comparing before and after the CdS production [93].

On the other hand, many other forms of bacteria have been shown to excrete enzymes and synthesize QDs and nanocrystals extracellularly. *Klebsiella pneumoniae* generated CdS quantum particles extracellularly [94]. The bacteria were placed in a nutrient-rich broth that acted as a source for sulfate ions. The sulfate ions were absorbed by the microbe and converted to adenosine phosphosulfate with the ATP sulfurylase secreted by the microbe. The created adenosine phosphosulfate was then phosphorylated into 3'phosphoadinosine phosphosulfate. Then, with the help of another enzyme provided by the microbe (phosphoadinosine phosphosulfate reductase), the 3'phosphoadinosine phosphosulfate was reduced into sulfite, which was further reduced to sulfide [94]. These sulfide ions could then react with metal ions like Cd^{2+} to form CdS NC extracellularly [94]. This biological method comes with inherent advantages that bypass intracellular synthesis drawbacks, using much less energy in the process.

In general terms, control over the nanoparticles size and shape is also seen as a drawback in regard to bacteria and microbes because of the many variables to be controlled in the synthesis. That being said, differences in pH, metal-salt solution concentration, and temperature may all play a key role in dictating the size and shape of the nanoparticles. These conditions have also been found to dictate whether the synthesis occurs intra or extracellularly, both responding to different factors.

Few known species can reliably create specifically shaped nanostructures. For instance, *Streptococcus thermophilus (Str. Thermophilus)* is known to produce ZnS and PbS hollow nano-spheres through a sonochemical process. During the respective processes, the ZnS and PbS nanoparticles cluster on the surface of the cell walls of the bacteria. The bacteria can then be removed using sonication, leaving just the hollow nanostructures built up by the cell walls [95].

The variety of bacteria available in the environment and described in the literature could help us take advantage of the biomineralization processes similar to the ones demonstrated by *Str. Thermophilus.*

Fungi are part of another important category of microorganisms that can be used in the preparation of chalcogenide nanoparticles. For instance, Tudu and coworkers prepared spherical CdS QDs using the *Termitomyces heimii* fungus extracted from mushrooms [96]. The synthesis was carried out in water, using $Cd(NO_3)_2$ and Na_2S as Cd^{2+} and S^{2-} sources, respectively. Then, the mushroom extract was added to the reaction mixture, which was further heated up at 60 to 80°C for 16 hours. Increasing the extract volume added led to a decrease in the particle size accompanied by an increase in the band gap energy, which varied between 2.5 and 2.8 eV. The infrared spectroscopy results revealed the presence of bands related to Cd-S bonds and also of proteins from the *Termitomyces heimii* fungus [96].

The rot fungus *Trametes versicolor* is another example of fungus used in the CdS preparation [97]. The fungus mycelium was added to an aqueous mixture of $Cd(NO_3)_2.4H_2O$, thioacetamide, and mercaptoacetic acid. The reaction mixture pH was raised to 10, and the mixture was shaken at 28°C for 24 hours. In the end of the process, spherical nanoparticles, with an average diameter of 6 nm were formed, according to TEM results. The XRD confirmed the presence of the CdS cubic phase. Similar to other works using fungal biosynthesis, the nanoparticle surface indicated the presence of fungal proteins. The authors hypothesized that the fungal proteins attached to the nanoparticle surface due to a defense mechanism of the mycelium from the Cd^{2+} presence. Then, these proteins chelated to Cd^{2+}, which further reacted with S^{2-} ions, producing the CdS nuclei [97].

Table 3 summarizes the main papers some advance towards the Green Chemistry application of biological structures to the synthesis of transition metal chalcogenides nanoparticles.

3.3 Mechanochemistry synthesis of transition metal chalcogenides

Mechanochemistry is the result of chemical transformations from grinding, milling, and similar changes in mechanical force, and this technique is often conducted without solvents [98].

Material	Size and morphology	Microorganism	Method and Temperature	Reference
CdS	Nanoparticles around 8 nm	*E. coli*	Intracellular. *E. coli* cells incubation and genetic expression at room temperature	[92]
CdS	Nanoparticles around 10 nm	*E. coli*	Intracellular. *E. coli* incubation with $CdCl_2$ in the dark at 37°C for different days	[93]
CdS	Particles ranging from 5 to 200 nm	*Klebsiella pneumoniae*	Extracellular. Biotransformation of Cd^{+2} ions into *K. pneumoniae* cells surface	[94]
PbS	Hollow nanostructures and hollow nanotubes	*Streptococcus thermophilus* (for nanospheres) and *L. bulgaricus* (for nanotubes)	Pb^{+2} ions and thioacetamide were dispersed in the microorganism suspension and sonicated at room temperature for 6 h	[95]
ZnS	Hollow nanostructures and hollow nanotubes	*S. thermophilus* (for nanospheres) and *Lactobacillus acidophilus* (for nanotubes)	Zn^{+2} ions and thioacetamide were dispersed in the microorganism suspension and sonicated at room temperature for 6 h	[95]
CdS	Nanoparticles in the 3 to 5 nm range	*Termitomyces heimii*	Cd^{+2}, microorganism extract, and Na_2S were added in aqueous solution and heated at 60–80°C for 16 in the dark	[96]
CdS	Nanoparticles with size below 10 nm	*Trametes versicolor*	Cd^{+2}, MPA, thioacetamide, and the microorganisms were added to water, pH adjusted to 10. The mixture was incubated at 28°C for 24 h	[97]

Table 3.
Papers presenting some advances towards the green chemistry application of biological organisms and structures for the synthesis of transition metal chalcogenides nanoparticles.

Mechanochemistry has been more recently revived in part because of its alignment with the principles of *Green Chemistry* [99–102]. A typical mechanochemistry application is the use of advanced milling methods, including high-energy and 'online' or real-time analyses. These methods allow for the assessment, and stepwise formation of specific nanoparticle materials and metal organic framework (MOF) polymorphs inaccessible through solution-based techniques [103–105]. Mechanochemistry also has demonstrated to perform various catalytic transformations making it an attractive alternative for carefully controlling the stoichiometry of nanoparticle products and their activity [101].

Mechanochemical syntheses with chalcogenides typically follow one of two routes in either a dry or wet mode [106]. In a dry reaction mode, metal and chalcogenide are milled together, while in wet systems, salts including acetates and sodium sulfide are used [107]. After completion, the reaction mixture is washed, and the sodium acetate or other salts are removed before drying. The reduction or elimination of solvent can have a cascade of impacts by increasing reaction efficiency and reducing costs related to waste disposal and treatment [108].

Mechanochemical methods have been leveraged to synthesize binary and complex metal sulfides and other chalcogenide nanoparticles [106, 109]. By controlling the mechanical energy of a system, chemists and engineers can tailor chemical and structural changes, polymorphic structures, and the materials resulting properties [101, 110, 111]. Short milling times on the order of seconds have yielded Cu_3Se_2 from copper selenide blends at room temperature [112].

Liquid-assisted milling systems have been demonstrated to run at low temperatures reducing energy consumption compared to reactions run at high temperatures [113]. Sonochemical methods represent different ways to promote and control reactions using sound energy in solutions where intense local heating and pressure events occur with short lifetimes [114–116]. These methods may be expanded to address how catalysts are used in fixed bed reactors in industry [117, 118].

The synthesis and use of metal chalcogenides present a green chemistry challenge from both the process safety and the environmental toxicology perspectives [119, 120]. Mechanochemical methods have been used to study the use of inert additives to synthesize copper sulfides in a non-explosive regime. This method allowed to control the obtained polymorph as covellite (CuS) or chalcocite (Cu_2S) [121].

Ohtani *et al*. have used mechanical alloying which is a solid-state technique performed with a high energy planetary ball mill to produce homogenous powder silver, samarium, or copper-based sulfides, selenides, or tellurides, allowing to control the stoichiometry and polymorph obtained [112, 122].

It is essential to highlight the additional use of complementary methods used alongside mechanochemical techniques such as thermal and microwave setups or electrical discharge milling (EDAMM) [106, 123]. All of these serve to extend the range and scope of transition metal chalcogenides and their engineered nanostructures possible to be obtained. Thus, providing accessible methods that spawn compelling functional products such as superconductive $In_XNb_3Te_4$ [109].

In situ monitoring of mechanochemical reactions may improve methods for the real-time detections of polymorphs, products, and potential toxins, in line with the 12th principle of green chemistry. For more comprehensive information about the mechanochemistry scope on preparing TMCs, readers are welcome to check the review article by Baláž *et al*. [106].

4. Applications of design of experiments for transition metal chalcogenides

The DoE can be applied in many situations involving TMCs preparation and performance. For instance, Ribeiro *et al*. used DoE of experiments to prepare CdTe quantum dots aiming to maximize the photoluminescence quantum yield (QY) [124]. These quantum dots were prepared in water, using microwaves as the heating method, fixing the temperature at 100°C for 30 minutes. The water as the solvent, the relatively low temperature, and the short reaction time are features that would be enough to consider this method as green.

However, in addition to these features, a 2^3 full factorial design followed by a central composite design (CCD) was carried out for each capping agent, which corresponds to eighteen experiments with four replicates in the center point. Three capping agents were studied, they were: 3-mercaptopropionic acid, L-glutathione, and 2-mercaptoethanesulfonate. The factors studied were the mole ratio between Cd^{2+} and Te^{2-}, the mole ratio between Cd^{2+} and the capping agent, and the pH. The factorial design allowed them to obtain a quadratic model for the set of experiments for each one of the capping agents. Interestingly, the statistically significant terms vary according to the capping agent used. The response surface graphs obtained for each model revealed that the QY was maximized when the L-glutathione was used as the capping agent, in the following conditions: pH of 9.8, Cd^{2+}/Te^{2-} mole ratio of 1:0.2, and Cd^{2+}/glutathione mole ratio of 1:0.77.

Copper antimony sulfides ($Cu_xSb_yS_z$) are a promising alternative for toxic chalcogenides containing cadmium or lead and for the indium-based semiconductors like $CuInSe_2$ due to the indium scarcity and consequent high price [125]. The $Cu_xSb_yS_z$ has four main polymorphs chalcostibite ($CuSbS_2$), tetrahedrite ($Cu_{12}Sb_4S_{13}$), skinnerite (Cu_3SbS_3), and famatinite (Cu_3SbS_4) [126]. To better understand the formation of each one of these phases, Pretto *et al.* performed a 2^4 factorial design to study the influence of the time (1 or 5 minutes), temperature (200 or 250°C), type of solvent (oleylamine or a mixture oleylamine and diphenyl ether), and the heating method (hot injection or heat up) [127]. The 2^4 factorial design required 16 independent experiments. The factorial design revealed that the heat up method led to the $CuSbS_2$ formation, whereas the hot injection led to the formation of Cu_3SbS_4 phase. The temperature as high as 250°C also led to the selective formation of $CuSbS_2$.

Another promising alternative to the binary Cd or Pb-containing semiconductor quantum dots is the ternary semiconductor $AgInS_2$. The $AgInS_2$ has photoluminescence QY higher than 50%, all over the visible and near infrared range of the electromagnetic spectrum, when passivated with a ZnS shell [128]. With the goal to prepare $AgInS_2$ based semiconductor quantum dots, Soares *et al.* prepared $AgInS_2$/ZnS capped by mercaptopropionic acid (MPA) [129]. The factors studied were the reaction time, temperature, Ag:In ratio, S:In ratio, Zn:In ratio, MPA:In ratio, and pH of the solution. The response factors studied were the PL maximum wavelength, the PL lifetime, and QY.

Due to the large number of factors studied, the authors had to initially perform a fractional factorial design of the 2^{5-1} type, which initially corresponded to 16 independent experiments. This initial design revealed that the significant factors were Ag:In ratio, MPA:In ratio, and the solution pH. Then, a 2^3 factorial design with central composite design was performed considering only the optimization of the three significant factors. The highest QY, which was around 0.46, was obtained for Ag:In equal to 0.1, pH 8.5, MPA:In equal 6. Besides the QY = 0.46, this set of conditions had an emission wavelength maximum at 625 nm, a lifetime emission of 400 ns, a Ag:In:Zn proportion about 1:3:5.

The application of semiconductor nanocrystals in solar cell devices requires the deposition of continuous thin films from the semiconductor materials. To obtain an appropriate thin film demands the optimization of several factors. In this sense, Ramírez *et al.* studied CZTS thin films' deposition by spray pyrolysis, starting from a mixture of the Cu^{2+}, Zn^{2+}, and Sn^{4+} salts in a mixture of acetone and DMSO [130]. The factors studied in this deposition process were the substrate temperature (350 and 450°C), the carrier gas pressure (2 and 4 bar), and the spray pulse time (0.4 to 1.2 s). The response factor studied was the film resistivity, which should be minimized. By performing a 2^3 factorial design with a face-centered central composite design, obtaining an empirical equation quadratic for the substrate temperature, linear for the other terms, and considering significant the interaction between the substrate temperature and pulse time. The lowest value for resistivity was obtained for films deposited at 400°C, carrier gas pressure lower than 0.3 bar, and spray pulse time lower than 0.8 s. For those conditions, the resistivity values between 10 to 52 Ω were obtained.

The examples presented in this section showed how the DoE could be used to decrease the number of experiments performed vastly in different contexts and stages of production of the TMCs semiconductor nanoparticles and thin films. Furthermore, the DoE provides statistical justification for each decision taken and for the empirical models developed, which can not be obtained through the OFAT approach.

5. Applications of life cycle assessment for transition metal chalcogenides

As commented in Strategy 7 in Section 2.7, life cycle assessments (LCA) can contribute to developing a greener process by analyzing the resources and energy input, the cost-effectiveness, and the social and economic impacts involved in that specific process.

To illustrate the utility of life cycle assessments, a simplified LCA was performed by aggregating and reframing data from several different routes to distinctive copper sulfide nanosystems [22, 25]. Five different synthetic protocols were selected from the literature and with open-access resources these data are used to calculate metrics related to economic, societal, and environmental impacts associated to the chemicals and relative amounts in the protocols [131–135].

The selected synthetic protocols for copper sulfides share similarities. Most use different amounts of dodecanethiol as a source of sulfur and capping agent, along with solvents such as ethanol and chloroform for workup [131–135]. Despite common reagents, the methods and routes vary across the protocols allowing for access to downstream heterostructured nanoparticles [131], generalized procedures (towards an array of TMCs nanosystems) [132], a one-step synthesis and assembly [133], control over product shape [134], and solventless techniques [135]. These pieces of information from these five protocols were extracted from the literature and using physicochemical reference data [136, 137], SDS, and online databases [138]. The above metrics were calculated for the different sets of methods (**Table 4**).

For this analysis, five metrics were selected to demonstrate the flexibility of this technique: process mass intensity, material cost, global warming potential, smog

Life Cycle Assessment of Copper Sulfide Nanosynthesis						
Procedure	**Process Mass Intensity**	**Material Cost**	**Global Warming Potential**	**Smog Formation Potential**	**Ingestion Toxicity**	**Product**
1	1.0	1.0	—	1.3	8.7	Cu_2S Hexagonal Nanoplatelets
2	9.3	1.9	—	4.9	50	Cu_9S_5 Digenite phase Nanowire Aggregates
3	5.9	33	3.7	3.7	8.1	Cu_2S Tunable Nanoparticles
4	1.3	8.4	1.0	1.0	1.2	Cu_2S 2D-Nanosheets
5	6.8	49	6.0	—	1.0	Cu_2S Hexagonal Nanoplatelets

Table 4.
Life cycle assessment of copper sulfide nanosystems. Metrics are normalized by column such that the least impact and most benign has a value of 1.

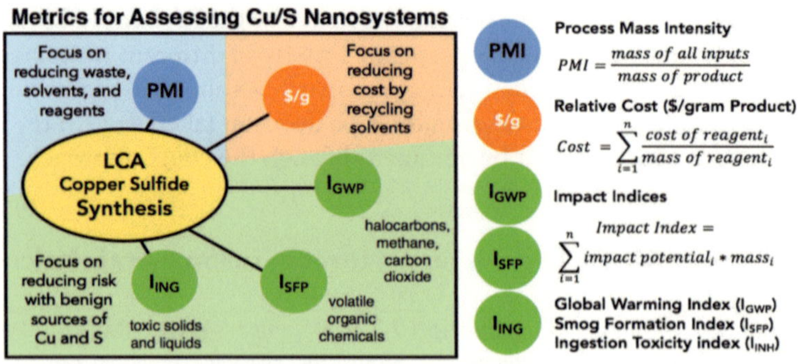

Figure 2.
Metrics used to assess the synthesis of copper sulfide nanosystems and the information used to calculate these metrics.

formation potential, and ingestion toxicity (**Figure 2**) [139–141]. While LCAs typically focus on environmental and energy impacts they may be broadened or adapted to address a variety of economic and societal impacts as with techno-economic assessments which have been performed for evaluating downstream copper recovery from spent electronics [20, 142–144]. An additional metric not used here which is relevant when screening metals with TMCs is the abiotic depletion potential, which examines the impacts of elemental scarcity in products including inorganic materials [26].

Process mass intensity (PMI) is a ratio of the masses of the chemical inputs for a process over the products [139]. Similar to E-Factor and related to atom economy, process mass intensity accounts for byproducts and wastes, including solvents [145]. This may serve to help identify materially efficient transformations contributing to the prevention of waste [146]. Cost is determined as a sum of the scaled costs of each reagent used to synthesize 1 gram of product and does not include operational costs related to energy, nor capital costs for equipment, which are the focus of other LCAs [23, 27].

The final three metrics are related to human and environmental health. They are calculated using available potentials and safety data related to the specific impact of a given chemical in a procedure to a reference chemical such as toluene [141, 145, 147]. The potential for each chemical is multiplied by the relative mass per product and the sum of each potential generates an impact index. The relative magnitude of these indices identifies the expected impact of the procedure on a model environment [145]. Global warming potential (GWP) accounts for the incorporation of greenhouse gases, such as carbon dioxide, chloro- and hydrochlorofluorocarbons [17, 26]. Smog formation potential (SFP) assesses a variety of volatile organic chemicals such as ethanol and acetone, that can partake in reactions with pollutant oxidants at ground level [147, 148]. Ingestion toxicity (ING) models how chemicals may disperse in an external environment and their relative potential to harm living systems [21, 119].

Collecting these metrics, we can get a more holistic perspective of how a specific synthesis procedure can have both up- and down-stream impacts (**Table 4**). Here no one metric is any more important than another, and we quickly realize that there is no one best protocol. This allows us to look back at our work and examine why a specific value is high or low and why and how another procedure has a better or worse value and perhaps adopt a chemical or method from this protocol.

Templates and other resources for this type of LCA are available online with the full set of calculations available of this LCA in the supplementary information [23, 147, 149]. The reader should note the assumptions that accompany this analysis: primarily using a multi-compartmental model to represent an external environment, the standard estimation of amounts for techniques such as washing and filtering, and neglecting to consider the cost of water [141]. The functional unit for this assessment represents a way to standardize the analysis across the procedures for differently scaled references and here was set as 1 gram of product. After calculations are performed the data is normalized per metric such that the most benign or least impactful value is scaled to one.

Assessing a synthesis or comparing several goes beyond setting any one metric. Here one may make the distinction between the cost effective and efficient methods in the first procedure and the more costly, yes less environmentally impactful methods in the fourth protocol. Examining protocol one's high ingestion toxicity parameter we can look back at the calculation to reveal it stems from a large excess of dodecanethiol. As this serves as solvent, a source of sulfur, and a capping agent, one may deem it a necessary hazard though they may also explore in the future how to better balance the stoichiometry of reagents.

Interestingly the solvent-free synthesis, method five, has a high process mass intensity and cost due to solvents such as chloroform that are used in the workup. LCA methods may be used to better model whether a system is greener than alternative methods by leveraging common and novel sustainability metrics [139, 140].

In contrast to the narrow focus on synthesis in the above example, LCAs ideally have a wide scope, including a look at sourcing reagents all the way to recovery or disposal [150]. A fundamental goal of green chemistry and systems thinking is the use of loops to close gaps between wastes and resources. LCAs may be used to address these considerations [151].

Although their power and flexibility LCAs are limited by the quality and accessibility of their source data. Even rudimentary assessments may prove inconclusive for certain values and metrics, note in **Table 4** for dashed entries without a calculated value. Empty values in an LCA may indicate a lack of literature information or imperfect metric, which also provides researchers with a focus on what information is missing from the literature. As LCAs are inherently complex many choose to use proprietary software such as GaBi or ASPEN though researchers must be careful to avoid using LCAs as a black box. To help steer and assess the synthesis, application, scale-up, and recycling of transition metal chalcogenides life cycle assessments can be used to systematically evaluate procedures for green chemists and engineers.

6. Applications of machine learning for transition metal chalcogenides

Proposing new materials requires extensive sustainability, availability, and economic metrics analysis. While reports of individual compounds focus on specific composition metrics, typically considering economic factors. Only within the most recent years, in the extensive review work, it is important to compare the proposed class of compounds (chalcogenides in our case) with other classes, taking into account trends in economics.

In this section, we look at machine learning applied to the chalcogenide materials from three main aspects: First, we present a study showing how machine learning was used to catalog the band gap diversity among materials containing p-block elements. Second, we present and examples from the recent literature where machine learning was applied to predict properties and trends related to TMCs.

6.1 Band gap diversity among materials containing p-block elements

To estimate the fraction of chalcogenide materials among all compounds with report or calculated band gap or absence of the band gap, 6031 reports were analyzed. The band gap values obtained experimentally were summarized in the literature reports [152–155]. For the compounds without band gaps, the data was extracted from the Materials Project database [156]. For this work, only a fraction of unique reports were used, and the full dataset is listed in the supporting information of the manuscript by Zhou *et al.* [157]. Out of 6031 reports, the compounds with elements from Group VII—452, Group VI—3801, Group V—1905, Group IV—1441, and Group III—1676. Out of all p-block elements, Group VI (O, S, Se, Te), mainly consisted of chalcogenides is the most frequently reported (**Figure 3**).

A detailed analysis of the band gap reports for the Group VI elements, revealed that the element are distributed as the following: O—1014, S—1416, Se—1081, Te—614. Interestingly, the reported band gap value follows the trend of shifting the distribution of band gaps from insulating to metallic character for the elements of the Group VI, when going from lower principal quantum number to higher, similarly to the periodic table property change from non-metals to metals (**Figure 4**).

This is the most diverse distribution of the band gaps among all other p-block elements, allowing a band gap engineering for tailoring materials for a specific application need. From the average of 3.5 eV for oxygen 2.0 eV for S, 1.8 eV for Se, and 0.6 eV for Te, chalcogenide materials are the most suitable candidates for semi-conductor synthesis and study.

6.2 Property prediction using machine learning for transition metal chalcogenides

The number of experimentally-confirmed predictions in the field of machine learning chemistry is limited [158]. Commonly, physics-based simulations (molecular

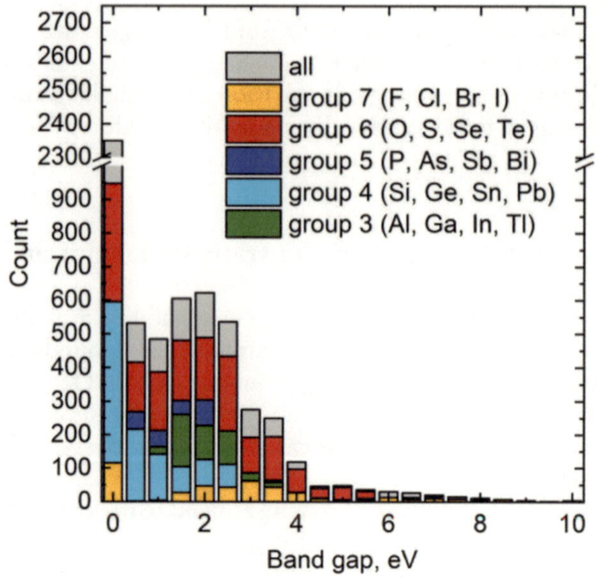

Figure 3.
Absolute number of reports of materials with band gap and with metallic character.

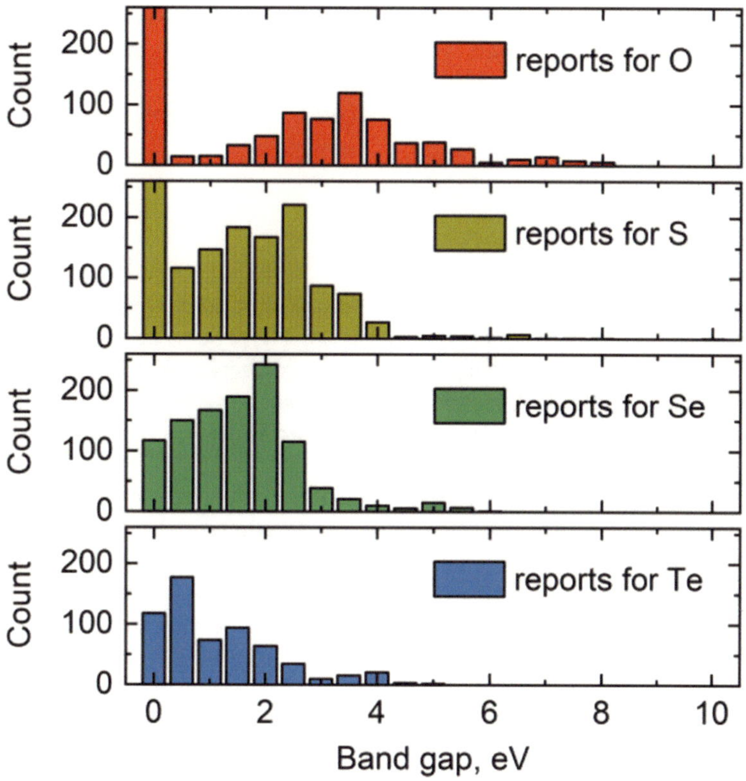

Figure 4.
Band gap energy distribution for different oxides, sulfides, selenides, and telurides.

dynamics or density functional theory) are regarded as experimental validation of a machine-learning model [159]. Developing a functional informatics infrastructure with training data pipeline, selection of appropriate algorithm, and assessment of model performance, requires expertise in both domain knowledge and informatics. This synergy is what converts model's prediction to a real sample of a lab bench.

Combination for ML and DoE makes exploration of optimal synthesis conditions, faster and more efficient. For example, organic solar cell efficiency was increased substantially, increasing the efficiency of solar energy conversion from 6–8%, confirmed with a real working device [160]. Given that solar cell is also an area where chalcogenides are viewed as promising candidates, we can expect similar performance boost with application of machine learning methods [161].

Band gap prediction (fundamental for semiconductor, lighting, sensor, and solar cell areas) is one of the most common property prediction that was tackled with machine-learning methods [157, 162]. Chalcogenides, especially the binary ones with ZnS-type structure, are promising candidates. Transition metals (most common components in binary ZnS-type chalcogenides) are prone to statistical mixing, which allows for a precise control of the band structure in these materials [163]. Ternary chalcogenides were predicted to be promising p-type transparent conductors [162]. The compositional diversity of the chalcogenide-rich candidates is represented by two transition metal elements present (VCu_3S_4), or TM with p-block element (IrSbS), or alkaline-earth with metalloid (Ba_2SiSe_4). Double perovskites of chalcogenides are promising photovoltaics. The formation of perovskites is governed by strict structural geometrical rules along and charge-balanced composition. This puts limitation on the list of possible candidates, however, given that for double perovskites quaternary systems have to be explored, machine learning is essential, since the unexplored chemical white space in quaternary phase diagrams is impossible to explore experimentally [164].

7. Conclusions

This chapter presented many views and examples of how Green Chemistry can be used to benefit TMCs in different scientific fields. These benefits can be explored in many different stages of the TMCs planning and production. For instance, there are different approaches to make TMCs syntetic processes greener, such as using more benign reagents and solvents, milder synthetic conditions. Or using biological media or solventless methods like mechanochemistry.

Additionally, DoE can be used to plan more efficiently the number of experiments necessary to draw certain conclusions and obtain models allowing to predict initially untested synthetic conditions. LCA can be used to predict the risks, benefits, and environmental impacts involved in the production, use, and disposal of TMCs. Machine Learning is important in predicting TMCs properties, which offers useful guidelines for the synthesis of known TMCs. It is also valuable for predicting the structural features and properties of the materials never synthesized, opening up possibilities for discovering new TMCs.

We hope this chapter can be a resource for scientists aiming to make their nanoparticle synthetic processes more benign and environmentally friendly.

Acknowledgements

The authors would like to thank Manhattan College Department of Chemistry & Biochemistry and Manhattan College School of Science for their support.

Conflict of interest

The authors declare no conflict of interest.

Author details

Alexandre H. Pinto*, Dylan R. Cho, Anton O. Oliynyk and Julian R. Silverman
Department of Chemistry and Biochemistry, Manhattan College, Riverdale, NY, USA

*Address all correspondence to: alex.pinto@manhattan.edu

References

[1] Anastas PT, Kirchhoff MM. Origins, current status, and future challenges of green chemistry. Anccounts of Chemical Research. 2002;**35**(9):686-694. DOI: 10.1021/ar010065m

[2] U. S. C. Pollution Prevention Act of 1990. Section 13101-13109. 1990. Available from: http://www.epw.senate.gov/PPA90.pdf%5Cnhttp://www.library.wisc.edu/selectedtocs/bb656.pdf%5Cnhttp://scholar.google.com/scholar?hl=en&btnG=Search&q=intitle:Pollution+prevention+act+of+1990#2

[3] Anastas PT, Warner JH. Green Chemistry: Theory and Practice. New York: Oxford University Press; 1998

[4] Hutchison JE. The road to sustainable nanotechnology: Challenges, progress and opportunities. ACS Sustainable Chemistry & Engineering. 2016;**4**(11):5907-5914. DOI: 10.1021/acssuschemeng.6b02121

[5] Zhuang Z, Peng Q, Li Y. Controlled synthesis of semiconductor nanostructures in the liquid phase. Chemical Society Reviews. 2011;**40**(11):5492. DOI: 10.1039/c1cs15095b

[6] Aldakov D, Lefrançois A, Reiss P. Ternary and quaternary metal chalcogenide nanocrystals: Synthesis, properties and applications. Journal of Materials Chemistry C. 2013;**1**(24): 3756-3776. DOI: 10.1039/c3tc30273c

[7] Talapin DV, Lee J-S, Kovalenko MV, Shevchenko EV. Prospects of colloidal nanocrystals for electronic and optoelectronic applications. Chemical Reviews. 2010;**110**(1):389-458. DOI: 10.1021/cr900137k

[8] National Fire Protection Agency (NFPA). NFPA 704 Standard System for the Identification of the Hazards of Materials and Emergency Response. NFPA 704 Standard System for the Identification of the Hazards of Materials and Emergency Response. 1960. Available from: http://www.nfpa.org/codes-and-standards/all-codes-and-standards/list-of-codes-and-standards?mode=code&code=704

[9] Dahl JA, Maddux BLS, Hutchison JE. Toward greener nanosynthesis. Chemical Reviews. 2007;**107**(6):2228-2269. DOI: 10.1021/cr050943k

[10] Baig RBN, Varma RS. Alternative energy input: Mechanochemical, microwave and ultrasound-assisted organic synthesis. Chemical Society Reviews. 2012;**41**(4):1559-1584. DOI: 10.1039/c1cs15204a

[11] Pinto AH. The importance of factorial design of experiments in functional nanomaterials preparation and performance. In: La Porta FA, Taft CA, editors. Functional Properties of Advanced Engineering Materials and Biomolecules. Switzerland: Springer; 2021. pp. 387-438. DOI: 10.1007/978-3-030-62226-8_14

[12] Behera JK, Zhou X, Ranjan A, Simpson RE. Sb2Te3 and its Superlattices: Optimization by statistical design. ACS Applied Materials & Interfaces. 2018;**10**(17):15040-15050. DOI: 10.1021/acsami.8b02100

[13] Sodipo BK, Aziz AA. Optimization of sonochemical method of functionalizing amino-Silane on superparamagnetic iron oxide nanoparticles using central composite design. Ultrasonics Sonochemistry. 2020;**64**:104856. DOI: 10.1016/j.ultsonch.2019.104856

[14] Lundstedt T, Seifert E, Abramo L, Thelin B, Nyström Å, Pettersen J, et al. Experimental design and optimization. Chemometrics and Intelligent Laboratory Systems. 1998;**42**(1-2):3-40. DOI: 10.1016/S0169-7439(98)00065-3

[15] Montgomery DC. Experimental design for product and process design and development. Journal of the Royal Statistical Society: Series D. 1999;**48**(2):159-177. DOI: 10.1111/1467-9884.00179

[16] Mee R. A Comprehensive Guide to Factorial Two-Level Experimentation. New York, NY: Springer; 2009. p. 545. DOI: 10.1007/b105081

[17] van der Harst E, Potting J. A critical comparison of ten disposable cup LCAs. Environmental Impact Assessment Review. 2013;**43**:86-96

[18] Amarakoon S, Vallet C, Curran MA, Haldar P, Metacarpa D, Fobare D, et al. Life cycle assessment of photovoltaic manufacturing consortium (PVMC) copper indium gallium (di)selenide (CIGS) modules. International Journal of Life Cycle Assessment. 2018;**23**(4):851-866. DOI: 10.1007/s11367-017-1345-4

[19] Guinée JB. Handbook on Life Cycle Assessment: Operational Guide to the ISO Standards. Dordrecht, Netherlands: Kluwer Academic Publishers; 2002. p. 692

[20] Zheng J-L, Zhu Y-H, Zhu M-Q, Sun G-T, Sun R-C. Life-cycle assessment and techno-economic analysis of the utilization of bio-oil components for the production of three chemicals. Green Chemistry. 2018;**20**(14):3287-3301

[21] Hischier R, Walser T. Life cycle assessment of engineered nanomaterials: State of the art and strategies to overcome existing gaps. Science of the Total Environment. 2012;**425**:271-282

[22] Pati P, McGinnis S, Vikesland PJ. Life cycle assessment of "green" nanoparticle synthesis methods. Environmental Engineering Science. 2014;**31**(7):410-420. DOI: 10.1089/ees.2013.0444

[23] Silverman JR, Hudson R. Evaluating Feedstocks, processes, and products in the teaching laboratory: A framework for students to use metrics to design greener chemistry experiments. Journal of Chemical Education. 2020;**97**(2):390-401

[24] Pourzahedi L, Eckelman MJ. Comparative life cycle assessment of silver nanoparticle synthesis routes. Environmental Science. Nano. 2015;**2**(4):361-369

[25] Bafana A, Kumar SV, Temizel-Sekeryan S, Dahoumane SA, Haselbach L, Jeffryes CS. Evaluating microwave-synthesized silver nanoparticles from silver nitrate with life cycle assessment techniques. Science of the Total Environment. 2018;**636**:936-943. DOI: 10.1016/j.scitotenv.2018.04.345

[26] Lunardi MM, Moore S, Alvarez-Gaitan JP, Yan C, Hao X, Corkish R. A comparative life cycle assessment of chalcogenide/Si tandem solar modules. Energy. 2018;**145**:700-709

[27] Zabalza Bribián I, Valero Capilla A, Aranda UA. Life cycle assessment of building materials: Comparative analysis of energy and environmental impacts and evaluation of the eco-efficiency improvement potential. Building and Environment. 2011;**46**(5):1133-1140

[28] Vink ETH, Rábago KR, Glassner DA, Gruber PR. Applications of life cycle assessment to NatureWorks™ polylactide (PLA) production. Polymer Degradation and Stability. 2003;**80**(3):403-419

[29] Grimaud G, Laratte B, Perry N. To transport waste or transport recycling plant: Insights from life-cycle analysis. Matériaux & Techniques. 2018;**105**(5/6):13

[30] Sparks TD, Oliynyk AO, Romaka V. Preface to the special issue on machine learning and data-driven design of materials issue in computational materials science. Computational Materials Science. 2021;**195**:110452. DOI: 10.1016/j.commatsci.2021.110452

[31] Pollice R, Dos Passos GG, Aldeghi M, Hickman RJ, Krenn M, Lavigne C, et al. Data-driven strategies for accelerated materials design. Accounts of Chemical Research. 2021;**54**(4):849-860

[32] Raccuglia P, Elbert KC, Adler PDF, Falk C, Wenny MB, Mollo A, et al. Machine-learning-assisted materials discovery using failed experiments. Nature. 2016;**533**(7601):73-76

[33] Oliynyk AO, Buriak JM. Virtual issue on machine-learning discoveries in materials science. Chemistry of Materials. 2019;**31**(20):8243-8247

[34] Zhang Z, Mansouri Tehrani A, Oliynyk AO, Day B, Brgoch J. Finding the next Superhard material through ensemble learning. Advanced Materials. 2021;**33**(5):1-8

[35] Mansouri Tehrani A, Oliynyk AO, Parry M, Rizvi Z, Couper S, Lin F, et al. Machine learning directed search for Ultraincompressible, Superhard Materials. Journal of the American Chemical Society. 2018;**140**(31):9844-9853

[36] Zhuo, Y., Mansouri Tehrani, A., Oliynyk, A.O. et al. Identifying an efficient, thermally robust inorganic phosphor host via machine learning. Nat Commun. 2018;**9**:4377. https://doi.org/10.1038/s41467-018-06625-z

[37] Gzyl AS, Oliynyk AO, Mar A. Half-Heusler structures with full-Heusler counterparts: Machine-learning predictions and experimental validation. Crystal Growth & Design. 2020;**20**(10):6469-6477. DOI: 10.1021/acs.cgd.0c00646

[38] Gzyl AS, Oliynyk AO, Adutwum LA, Mar A. Solving the Coloring problem in half-Heusler structures: Machine-learning predictions and experimental validation. Inorganic Chemistry. 2019;**58**(14):9280-9289. DOI: 10.1021/acs.inorgchem.9b00987

[39] Oliynyk AO, Mar A. Discovery of intermetallic compounds from traditional to machine-learning approaches. Accounts of Chemical Research. 2018;**51**(1):59-68. DOI: 10.1021/acs.accounts.7b00490

[40] Oliynyk AO, Antono E, Sparks TD, Ghadbeigi L, Gaultois MW, Meredig B, et al. High-throughput machine-learning-driven synthesis of full-Heusler compounds. Chemistry of Materials. 2016;**28**(20):7324-7331

[41] Oliynyk AO, Adutwum LA, Harynuk JJ, Mar A. Classifying crystal structures of binary compounds AB through cluster resolution feature selection and support vector machine analysis. Chemistry of Materials. 2016;**28**(18):6672-6681

[42] Wang AYT, Murdock RJ, Kauwe SK, Oliynyk AO, Gurlo A, Brgoch J, et al. Machine learning for materials scientists: An introductory guide toward best practices. Chemistry of Materials. 2020;**32**(12):4954-4965

[43] Gao M-R, Xu Y-F, Jiang J, Yu S-H. Nanostructured metal chalcogenides: Synthesis, modification, and applications

in energy conversion and storage devices. Chemical Society Reviews. 2013;**42**(7):2986-3017

[44] Trindade T, O'Brien P, Pickett NL. Nanocrystalline semiconductors: Synthesis, properties, and perspectives. Chemistry of Materials. 2001;**13**(11):3843-3858

[45] Smith AM, Nie S. Semiconductor nanocrystals: Structure, properties, and band gap engineering. Accounts of Chemical Research. 2010;**43**(2):190-200

[46] Murray CB, Norris D, Bawendi MG. Synthesis and characterization of nearly monodisperse CdE (E = S, Se, Te) semiconductor nanocrystallites. Journal of the American Chemical Society. 1993;**115**(4):8706-8715

[47] Li Y, Li X, Yang C, Li Y. Ligand-controlling synthesis and ordered assembly of ZnS Nanorods and Nanodots. The Journal of Physical Chemistry. B. 2004 Oct;**108**(41):16002-16011

[48] Azad Malik M, O'Brien P, Revaprasadu N. Synthesis of TOPO-capped Mn-doped ZnS and CdS quantum dots. Journal of Materials Chemistry. 2001;**11**:2382-2386

[49] Jun Y, Koo J-E, Cheon J. One-step synthesis of size tuned zinc selenide quantum dots via a temperature controlled molecular precursor approach. Chemical Communications. 2000;**14**:1243-1244

[50] Talapin DV, Rogach AL, Mekis I, Haubold S, Kornowski A, Haase M, et al. Synthesis and surface modification of amino-stabilized CdSe, CdTe and InP nanocrystals. Colloids and Surfaces A: Physicochemical and Engineering Aspects. 2002;**202**(2-3):145-154

[51] Trioctylphosphine. Safety Data Sheet (SDS). Safety Data Sheet (SDS)–Sigma-Aldrich. http://www.sigmaaldrich.com/catalog/product/aldric. Available from: http://www.sigmaaldrich.com/catalog/product/aldrich/117854?lang=en&region=US

[52] Sigma-Aldrich. Trioctylphosphine Oxide. 2017: pp. 1-7

[53] Gelest. Dimethylcadmium. 2015: pp. 2-9

[54] Peng ZA, Peng X. Mechanisms of the shape evolution of CdSe nanocrystals. Journal of the American Chemical Society. 2001;**123**(7):1389-1395

[55] Peng ZA, Peng X. Formation of high-quality CdTe, CdSe, and CdS nanocrystals using CdO as precursor. Journal of the American Chemical Society. 2001;**123**(1):183-184

[56] Sigma-Aldrich. Hexylphosphonic Acid. 2017: pp. 1-6

[57] Sigma-Aldrich. Tetradecylphosphonic Acid. 2017: pp. 1-6

[58] Unni C, Philip D, Smitha SL, Nissamudeen KM, Gopchandran KG. Aqueous synthesis and characterization of CdS, CdS:Zn2+ and CdS:Cu2+ quantum dots. Spectrochimica Acta, Part A: Molecular and Biomolecular Spectroscopy. 2009;**72**(4):827-832

[59] Hu J, Lu Q, Tang K, Qian Y, Zhou G, Cheng H, et al. Hydrothermal growth of β-Ag2Se tubular crystals. Chemical Communications. 2000;**8**:715-716

[60] Sweat DP, Stephens CE. A modified synthesis of tellurophene using NaBH4 to generate sodium telluride. Journal of Organometallic Chemistry. 2008;**693**(14):2463-2464

[61] Manna A, Bhattacharya R, Das TK, Saha S. Effect of reducing agent in the formation of CdSe nanoparticles by

chemical reduction route. Physica B: Condensed Matter. 2011;**406**(4): 981-984

[62] Pinto AH, Leite ER, Longo E, De Camargo ER. Crystallization at room temperature from amorphous to trigonal selenium as a byproduct of the synthesis of water dispersible zinc selenide. Materials Letters. 2012;**87**:62-65. DOI: 10.1016/j.matlet.2012.07.067

[63] Zuala L, Agarwal P. Thermal and structural studies of CdSe nanorods synthesized by solvothermal process. Physica Status Solidi. 2016;**213**(7): 1885-1893. DOI: 10.1002/pssa.201532962

[64] Gong H, Hao X, Gao C, Wu Y, Du J, Xu X, et al. Facile aqueous synthesis and growth mechanism of CdTe nanorods. Nanotechnology. 2008;**19**(44):445603

[65] Cao F, Shi W, Zhao L, Song S, Yang J, Lei Y, et al. Hydrothermal synthesis and high photocatalytic activity of 3D Wurtzite ZnSe hierarchical nanostructures. Journal of Physical Chemistry C. 2008;**112**(44):17095-17101

[66] Jang YJ, Jang J-W, Lee J, Kim JH, Kumagai H, Lee J, et al. Selective CO production by Au coupled ZnTe/ZnO in the photoelectrochemical CO2 reduction system. Energy & Environmental Science. 2015;**8**(12):3597-3604

[67] Sobhani A, Davar F, Salavati-Niasari M. Synthesis and characterization of hexagonal nano-sized nickel selenide by simple hydrothermal method assisted by CTAB. Applied Surface Science. 2011;**257**(18):7982-7987

[68] Peng Q, Dong Y, Li Y. Synthesis of uniform CoTe and NiTe semiconductor nanocluster wires through a novel Coreduction method. Inorganic Chemistry. 2003;**42**(7):2174-2175

[69] Wang Q, Chen G, Shi X, Jin R, Wang L, Chen D. Controllable synthesis of Cu7Te4 nanoparticles and sheet-like particles through the delayed reaction and their thermal stability. Powder Technology. 2011;**207**(1-3):192-198

[70] Wu M, Xiong Y, Jiang N, Ning M, Chen Q. Hydrothermal preparation of α-MnSe and MnSe2 nanorods. Journal of Crystal Growth. 2004;**262**(1-4):567-571

[71] Yuan M, Mitzi DB. Solvent properties of hydrazine in the preparation of metal chalcogenide bulk materials and films. Dalton Transactions. 2009;**31**:6078-6088

[72] Sigma-Aldrich. Hydrazine Safety Data Sheet. 2017: pp. 1-9. Available from: www.sigmaaldrich.com

[73] Xia Y, Zhu C. Aqueous synthesis of type-II core/shell CdTe/CdSe quantum dots for near-infrared fluorescent sensing of copper(II). The Analyst. 2008;**133**(7):928-932

[74] Barton DHR, McCombie SW. A new method for the deoxygenation of secondary alcohols. Journal of the Chemical Society, Perkin Transactions 1. 1975;(16):1574-1585. DOI https://doi.org/10.1002/anie.201906755

[75] Zhang H, Wang D, Yang B, Möhwald H. Manipulation of aqueous growth of CdTe nanocrystals to fabricate Colloidally stable one-dimensional nanostructures. Journal of the American Chemical Society. 2006;**128**(31):10171-10180

[76] Xu S, Wang C, Xu Q, Zhang H, Li R, Shao H, et al. Key roles of solution pH and ligands in the synthesis of aqueous ZnTe nanoparticles. Chemistry of Materials. 2010;**22**(21):5838-5844

[77] Chen C, He X, Gao L, Ma N. Cation exchange-based facile aqueous synthesis of small, stable, and

nontoxic near-infrared Ag2Te/ZnS Core/Shell quantum dots emitting in the second biological window. ACS Applied Materials & Interfaces. 2013;**5**(3):1149-1155

[78] Xiao Q, Weng D, Yang Z, Garay J, Zhang M, Lu Y. Efficient synthesis of PbTe nanoparticle networks. Nano Research. 2010;**3**(10):685-693

[79] Fowler BA. Monitoring of human populations for early markers of cadmium toxicity: A review. Toxicology and Applied Pharmacology. 2009;**238**(3):294-300

[80] Järup L, Akesson A. Current status of cadmium as an environmental health problem. Toxicology and Applied Pharmacology. 2009;**238**(3):201-208

[81] Fthenakis VM. Life cycle impact analysis of cadmium in CdTe PV production. Renewable and Sustainable Energy Reviews. 2004;**8**:303-334

[82] Henry CH. Limiting efficiencies of ideal single and multiple energy gap terrestrial solar cells. Journal of Applied Physics. 1980;**51**(8):4494-4500

[83] Du Y-F, Fan J-Q, Zhou W-H, Zhou Z-J, Jiao J, Wu S-X. One-step synthesis of stoichiometric Cu2ZnSnSe4 as counter electrode for dye-sensitized solar cells. ACS Applied Materials & Interfaces. 2012;**4**(3):1796-1802

[84] Kush P, Deka S. Anisotropic kesterite Cu2ZnSnSe4 colloidal nanoparticles: Photoelectrical and photocatalytic properties. Materials Chemistry and Physics. 2015;**162**:608-616

[85] Yu X, Shavel A, An X, Luo Z, Ibáñez M, Cabot A. Cu2ZnSnS4-Pt and Cu2ZnSnS4-Au heterostructured nanoparticles for photocatalytic water splitting and pollutant degradation. Journal of the American Chemical Society. 2014;**136**(26):9236-9239

[86] Kappe CO. Controlled microwave heating in modern organic. Angewandte Chemie, International Edition. 2004;**43**(46):6250-6284

[87] Sigma-Aldrich. Ethylene Glycol Safety Data Sheet. 2017: pp. 1-8. Available from: www.sigmaaldrich.com

[88] Pinto AH, Shin SW, Isaac E, Knutson TR, Aydil ES, Penn RL. Controlling Cu 2 ZnSnS 4 (CZTS) phase in microwave solvothermal synthesis. Journal of Materials Chemistry A. 2017;**5**(44):23179-23189

[89] Pinto AH, Shin SW, Sharma A, Penn RL, Aydil ES. Synthesis of Cu2(Zn1-xCox)SnS4 nanocrystals and formation of polycrystalline thin films from their aqueous dispersions. Journal of Materials Chemistry A. 2018;**6**(3):999-1008

[90] Dahoumane SA, Wujcik EK, Jeffryes C. Noble metal, oxide and chalcogenide-based nanomaterials from scalable phototrophic culture systems. Enzyme and Microbial Technology. 2016;**95**:13-27. DOI: 10.1016/j.enzmictec.2016.06.008

[91] Mahmoud WE, El-Mallah HM. Journal of Physics D: Applied Physics. 2009;**42**:035502, (5pp). https://doi.org/10.1088/0022-3727/42/3/035502

[92] Mi C, Wang Y, Zhang J, Huang H, Xu L, Wang S, et al. Biosynthesis and characterization of CdS quantum dots in genetically engineered Escherichia coli. Journal of Biotechnology. 2011;**153**(3-4):125-132. DOI: 10.1016/j.jbiotec.2011.03.014

[93] Yan ZY, Du QQ, Qian J, Wan DY, Wu SM. Eco-friendly intracellular biosynthesis of CdS quantum dots without changing Escherichia coli's antibiotic resistance. Enzyme and

Microbial Technology. 2017;**96**:96-102. DOI: 10.1016/j.enzmictec.2016.09.017

[94] Holmes JD, Richardson DJ, Saed S, Evans-Gowing R, Russell DA, Sodeau JR. Cadmium-specific formation of metal sulfide “Q-particles” by Klebsiella pneumoniae. Microbiology. 1997;**143**(8):2521-2530

[95] Zhou H et al. Nanotechnology. 2009;**20**: 085603, (10pp). https://doi.org/10.1088/0957-4484/20/8/085603

[96] Tudu SC, Zubko M, Kusz J, Bhattacharjee A. CdS nanoparticles (< 5 nm): Green synthesized using Termitomyces heimii mushroom–structural, optical and morphological studies. Applied Physics A: Materials Science & Processing. 2021;**127**(2):1-9. DOI: 10.1007/s00339-020-04245-3

[97] Qin Z, Yue Q, Liang Y, Zhang J, Zhou L, Hidalgo OB, et al. Extracellular biosynthesis of biocompatible cadmium sulfide quantum dots using Trametes versicolor. Journal of Biotechnology. 2018;**284**:52-56. DOI: 10.1016/j.jbiotec.2018.08.004

[98] Do JL, Friščić T. Mechanochemistry: A force of synthesis. ACS Central Science. 2017;**3**(1):13-19

[99] James SL, Adams CJ, Bolm C, Braga D, Collier P, Friščić T, et al. Mechanochemistry: Opportunities for new and cleaner synthesis. Chemical Society Reviews. 2012;**41**(1):413-447

[100] Gorrasi G, Sorrentino A. Mechanical milling as a technology to produce structural and functional bio-nanocomposites. Green Chemistry. 2015;**17**(5):2610-2625

[101] Friščić T, Mottillo C, Titi HM. Mechanochemistry for synthesis. Angewandte Chemie, International Edition. 2020;**59**(3):1018-1029. DOI:https://doi.org/10.1002/anie.201906755

[102] Baláž P. Mechanochemistry in minerals engineering. In: Mechanochemistry in Nanoscience and Minerals Engineering. Berlin Heidelberg: Springer; 2008. pp. 257-296

[103] Julien PA, Užarević K, Katsenis AD, Kimber SAJ, Wang T, Farha OK, et al. In situ monitoring and mechanism of the Mechanochemical formation of a microporous MOF-74 framework. Journal of the American Chemical Society. 2016;**138**(9):2929-2932

[104] Halasz I, Kimber SAJ, Beldon PJ, Belenguer AM, Adams F, Honkimäki V, et al. In situ and real-time monitoring of mechanochemical milling reactions using synchrotron X-ray diffraction. Nature Protocols. 2013;**8**(9):1718-1729

[105] Friščić T, Halasz I, Beldon PJ, Belenguer AM, Adams F, Kimber SAJ, et al. Real-time and in situ monitoring of mechanochemical milling reactions. Nature Chemistry. 2013;**5**(1):66-73

[106] Baláž P, Baláž M, Achimovičová M, Bujňáková Z, Dutková E. Chalcogenide mechanochemistry in materials science: Insight into synthesis and applications (a review). Journal of Materials Science. 2017;**52**(20):11851-11890

[107] Rusanov V, Chakurov C. Percolation phenomena in explosive mechanochemical synthesis of some metal chalcogenides. Journal of Solid State Chemistry. 1990;**89**(1):1-9

[108] Tan D, Friščić T. Mechanochemistry for organic chemists: An update. European Journal of Organic Chemistry. 2018;**2018**(1):18-33

[109] Ueda, Yutaka, and Tsukio Ohtani. “Mechanochemical synthesis, vacancy-ordered structures and low-dimensional properties of transition metal chalcogenides.” Handbook of Solid State

Chemistry, 6 Volume Set (2017). pp 383-433

[110] Takacs L. The historical development of mechanochemistry. Chemical Society Reviews. 2013;**42**(18): 7649-7659

[111] Moores A. Bottom up, solid-phase syntheses of inorganic nanomaterials by mechanochemistry and aging. Current Opinion in Green and Sustainable Chemistry. 2018;**12**:33-37

[112] Ohtani T, Motoki M, Koh K, Ohshima K. Synthesis of binary copper chalcogenides by mechanical alloying. Materials Research Bulletin. 1995;**30**(12):1495-1504

[113] Balema V, Hlova I, Pecharsky VK. Preparation of metal chalcogenides. United States; US20190039913A1. 2018

[114] Ohtani T, Araki M, Shohno M. Formation of Cu3Se2 and Ag2Se by sonochemical solid-state reactions at room temperature. In: Solid State Ionics. Elsevier; 2004. pp. 197-203. https://doi.org/10.1016/j.ssi.2004.05.017

[115] Jung JH, Lee JH, Silverman JR, John G. Coordination polymer gels with important environmental and biological applications. Chemical Society Reviews. 2013;**42**(3):924-936

[116] Suslick KS, Fang MM, Hyeon T, Mdleleni MM. Applications of Sonochemistry to materials synthesis. In: Sonochemistry and Sonoluminescence. Netherlands: Springer; 1999. pp. 291-320

[117] Rashmi Pradhan S, Colmenares-Quintero RF, Colmenares Quintero JC. Designing Microflowreactors for Photocatalysis using Sonochemistry: A systematic review article. Molecules. 2019;**24**(18):3315

[118] Mosleh S, Rahimi MR, Ghaedi M, Dashtian K, Hajati S. Sonochemical-assisted synthesis of CuO/Cu2O/Cu nanoparticles as efficient photocatalyst for simultaneous degradation of pollutant dyes in rotating packed bed reactor: LED illumination and central composite design optimization. Ultrasonics Sonochemistry. 2018;**40**: 601-610

[119] Latiff N, Teo WZ, Sofer Z, Huber Š, Fisher AC, Pumera M. Toxicity of layered semiconductor chalcogenides: Beware of interferences. RSC Advances. 2015;**5**(83):67485-67492

[120] Takebe H, Kitagawa R, Hewak DW. Non-toxic Sulfide glasses and thin films for optical applications. Journal of the Ceramic Society of Japan. 2005;**113**(1313):37-43

[121] Baláž M, Zorkovská A, Urakaev F, Baláž P, Briančin J, Bujňáková Z, et al. Ultrafast mechanochemical synthesis of copper sulfides. RSC Advances. 2016;**6**(91):87836-87842

[122] Ohtani T, Maruyama K, Ohshima K. Synthesis of copper, silver, and samarium chalcogenides by mechanical alloying. Materials Research Bulletin. 1997;**32**(3):343-350

[123] Užarević K, Halasz I, Friščić T. Real-time and In situ monitoring of Mechanochemical reactions: A new playground for all chemists. Journal of Physical Chemistry Letters. 2015;**6**(20):4129-4140

[124] Ribeiro DSM, de Souza GCS, Melo A, Soares JX, Rodrigues SSM, Araújo AN, et al. Synthesis of distinctly thiol-capped CdTe quantum dots under microwave heating: Multivariate optimization and characterization. Journal of Materials Science. 2017;**52**(6):3208-3224

[125] Krishnan B, Shaji S, Ernesto OR. Progress in development of copper antimony sulfide thin films as an alternative material for solar energy harvesting. Journal of Materials Science: Materials in Electronics. 2015;**26**(7):4770-4781

[126] Ramasamy K, Sims H, Butler WH, Gupta A. Selective nanocrystal synthesis and calculated electronic structure of all four phases of copper-antimony-sulfide. Chemistry of Materials. 2014;**26**(9):2891-2899

[127] Pretto T, Baum F, Fernandes Souza Andrade G, Leite Santos MJ. Design of experiments a powerful tool to improve the selectivity of copper antimony sulfide nanoparticles synthesis. CrystEngComm, 2021;**23**:397-403. DOI https://doi.org/10.1039/D0CE01563F

[128] Martynenko IV, Baimuratov AS, Weigert F, Soares JX, Dhamo L, Nickl P, et al. Photoluminescence of Ag-In-S/ZnS quantum dots: Excitation energy dependence and low-energy electronic structure. Nano Research. 2019;**12**(7):1595-1603

[129] Soares, J.X., Wegner, K.D., Ribeiro, D.S.M. et al. Rationally designed synthesis of bright AgInS2/ZnS quantum dots with emission control. Nano Res. 2020;**13**:2438-2450. https://doi.org/10.1007/s12274-020-2876-8

[130] Ramírez EA, Ramírez A, Gordillo G. Cu2ZnSnS4 films grown in one-step process by spray pyrolysis with improved properties. Materials Science in Semiconductor Processing. 2017;**67**:110-117

[131] Han W, Yi L, Zhao N, Tang A, Gao M, Tang Z. Synthesis and shape-tailoring of copper sulfide/indium sulfide-based nanocrystals. Journal of the American Chemical Society. 2008;**130**(39):13152-13161. DOI: 10.1021/ja8046393

[132] Saldanha PL, Brescia R, Prato M, Li H, Povia M, Manna L, et al. Generalized one-pot synthesis of copper Sulfide, selenide-Sulfide, and telluride-Sulfide nanoparticles. Chemistry of Materials. 2014;**26**(3):1442-1449. DOI: 10.1021/cm4035598

[133] Lu Q, Gao F, Zhao D. One-step synthesis and assembly of copper Sulfide nanoparticles to nanowires, nanotubes, and Nanovesicles by a simple organic amine-assisted hydrothermal process. Nano Letters. 2002;**2**(7):725-728. DOI: 10.1021/nl025551x

[134] Du XS, Mo M, Zheng R, Lim SH, Meng Y, Mai YW. Shape-controlled synthesis and assembly of copper sulfide nanoparticles. Crystal Growth & Design. 2008;**8**(6):2032-2035. DOI: 10.1021/cg701145q

[135] Sigman MB, Ghezelbash A, Hanrath T, Saunders AE, Lee F, Korgel BA. Solventless synthesis of monodisperse Cu2S Nanorods, Nanodisks, and Nanoplatelets. Journal of the American Chemical Society. 2003;**125**(51):16050-16057. DOI: 10.1021/ja037688a

[136] National Institute of Standards and Technology. NIST Chemistry WebBook. 2017. Accessed at: https://webbook.nist.gov/

[137] Kim S, Thiessen PA, Bolton EE, Chen J, Fu G, Gindulyte A, et al. PubChem substance and compound databases. Nucleic Acids Research. 2016 Jan;**44**:D1202-D1213. DOI: 10.1093/nar/gkv951

[138] Wuxi AppTec. WuXi LabNetwork. 2020. Accessed at: https://www.labnetwork.com/frontend-app/p/#!/.

[139] Dicks AP, Hent A. Green Chemistry Metrics :A Guide to Determining and Evaluating Process Greenness. Chemistry and Materials Science. Cham: Springer; 2014.

[140] Lapkin A, Constable DJC. Green Chemistry Metrics: Measuring and Monitoring Sustainable Processes. Ames, Iowa: Wiley; 2008

[141] Mercer SM, Andraos J, Jessop PG. Choosing the greenest synthesis: A multivariate metric green chemistry exercise. Journal of Chemical Education. 2012;**89**(2):215-220. DOI: 10.1021/ed200249v

[142] Kazi FK, Fortman JA, Anex RP, Hsu DD, Aden A, Dutta A, et al. Techno-economic comparison of process technologies for biochemical ethanol production from corn Stover. Fuel. 2010;**89**:S20-S28. DOI: 10.1016/J.FUEL.2010.01.001

[143] Behnamfard A, Salarirad MM, Veglio F. Process development for recovery of copper and precious metals from waste printed circuit boards with emphasize on palladium and gold leaching and precipitation. Waste Management. 2013;**33**(11):2354-2363. DOI: 10.1016/j.wasman.2013.07.017

[144] Ghodrat M, Rhamdhani MA, Brooks G, Masood S, Corder G. Techno economic analysis of electronic waste processing through black copper smelting route. Journal of Cleaner Production. 2016;**126**:178-190. DOI: 10.1016/j.jclepro.2016.03.033

[145] Allen DT, Shonnard DR. Green Engineering: Environmentally Conscious Design of Chemical Processes. Upper Saddle River, NJ: Prentice Hall; 2002. p. 552

[146] Mooney D. Effectively minimizing hazardous waste in academia: The green chemistry approach. Chemical Health and Safety. Lawrence, Kansas, USA. 2004;**11**(3):24-28

[147] Silverman J. Open-Access Life Cycle Assessments: ExceLCA for Undergraduate Scholars. KU ScholarWorks. University of Kansas Libraries; 2018 Lawrence, Kansas, USA

[148] Wu W, Zhao B, Wang S, Hao J. Ozone and secondary organic aerosol formation potential from anthropogenic volatile organic compounds emissions in China. Journal of Environmental Sciences. 2017;**53**:224-237

[149] Silverman JR, Bode C, Subramaniam B. Open-access chemical assessments for students and educators: A case study for evaluating aspirin synthesis. ACS Symposium Series. 2019;**1318**:119-127

[150] Guinee JB. Handbook on life cycle assessment operational guide to the ISO standards. Int J LCA 7, 311 (2002). https://doi.org/10.1007/BF02978897

[151] McDonough W, Braungart M. Cradle to Cradle: Remaking the Way we Make Things. New York: North Point Press; 2002

[152] Kiselyova NN, Dudarev VA, Korzhuyev MA. Database on the bandgap of inorganic substances and materials. Inorganic Materials: Applied Research. 2016;7(1):34-39

[153] Strehlow WH, Cook EL. Compilation of energy band gaps in elemental and binary compound semiconductors and insulators. Journal of Physical and Chemical Reference Data. 1973;**2**(1):163-200

[154] Joshi NV. In: Joshi NV, editor. Photoconductivity. Photoconductivity Art, Science, and Technology. 1st ed.

New York: Routledge; 2017. DOI: 10.1201/9780203743522

[155] Madelung O. Semiconductors: Data Handbook. Berlin, Heidelberg: Springer Berlin Heidelberg; 2004. DOI: 10.1007/978-3-642-18865-7

[156] Jain A, Ong SP, Hautier G, Chen W, Richards WD, Dacek S, et al. Commentary: The materials project: A materials genome approach to accelerating materials innovation. APL Materials. 2013;**1**(1):1-11

[157] Zhuo Y, Mansouri Tehrani A, Brgoch J. Predicting the band gaps of inorganic solids by machine learning. Journal of Physical Chemistry Letters. 2018;**9**(7):1668-1673. DOI: 10.1021/acs.jpclett.8b00124

[158] Saal JE, Oliynyk AO, Meredig B. Machine learning in materials discovery: Confirmed predictions and their underlying approaches. Annual Review of Materials Research. 2020;**50**(1):49-69. DOI: 10.1146/annurev-matsci-090319-010954

[159] Xi L, Pan S, Li X, Xu Y, Ni J, Sun X, et al. Discovery of high-performance thermoelectric chalcogenides through reliable high-throughput material screening. Journal of the American Chemical Society. 2018;**140**(34):10785-10793. DOI: 10.1021/jacs.8b04704

[160] Cao B, Adutwum LA, Oliynyk AO, Luber EJ, Olsen BC, Mar A, et al. How to optimize materials and devices via design of experiments and machine learning: Demonstration using organic photovoltaics. ACS Nano. 2018;**12**(8):7434-7444. DOI: 10.1021/acsnano.8b04726

[161] Choudhary K, Bercx M, Jiang J, Pachter R, Lamoen D, Tavazza F. Accelerated discovery of efficient solar cell materials using quantum and machine-learning methods. Chemistry of Materials. 2019;**31**(15):5900-5908. DOI: 10.1021/acs.chemmater.9b02166

[162] Kormath Madam Raghupathy R, Wiebeler H, Kühne TD, Felser C, Mirhosseini H. Database screening of ternary chalcogenides for P-type transparent conductors. Chemistry of Materials. 2018;**30**(19):6794-6800. DOI: 10.1021/acs.chemmater.8b02719

[163] Wang X, Xu Y, Yang J, Ni J, Zhang W, Zhu W. ThermoEPred-EL: Robust bandgap predictions of chalcogenides with diamond-like structure via feature cross-based stacked ensemble learning. Computational Materials Science. 2019;**169**:109117. DOI: 10.1016/j.commatsci.2019.109117

[164] Agiorgousis L, Michael, Sun YY, Choe DH, West D, Zhang S. Machine learning augmented discovery of chalcogenide double perovskites for photovoltaics. Advanced Theory and Simulations. 2019;**2**(5):1-9. DOI: 10.1002/adts.201800173

Section 3

Green Chemistry - Miscellenous

Chapter 9

Biopolymers

Ioana Stanciu

Abstract

Significant progress has been made on biopolymers in recent years. Biopolymers are preferred to other materials because they have specific physical, chemical, biological, biomechanical, and degradation properties. Many natural or synthetic biopolymers can degrade hydrolytically or enzymatically and are used for many applications.

Keywords: biopolymers, classification, structure

1. Introduction

Biopolymers are a large group of biomaterials, for which it is important that they are biocompatible and noncytotoxic and produce decomposition products that are themselves nontoxic. They should support cell adhesion and proliferation and be an active participant in the process of creating new tissues, ensuring that the dynamics of formation and resorption of these tissues take place efficiently, with a constant balance between the two processes. Biopolymers by their properties, although they include different types of hydrophilic and hydrophobic polymers (which include different types of complex polymers, such as grafted and block copolymers, hydrogels, different types of thermoplastic biodegradable polymers, and polypeptide polymers, which may be of natural origin or artificial, different functional properties, network kinetics, and biodegradation), mostly belong to the so-called family smart polymers, due to their ability showing significant changes in their structural properties when the environmental conditions in which they are changing.

Among the best-known polymers are polyethylene, polyvinyl chloride, polystyrene, polypropylene, polyvinyl alcohol, etc. Their basic properties are satisfactory strength, flexibility, ease of painting and modeling, and hardness. What is the basic specificity of smart polymers is their extreme response to slight changes in environmental conditions, such as changes in temperature, pH, water, and light. These changes cause changes in the structural characteristics of these polymers. Therefore, they have significant potential for different types of applications in the field of biotechnology and biomedicine. Knowledge of the chemistry of conformational changes in polymers, which are conditioned by changes in environmental conditions, offers the possibility of the synthesis of new smart polymers, whose changes can be controlled by changes in environmental factors. This is extremely important for biological systems, as it allows the efficient use of polymers for the delivery of drugs or other substances important for the control of metabolic mechanisms. Smart polymers can be designed in the form of hydrogels, patches, bags degradation of plastic, chewing gum, blood glucose detectors, and polymers for the targeted release of insulin.

Graft and block copolymers are a special group of polymers, which consist of two different polymers, which are linked together in a graft. There are many patents that offer different combinations of such polymers, which have different reactive groups. The products are a combination that has properties characteristic of both components involved in the construction of such a polymer, which gives a new dimension to the structures of smart polymers, thus creating conditions for increasingly diverse applications. The crosslinked hydrophobic and hydrophilic polymers form micelle-like structures in which the safe encapsulation and protection of the drug can be related to the delivery of the drug through an aqueous medium until it reaches the target location when degradation occurs by breaking the bonds between the two polymers. An example of such a polymer is a polyacrylic acid PAAc bioadhesive polymer. The polyacrylic acid adheres, swells rapidly, and degrades to pH = 7.4, resulting in the rapid release of the drug trapped in the matrix. The combination of polyacrylic acid with other polymers, which are less susceptible to changes in neutral pH, increases the retention time of the drug and slows its release, while showing an increase in its bioavailability and efficacy.

Hydrogels are networks of polymers that do not dissolve in water, but swell and collapse when the aquatic environment changes. They are used in biotechnology for phase separation because such polymers are suitable for recycling. Highly specialized hydrogels have also been developed for the delivery and release of drugs into specific tissues. Polyacrylic acid hydrogels have very pronounced bioadhesive properties and exceptional absorption.

Enzyme immobilization in hydrogels is a highly developed technique for introducing enzymes into biological systems. In such an enzyme application, the body's response depends on the product of the enzyme reaction. The process of introducing enzymes, receptors, and antibodies takes place by establishing the connection of the given entities with the selected molecule, within the hydrogel. After the connection is established, a targeted chemical reaction takes place in the hydrogel, the product of which may be, for example, oxygen, due to the sensitivity of the system to the presence of redox enzymes or pH changes of hydrogels that are sensitive to pH changes. Thus, for example, combinations of glucose oxidase and insulin packaging are converted to pH-sensitive hydrogels. Namely, in the presence of glucose, the formation of gluconic acid by enzymes leads to the targeted release of insulin from the hydrogel. The two basic criteria for the efficient operation of this technology are stability enzymatic and rapid kinetics (rapid response to change activator and recovery after cessation of its action).

Smart polymers are not yet suitable for drug delivery, although their properties are suitable for bioseparation. The time and cost involved in protein purification can be reduced with the significant use of smart proteins, which undergo rapid reversible changes in response to changes in average properties. Conjugate systems have long been used in physical and chemical separations and immunoassays. Microscopic changes in the structure of the polymer are manifested by the formation of a precipitate, which can be used to help separate the trapped proteins from the solution. These systems work in such a way that if a protein or other molecule is separated from the mixture forms a biconjugate with the polymer, then it precipitates with the polymer when the properties of the medium change. The precipitate is removed from the medium, separating the desired conjugate component from the rest of the mixture. The removal of these components from the conjugate depends on the speed of recovery of the polymer and its return to baseline. Another possibility of controlling biological reactions using smart polymers is the preparation of recombinant proteins, with characteristic binding sites embedded in ligand-like proteins or binding sites in cells. This technique is used to control both ligands and cell binding activities, which are responsible for various stimuli of activity, such as temperature or light [1, 2]. Future applications of smart polymers will be related to the development of polymers that will be able to learn and correct their own behavior over time. In the near future, smart toilets will use such polymers for analysis, for example, urine, and provides assistance in identifying health problems. Many creative approaches to targeted drug delivery systems and their self-regulation based on their unique cellular environment are already the subject of intense research.

Intelligent natural polymers and their response to external stimulation.

Polymers found in living systems (proteins, carbohydrates, and nucleic acids) play a significant physiological role in biological systems. Smart polymers have become especially important after studying their chemistry and the conditions that serve as activators of their conformational changes. The new polymeric materials are formulated to "feel" specific changes in biological systems by adjusting to respond to them in a predictable way, allowing them to be used as a useful tool for the administration of drugs and the control of various metabolic mechanisms. The nonlinear response of smart polymers makes them unique and efficient. Significant changes in the structure and properties of smart polymers can be caused by a small change in stimuli. Once a change occurs, there are no changes, which means that it is predictable whether or not there will be a response. It is important to note that the change in the conformation of the polymer is homogeneous for the system as a whole. Smart polymers change their conformational, adhesive, or retention properties when present in water, with small changes in pH, ionic strength, temperature, or other change activators. Another factor that defines the response efficiency of smart polymers is their inherent nature. The response of molecules to stimulus-induced change occurs simultaneously in a number of individual monomeric units. Although the change, which is related to only one of the monomer molecules, is almost insignificant when it comes to the whole set of such molecules (because the total change is equal to the sum of these identical or similar minor changes and theirs), corresponding responses (hundreds or thousands) such a change has a significant value, conditioning the appearance of a very significant force of stimulation of the appropriate biological responses (processes).

2. Classification and basic applications of polymers

Today, the most common application of polymers in biomedicine is for the targeted delivery of drugs. With over-the-counter pharmaceuticals, the basic problem for scientists is to find an effective way to deliver drugs to a certain place in the body without first degrading it to very acidic stomach acid. Preventing side effects on healthy bones and tissues is also a significant issue. It is advisable to monitor the release of medicines until the delivery system is in the desired location [3].

The release of the drug is controlled by chemical or physiological activation. Intelligent linear or matrix polymers are in different varieties and may have different properties depending on the type of functional groups or side chains present in them. These groups may be sensitive to pH, temperature, ionic power, electric or magnetic field, and light. Some polymers crosslink with non-covalent bonds and can cleave and reform depending on external conditions. Dendrimers are known as typical particle-free polymers, which are most commonly used for drug delivery. Lactic acid polymers are used in the traditional encapsulation of medicines. The release of drugs from smart polymer matrices is achieved by chemical or physiological reactions, such as hydrolysis, in which the bonds are broken and the drug is released when the matrix is divided into biodegradable components. Artificially obtained polymers, such as polyanhydrides, polyesters, polyacrylic acids, poly (methyl methacrylates), and polyurethanes, are most often used for this purpose. Low-molecular-weight hydrophilic amorphous polymers containing heteroatoms (other atoms relative to carbon) have been shown to degrade most rapidly. By controlling the rate of degradation, today's scientists are trying to control the rate of drug delivery [3–5]. Another very important field of application of polymers is tissue engineering, where many polymers of natural and artificial origin can be part of such composite structures that allow better adhesion and attachment of suitable cells (stem cells), their differentiation according to a certain phenotype, and finally, of them proliferation to new tissue [3–5].

Due to the relevant economic and environmental aspects, the growing interest in natural polymers is becoming more important in natural polymers due to their biodegradability, low toxicity, low cost, low availability, and renewable costs. Moreover, they offer a wide range of benefits for tissue engineering applications, such as biological signaling, cell adhesion, degradation, and responsible cell remodeling. However, their inadequate physical properties, such as solubility, rapid degradation, and possible loss of biological properties during casting, often limit their use as scaffolding materials. In addition, the risk of immune rejection and transmission of diseases require further purification. Physical or chemical modification of such polymers is an effective way to improve the stability of the material and its physical characteristics [6].

3. Polymers of natural origin

Natural polymers used in tissue engineering are very diverse in origin and composition. Such polymers include proteins, collagen, silk fibroin, polysaccharides, chitosan, and its derivatives, hyaluronic acid, alginates, starch, cellulose, dextrans, and polyesters of microbial origin.

3.1 Proteins

There have been numerous studies on proteins that make up the extracellular matrix (ECM) of natural tissues, such as scaffolding in tissue engineering. The use of biomacromolecules, such as collagen and fibronectin, which transmit biological information and provide the necessary physicochemical functions, is attempting to mimic ECM in the development of targeted tissues. A group of collagen-based biomaterials has been investigated with a particular interest in their application in bone regeneration [7].

Despite attempts to replace natural fibrous protein materials with artificial polymeric materials, natural protein fibers, such as silk, wool, and leather fibers, have not yet been successfully reproduced as artificial materials. Bone tissue and cartilage engineering based on the use of silk fibroin in the construction of scaffolds based on biomaterials is a very current topic of modern research [7–9].

3.2 Collagen

Collagen is a biological protein with a high content of glycine (almost 33%) and other amino acids (almost 20%). It is rich in the ECM of many tissues (skin, bones, cartilage, blood vessels, and teeth) where it provides the basic structure and mechanical support of these systems. Type I collagen is found in bones, tendons, cartilage, and corneal fibers, type II in hyaline cartilage and the nucleus pulposus (the gelatinous substance in the central part of the spine), type III in the intestines and uterine walls, type IV in the endothelium tissues and epithelial membrane, and type V in the cornea, placenta, bones, and heart valves. In the structure of collagen fibers, glycine is the most abundant, which is in place of every third residue in the polypeptide chain of collagen, forming (Gly-XY) a motif repeatedly adapted to the helical structure oriented to the left (α chain) with a length of about 1400 amino acids and three residues per reason. The most common collagen sequences are composed of glycine (gly), proline (pro), and hydroxyproline (hyp) (Gly-Pro-Hyp). Three α chains are wrapped around each other, twisted to the right, and densely wrapped in a triple helix (**Figure 1**) [10, 11].

Collagen fibers are stabilized by specific covalent crosslinkers between collagen molecules. Biodegradability, low antigenicity, and cell binding properties make collagen a valuable material for use in tissue engineering. Collagen sponge accelerates the growth and growth of cells and tissues and improves bone formation by promoting the differentiation of osteoblasts.

Studies involving the seeding of mesenchymal stem cells (MSCs) in collagen gel, when implanted in osteochondral defects in rabbits, lead to bone and hyaline cartilage formation, although the mechanical properties of regenerated tissue are significantly lower than in normal tissues. It is important to note that, after the implantation process, no degeneration of the surrounding tissues was observed in the first 24 weeks. The main disadvantage of using collagen as a biomaterial for tissue repair is its high rate of degradation, which leads to the rapid loss of mechanical properties of scaffolds based on it. Many attempts have been made to overcome this problem by further processing to ensure that the collagen remains insoluble during a certain critical period, by adding suitable mineral crystals or a combination of collagen with other natural materials, such as glycosaminoglycans (GAGs) or synthetic polymers, derived from methacrylate, or by applying specific methods of collagen crosslinking [10–12]. Highly porous hydroxyapatite/collagen composite constructions seeded with chondrocytes showed an increase in the stability of the foam composite in the culture medium upon ECM deposition. A matrix made by crosslinking collagen fibers with modified hydroxyapatite, implanted in the cranial defects of mice, showed good biocompatibility and improved osteoconductivity in relation to the collagen material itself. The improvement of mechanical properties can be achieved by crosslinking the collagen chains by amino groups of lysine and hydroxylysine using glutaraldehyde or other agents, such as 1-ethyl-3-(3-dimethyl aminopropyl), carbodiimide (1-ethyl-3-(3)-dimethyl aminopropyl) carbodiimide, and hexamethylene diisocyanate. However, these treatments are not sufficiently cytocompatible due to the potential toxicity of any of the network agents used.

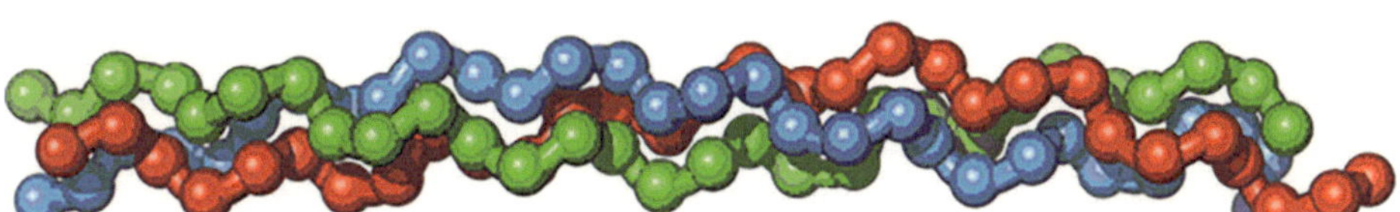

Figure 1.
Collagen triple helix [http://www.rcsb.org/].

Other strategies for modifying collagen fibers are related to the chemical modification of the side chains of the collagen fiber to form double bonds, which allow subsequent crosslinking with chemical or photoinitiated free radicals or crosslinking with the enzyme lysyl oxidase. Takayama and Mizumachi have shown that lactoferrin adsorbs type I collagen, thus accelerating the calcification of collagen *in vitro* and the differentiation of human osteoblastic cells. Biocompatible scaffolds derived from type I collagen electrospun and hydroxyapatite nanoparticles show a significant increase in tensile strength and modulus. The mechanical properties of the scaffold are further improved by crosslinking with the vapor phase glutaraldehyde.

A new method for obtaining a biostyclic-nanofibrocollagen nanocomposite, in the form of a membrane or macroporous scaffold, has recently been published. The nanocomposite matrix exhibits "*in vitro*," bioactivity induced by the rapid formation of bone-like mineral apatite on its surface when incubated in body fluid (SBF). Osteoblastic cells show a favored growth on nanocomposite scaffolds, and their alkaline phosphatase activity is significantly higher than in pure collagen scaffolds [13].

3.3 Silk fiber

Silk is a natural polymer that has been used clinically for centuries. Naturally extruded by insects or larvae, the silk took the form of a fiber consisting of a protein fiber called fibroin and a sticky coating made of sericin protein. Today, fibroin fibers are increasingly used as biomaterials for new biomedical applications, especially, in the field of tissue engineering, due to their biocompatibility, slow degradability, and exceptional mechanical properties, as well as better ability to control molecular structure and morphology through various forms of processing and surface modification.

Silk fiber in various forms (films, fibers, nets, membranes, sponges, etc.) supports the adhesion, proliferation, and differentiation of stem cells "*in vitro*" and promotes the repair of tissues "*in vivo*." 3D fibrous silk scaffolds have a special application in skeletal tissue engineering's, such as bones, ligaments, and cartilage, as well as in connective tissue engineering (skin tissues). The dominant source of silk-based materials to date is fibroin produced from Bombyx mori silkworm caterpillars. Silk fibroin from spiders and modified transgenic species is also a very important source. Such a silk fibroin shows the possibility of modifying certain fibroin sequences from natural silk and thus indirectly new possibilities for the production of materials inspired by silk fibroin, for various medical applications.

4. Structure and properties of silk

Silk belongs to a group of natural polymers that produce a wide variety of insects and spiders. In nature, silk has different structures and functions that are evolutionarily oriented toward the needs of the animal species that produce it in a given environment. The different functions of silk are classified as a series of tissue constructions, from tissue in the form of a victim trap (spider weaving), through safety rope (pull rope) to reproductive form (cocoon shape). Silk offers an excellent combination of lightweight (1.3 g/cm3), high tensile strength (over 4.8 GPa, as the strongest fiber known in nature), and outstanding hardness and elasticity (with a tightening length, before breaking the fiber by more than 35%). The tensile strength of silk yarn is comparable to most synthetic Kevlar-49 fibers, while the elasticity of silk fibers is 4–7 times higher than Kevlar-49 fibers, and the energy required to break the fibers is 3–4 times bigger. In addition to these remarkable mechanical properties, silk has thermal stability of up to about 250°C, which allows it to be processed over a wide temperature range [14].

The most commonly studied silk is derived from the larvae of Bombyx mori and the silk thread of the spider Nephila clavipes. Structurally, fibroin silks of these species are characterized by natural block copolymers, consisting of hydrophobic blocks with very regular repeated sequences, consisting of short chains of amino acids, such as glycine and alanine, and hydrophilic blocks with more complex sequences. Consisting of long chains of amino acids, such as charged amino acids (**Figures 2** and **3**) [15, 16].

Hydrophobic blocks tend to form β-plates or crystals bound by hydrogen bonds and hydrophobic interactions, forming the basis of the tensile strength of silk fibroin. These arranged hydrophobic blocks are responsible for the high elasticity and hardness of the silk fibers [17].

The method of processing fibroin solutions into fibers by various bodies is still insufficiently studied and still remains in the field of intensive research. The process of obtaining fibroin fibers involves spinning a highly concentrated aqueous solution of silk fibroin into a non-Newtonian liquid crystalline state, in which the silk fibroin is lubricated and stabilized with water and forms micelle-like structures by separating the phases due to the process, and internal hydrophilic and hydrophobic block structures. The process is also supported by the water content and location. During the natural process of producing fibroin fiber, the concentration of the solution of silk fibroin in the glands gradually increases to form mycelium, which further aggregates to form globules or gel-like structures. At this stage, the fibrous silk protein is organized into a metastable state that maintains enough water to avoid premature conversion to the structure of the β-plate. The tensile stress during the spinning of the fibers (movements of the head in a silk caterpillar or pushing of a spider's legs) determines the final assembly of the β-plate into crystal blocks [18].

In the final stage of spinning a silkworm, hydrophilic proteins, such as sericin, form a composite matrix with a silk fibroin core. The resulting silk fibers are insoluble in many solvents, such as water, ethanol, dilute acids and bases, hexafluoroisopropanol (hexafluoroisopropanol (HFIF)), calcium nitrate, or LiBr solution. The crystalline region of silk fibroin contains repetitive sequences rich in alanine or alanine glycine. These repetitive sequences are used as a basis for the genetic engineering of silk fibroin-like polymers in host systems, such as Escherichia coli, yeast, mammalian cells, and plants. Similar to natural fibroin fibers, most recombinant fibroin polymers have low water solubility due to their hydrophobicity [17, 19].

Strategies used to self-organize recombinant polymers similar to silk fibroin to increase solubility include, but are not limited to the following: i) the inclusion of molecular activators (triggers) of reactions, such as methionine reduction/oxidation

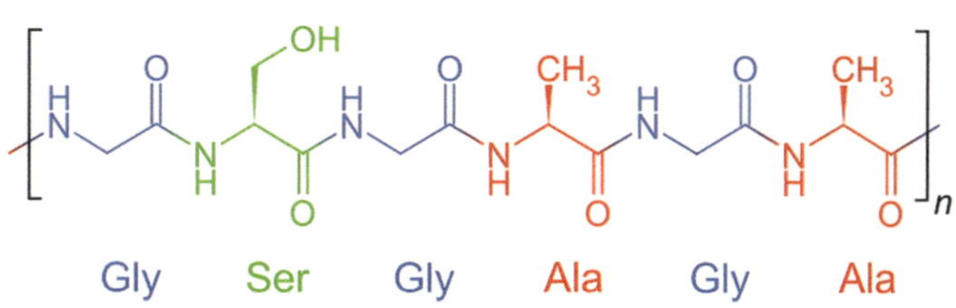

Figure 2.
The primary structure of silk fibroin [wikipedia.org].

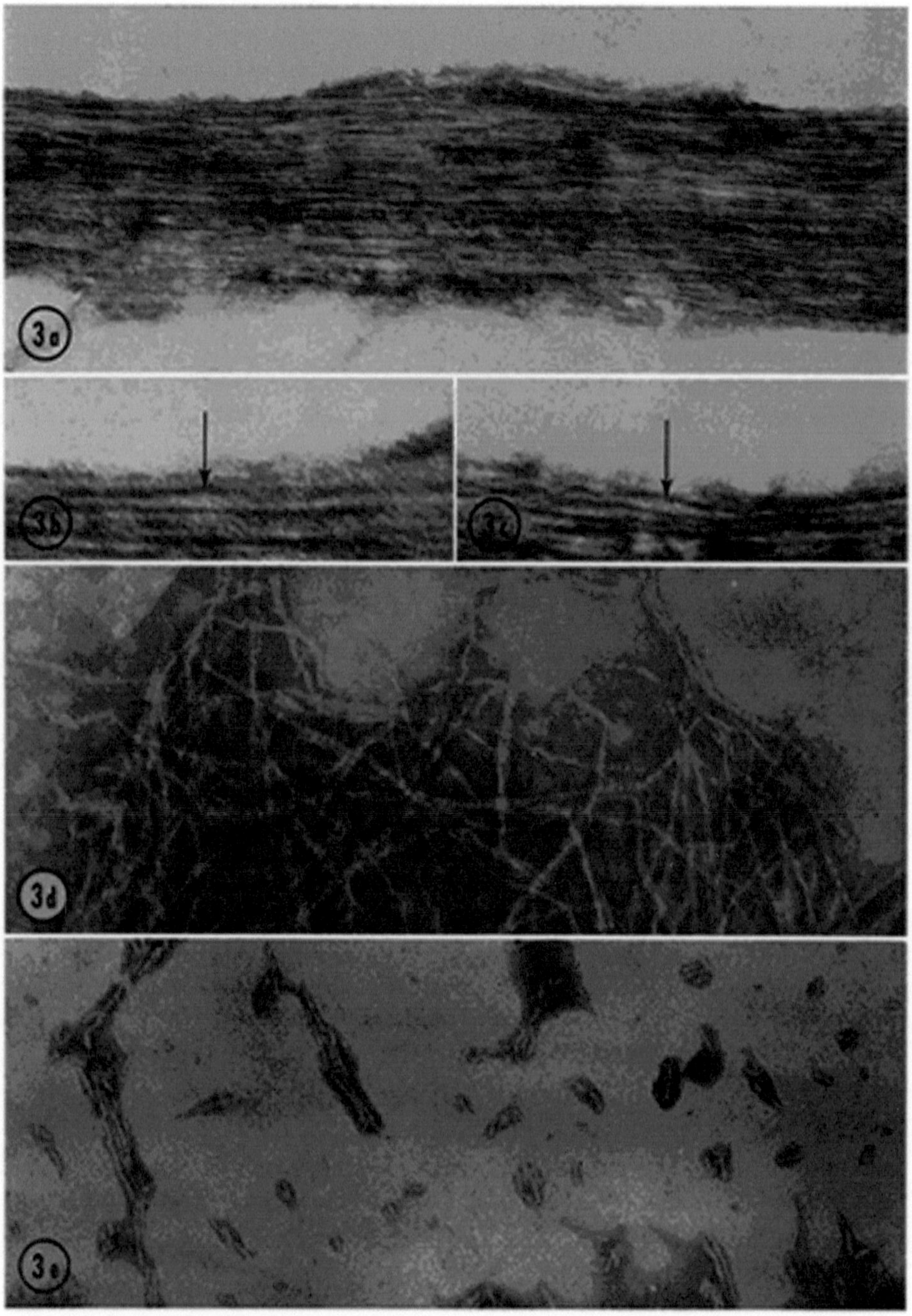

Figure 3.
a) Negatively charged disintegrated Anaphe moloneyi silk, dyed with sodium phospholphramate, with fibrous plaques, magnification 120.000x b) and c) fibers twisted by 900, fiber width, magnification 350,000x; d) a film poured from a fibroin solution of Bombyx silkworm obtained by treatment with cupriethylenediamine negatively charged, stained with sodium phospholphramate, magnification 120,000 x; e) fibers obtained from a sequentially charged polypeptide (alanine-glycerol-alanine-glycerol-serine) polypeptide, stained with sodium phospholphramate. Magnification 120.000x [J.Cell biol, 33, (1967), 289].

reactions to control β-plate formation or phosphorylation/phosphorylation kinase action, ii) construction of chimeric polymer fibers of fibroin-like silk to incorporate α helical structures, and iii) inclusion of GVGVP (glycine-valine-glycine-valine-proline) domain characteristic of elastin to reduce crystallinity. The latter approach, based on copolymers of fibroin silk and elastin, whose hydrogels are exposed to physiological conditions, makes this combination of copolymers attractive for use in injectable systems for the controlled administration of therapeutic agents.

5. Conclusions

This review detailed the classification, types, and some important applications of biomaterials used in different applications in our life. From the biomaterial's unique properties, these materials are perfect candidates for different bio-related applications. Biopolymers have attractive properties, particularly biodegradability, biocompatibility, selective permeability, and modifiable physical-mechanical properties. These properties find targeted applications in a variety of fields, particularly in the field of pharmacy and medicine. The use of natural biopolymers makes it possible to manufacture sustained-release forms that should avoid post-consumer peaks (diminish side and undesirable effects) and spread the effectiveness of this molecule over time (decrease the number of daily doses).

Author details

Ioana Stanciu
Faculty of Chemistry, Department of Physical Chemistry, University of Bucharest, Bucharest, Romania

*Address all correspondence to: istanciu75@yahoo.com

References

[1] Van de Velde K, Kiekens P. Biopolymers: Overview of several properties and consequences on their applications. Polymer Testing. 2002;**21**(4):433-442

[2] Endres HJ, Siebert-Raths A. Engineering biopolymers. Engineering Biopolymers. 2011;**71148**:3-15

[3] Niaounakis M. Biopolymers: Applications and Trends. William Andrew; 2015

[4] Kaplan DL. Introduction to biopolymers from renewable resources. In: Biopolymers from Renewable Resources. Berlin, Heidelberg: Springer; 1998. pp. 1-29

[5] Moradali MF, Rehm BH. Bacterial biopolymers: From pathogenesis to advanced materials. Nature Reviews Microbiology. 2020;**18**(4):195-210

[6] Honarkar H, Barikani M. Applications of biopolymers I: Chitosan. Monatshefte für Chemie-Chemical Monthly. 2009;**140**(12):1403-1420

[7] Mohanty AK, Misra M. In: Drzal LT, editor. Natural Fibers, Biopolymers, and Biocomposites. CRC Press; 2005

[8] Lifshitz IM. Some problems of the statistical theory of biopolymers. Soviet Physics JETP. 1969;**28**(6):1280-1286

[9] Schnepp Z. Biopolymers as a flexible resource for nanochemistry. Angewandte Chemie International Edition. 2013;**52**(4):1096-1108

[10] Rao MG, Bharathi P, Akila RM. A comprehensive review on biopolymers. Scientific Reviews and Chemical Communications. 2014;**4**(2):61-68

[11] http://www.rcsb.org/]

[12] Hassan MES, Bai J, Dou DQ. Biopolymers; definition, classification and applications. Egyptian Journal of Chemistry. 2019;**62**(9):1725-1737

[13] Knorr D, Heinz V, Buckow R. High pressure application for food biopolymers. Biochimica et Biophysica Acta (BBA)-Proteins and Proteomics. 2006;**1764**(3):619-631

[14] http://wikipedia.org/

[15] Revel JP, Karnovsky M. Hexagonal array of subunits in intercellular junctions of the mouse heart and liver. The Journal of Cell Biology. 1967;**33**(3):C7

[16] Sivakanthan S, Rajendran S, Gamage A, Madhujith T, Mani S. Antioxidant and antimicrobial applications of biopolymers: A review. Food Research International. 2020;**136**:109327

[17] Kalia S, Averous L. Biopolymers: Biomedical and Environmental Applications. Vol. 70. John Wiley & Sons; 2011

[18] Angell CA. Formation of glasses from liquids and biopolymers. Science. 1995;**267**(5206):1924-1935

[19] Fisher ME. Effect of excluded volume on phase transitions in biopolymers. The Journal of Chemical Physics. 1966;**45**(5):1469-1473

Chapter 10

New Frontier of Plant Breeding Using Gamma Irradiation and Biotechnology

Vichai Puripunyavanich, Lamai Maikaeo, Mayuree Limtiyayothin and Piyanuch Orpong

Abstract

Mutation is an underlying cause of evolution as a mutant, either natural or artificial, with a novel trait may be preferentially selected for nature because of its superior survival adaptive features. Because of the desirability of the novelty, mutation is the heritable change to an individual's genetic makeup, which is passed on from parent to offspring and thereby, drives evolution. In nature, mutations are spontaneously caused by errors in the DNA replication. Gamma radiation induced mutation in plant breeding is the one effective method that can cause DNA changes via direct and indirect actions. Many crop varieties have been created using gamma irradiation mutagenesis technology for trait improvement that enhance the characteristic or increase the abiotic and biotic stress tolerance. Plant breeding and genetics procedure usually start from mutation induction by gamma irradiation and work with the other modern enabling technologies, such as tissue culture or molecular genetics. Tissue culture and bioreactor techniques are used for synthesizing new plant varieties, while the molecular genetic technique is used for genetic analysis of the new varieties. The irradiation coupled with new modern tissue culture and molecular genetic technology is widely used to induce plant mutation breeding for creating new commercial plant varieties.

Keywords: plant mutation breeding, irradiation, biotechnology, tissue culture, molecular genetic

1. Introduction

Nowadays, agriculture has changed dramatically compared to the past. Technologies have played an important role that can be applied to use in almost every aspect of the agricultural value chain from production to consumption. The world population is currently around 7.9 billion and is expected to reach 10 billion in 2050 [1]. To feed the expected world population, the production of food will need to increase to meet the global food demand. Improved agricultural production through plant breeding, is one of the key solutions to sustainable development for supporting the growing global population. Plant breeding refers to the development of plant cultivars suitable for agricultural cultivation. In nature, natural selection is responsible for the survival of plants, based on their characteristics. Plants with low survival quality would be eliminated naturally. High-quality plants can be crossed and matched up using farmers' knowledge and expertise. Gregor Mendel, the modern genetics pioneer, did research based on the principles of genetic theories. Later, radiation and chemical mutagens were developed to help induce genetic mutation more rapidly than natural evolution.

2. Mechanism and types of mutation

2.1 Mutagenesis mechanism

Mutations involve changes in the genetic material (DNA) of cells and can be passed on to offspring through the cell division process. Mutations may be associated with the loss of a gene or structural changes within a gene, which is called gene mutation or point mutation. Chromosome-related mutations can involve changes in the number of chromosomes, exchange of chromosomes, and the disappearance/addition of a segment of a chromosome.

1. Gene Mutation

Gene mutation or point mutation is a chemical change in a gene involving only a few nucleotides bases in a gene that affects some genetic changes in that gene.

2. Chromosome Mutation

Chromosome mutation is the change in the number or the structure of the chromosomes.

a. Changes in the Structure of Chromosomes

A missing or duplication part of the chromosome can cause a number of genes to be changed as well. The effects of these changes vary greatly depending on the type of gene and the number of genes involved.

b. Changes in Chromosome Number

Usually, plants have diploid chromosome numbers, an increase or decrease can occur only on certain chromosomes called aneuploidy, for example, $2n + 1$ or $2n - 1$. In case of increase or decrease of chromosomes sets, it is called polyploidy.

2.2 Type of mutagens

Mutagen can be divided into three types, which are as follows:

1. Physical mutagen: Radiation such as X-rays, gamma rays, ion beams, electron beams, and neutron particles.

2. Chemical mutagen: Chemicals that cause mutations in genes or chromosomes. These chemical substances are ethyl methane sulphonate (EMS), diethyl sulfate (dES), ethyleneimine (EI), N-ethyl-N-Nitrosourea (ENU).

3. Biological mutagen: Genetic disruption might be the result of the virus and bacteria.

Comparatively, gamma and X-ray irradiation can be used to induce changes in seeds or other reproductive organs of the plant, such as branches, buds, stems, bulbs, and corm because they can penetrate into the internal tissues better than chemicals and also the method does not lead to chemical waste [2].

3. The use of radiation-induced mutations

It has been over 100 years since Darwin developed the theory of evolution by natural selection. Many scientists have learned and discovered the evidence of this theory that relates to a genetic mutation. Furthermore, they have found ways to induce and utilize mutations. The development of genomic study helps enhance the plant mutagenesis in crop improvement more efficiently. The mutant traits, such as dwarfing, sterility, disease and pest resistance, change of metabolites content, and many others, were shown in many species from crops to ornamentals by mutation induction. The need for plant breeding has been demonstrated usually by combining genetic modifiers to change the original mutant phenotype to become a significant breeding target. The mutagenesis technique is one of the apparatus that has resulted in many impressive mutant varieties. The number of such varieties is high across economic plant species, including vegetatively propagated species [3, 4].

Mutation techniques have been used widely in efforts to create new plant varieties that are more resistant to biotic and abiotic stress. The effects induced by physical and chemical mutagens can be similar to the natural mutation in major crops. The use of *in vitro* mutagenesis combined with *in vitro* selection with the parental lines has significantly improved the plant mutation breeding efficiency, especially in bananas and potatoes [5]. The success of mutation breeding depends on three main factors—mutagenesis efficiency, the starting plant material, and mutant screening. Identification of the genes associated with valuable traits would help in understanding the genetic background and choosing suitable starting material by molecular genetic technologies. Finally, the efficiency of mutant screening will impact the cost of mutation breeding, especially in plant species requiring long periods before the mutant traits can be evaluated [6].

Genetic variation is a prerequisite for plant breeding required to obtain useful traits for crop improvement. Often, the desired variation is either decreasing or disappearing over time. Moreover, spontaneous mutations occur at a very low rate that cannot be used in plant breeding for creating and developing new plant varieties. Therefore, the increase in mutation rate needs to be enhanced by the induction of genetic variability using mutagen treatments, such as gamma-ray, X-ray, fast neutron, ethyl methane sulfonate (EMS), and sodium azide, among others.

Plant breeders recognize the importance of ionizing radiation in plant breeding research programs. The technique has been tried out since 1930 and became accepted around 1950 because it solved some problems that standard conventional plant breeding methods could not. The two main advantages of irradiation are—1) some crops, especially cereal crops, have undergone extensive breeding that in some areas and could not be advanced from existing germplasm or strains, some characteristics, such as disease resistance, were rarely found in natural plants. 2) the increase in world population destroyed native plants because the farmer's plant only improved the cultivation of plant species that have a disadvantage in terms of having a narrow genetic background. The use of radiation induces mutations in plants is one way to solve the problem of variability in plant species by inducing the desired characteristics that are not found in genetic sources. Ionizing radiation, such as X-rays, gamma rays, and neutron rays, can cause damage to genetic material and lead to phenotypic changes. Any propagated parts of plants can be irradiated, including seeds, buds, shoots, and stolon. However, seeds are most commonly used and most convenient. Radiation-induced mutations occur randomly, but their rate of occurrence could be made higher than spontaneous mutations. Evolution and practical plant mutation breeding both depend on genetic variation. Desired characteristics are selected and the obtained mutants are investigated and screened criteria essentially include the direct or integrated study of morphological variability, alteration of physiological and biochemical parameters, gene expression analysis, etc., under different degrees of stress conditions. Before conducting any plant mutation breeding experiment, one has to define the desired characteristics. The screening for such characteristics can be done *in vitro* and/or *in vivo* depending on the species, trait/character (abiotic stresses) understudy, biotic/abiotic factors influencing the expression of the trait(s), etc.

4. Technology in mutation breeding and tissue culture

4.1 Induced mutation in the plant by irradiation

Plant breeding is the development of plant species to have better characteristics than the original variety and to meet human needs. Plant breeding can be used for many different purposes, such as to increase productivity, nutritional value, to withstand inappropriate cultivation conditions disease and insect resistance, for enhanced appearance, etc. Plant breeding methods are as follows.

4.2 Conventional plant breeding

Conventional plant breeding is the development or improvement of plant varieties by using conservative tools for manipulating plant genomes within the natural genetic boundaries of the species. The common methods for breeding self-pollinated species include mass selection, pure line selection, pedigree, bulk population, single seed descent, backcrossing, multiline and composite [7].

4.3 Induce mutation plant breeding

Natural mutations may occur, whereas at low rates and over a lengthy period. In this way, the mutation's induction takes less time than a spontaneous mutation. To induce mutations, mutagens must be used. The most commonly used mutagens are chemicals, but it is not safe for workers and the environment. Irradiation is another popular mutagen used to induce mutations.

4.4 Induced mutation in the plant by irradiation and tissue culture

Advantages of plant propagation by tissue culture are as follows:

1. It can increase the volume of plants in large quantities in less time than normal propagation. This saves space and labor used for propagation.

2. Disease-free plants can be obtained. Diseases are caused by fungi or bacteria infecting seeds or other propagation. In tissue culture, plant parts are sterilized before being cultured on a tissue culture medium. If there is mold and bacteria on it, it will show and can be removed.

The process of induced mutation in plants by irradiation and tissue culture

1. Study the appropriate tissue culture medium.

Plant tissue culture should be studied in a formula that is appropriate for the species and parts of the plants tested to enable the plants to grow well. and expand the number appropriately with strong plants.

2. Irradiation

Two methods of irradiation combined with tissue culture can be performed [8]

a. Irradiated plant parts such as seeds, lateral buds, shoots, and branches. Then the plant parts are sterilized and tissue cultured. The disadvantage of this method is that the plant becomes contaminated with microorganisms after tissue culture, resulting in less plant tissue than in reality.

b. Irradiated plant parts, such as seeds, lateral buds, shoots, and branches. Then the plant parts are sterilized and tissue cultured. The disadvantage of this method is that the plant becomes contaminated with microorganisms after tissue culture has fewer plants than the number of plants that should have been the chances of selecting a mutant will also be less.

Irradiation can be done in two ways: 1. Acute irradiation is the irradiation of a high dose in a short time or irradiation at a high dose rate, 2. Chronic irradiation is the irradiation of a low dose and uses a long time or irradiation at a low dose rate (**Figure 1**).

Plant tissue before irradiation is called M_0V_0, with M standing for meiotic and V for vegetative, when irradiated, it is called M_1V_1 generation. New plants or new shoots from M_1V_1 generation are called M_1V_2 generation. It can be subcultured to the next-generations, called M_1V_3, M_1V_4, M_1V_5, and so on (**Figure 2**).

3. Selection of mutations by tissue culture technique

After irradiation of plant tissue, the tissue is subcultured to M_1V_2 generation. In M_1V_2 generation, there can be genetic differences in new plants or new shoots that can

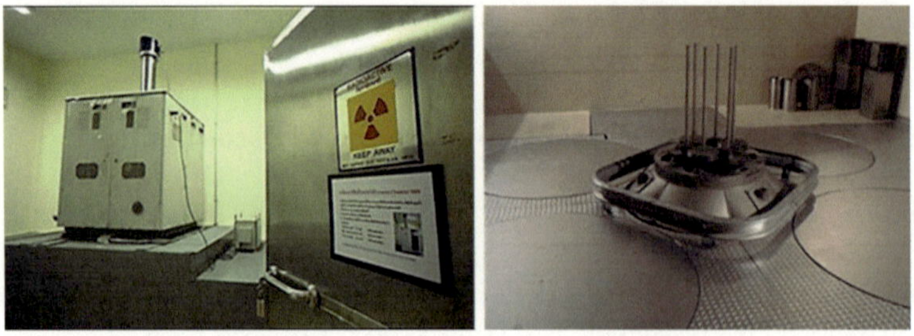

Figure 1.
(left) Gammator and (right) Gamma room at Thailand Institute of Nuclear Technology (Public Organization).

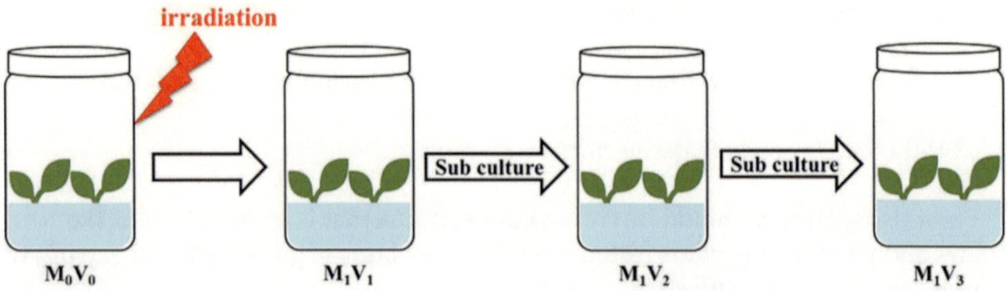

Figure 2.
M_1V_0 - M_1V_3 generation.

be divided into three groups: 1. The original species, 2. The group of tissues that are all mutated is called a solid mutant, and 3. The group in which the mutated and normal tissues are combined is called a chimera [8]. The occurrence of chimera in different layers of plant cells affects different plant phenotypes, as illustrated in **Figure 3**.

The size and frequency of sectoring are a function of transposon (or other mutagens) activity and the rate of cell division [9]. Tissue culture can help in the separation of mutated and normal sections by subculture. The mutated section is cultured, and the subculture continues until the solid mutant is obtained (**Figure 4**).

Examples of research on plant breeding by radiation in combination with tissue culture.

Study on the effect of gamma irradiation on tissue culture of aquatic plant "*Anubias nana*." It was found that the selected mutants in the M_1V_4 generation had variegated leaves, dwarfism, light green leaves, abnormal leaves, and albinism (**Figure 5**).

Mutation's breeding on tissue culture of Curcuma hybrid "Laddawan." The experiment revealed that morphological variations in the M_1V_2 generation were observed at 50 Gy gamma irradiation treatment samples with variegated leaves and light green leaves were observed (**Figure 6**).

The use of tissue culture techniques to assist in the selection of mutations in tissue culture of bromeliad "*Tillandsia cyanea*". The selected mutants in M_1V_3 generation had light green leaves, variegated leave, dwarf, and giant characteristics. The most common characteristic was light green leaves (**Figure 7**).

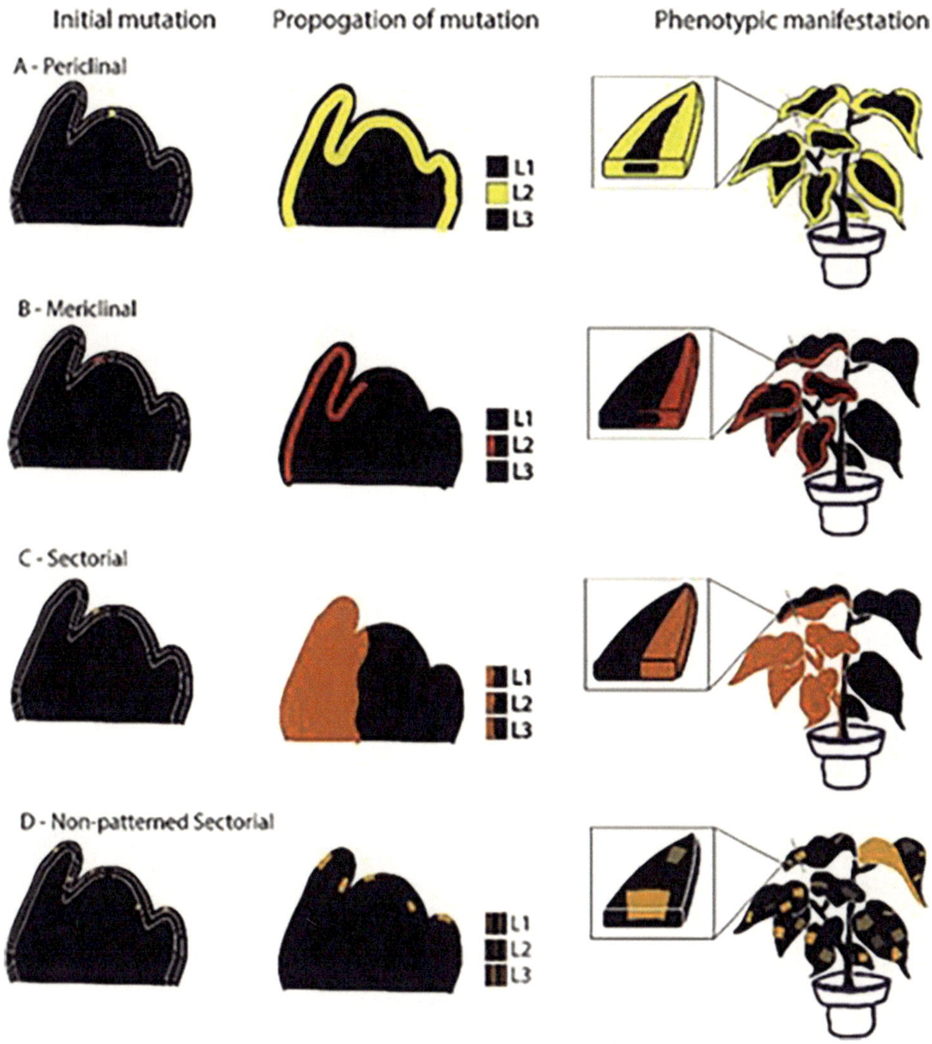

Figure 3.
Shows mutation characteristics and phenotypic expression. (A-B) has a mutation at the L2 layer (shown in yellow and red, respectively) causing phenotypic mutation along the leaf margins. (C-D) has a mutation at L1, L2, and L3 layers, which are often characterized as being unstable. (C) Sectorial are traverse all layers of the shoot meristem. (D) Non-patterned sectorial are variegated appearance.

Besides the traditional tissue culture method, another tissue culture method may increase the number of plant tissues in a shorter time. This is called the TIB (Temporary Immersion Bioreactor) System. This system has a container that separates the liquid tissue culture medium and plant parts into two parts, each with a tube connected to allow the liquid tissue culture medium to be pushed back and forth with air pressure from an air pump. The condition inside the bottle was aseptic by filtering the air entering the bioreactor with a filter with a pore size of 0.2 µm plant This prevents the plant from drowning in liquid food all the time [13].

The working principle is divided into four phases as follows [14]:

1. Stationary phase: Systemic tissue is normal in culture parts and media parts with liquid food.

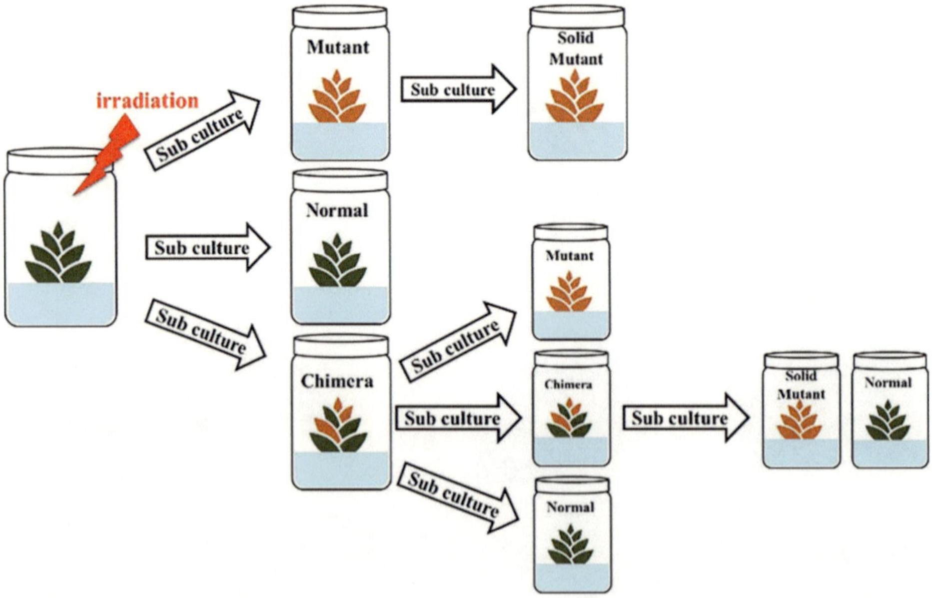

Figure 4.
Selection of mutations by tissue culture technique.

Figure 5.
Characteristics in the M_1V_4 generation of Anubias nana, (A) normal plant, (B) albinism, and (C) variegated leaves [10].

2. Immersion phase: It is the stage of air pressure to deliver liquid food to flood the plant tissues in the culture part at the specified time

3. Drain phase: Allowing liquid food to flow back into the vessels below by gravity.

4. Ventilation phase: The phase of air pressure into the vessels above (**Figure 8**).

The temporary immersion bioreactor system is a fast automated system. Therefore, the use of labor in the work is reduced. In addition, the container capacity of the bioreactor system temporarily sinks and can increase by 4–5 times. Thus, increasing the volume per area can also reduce the area of the tissue culture room down. Therefore, the temporary immersion bioreactor system is a new and suitable plant propagation system. It can replace the traditional plant tissue culture system (solid medium) and can be used in industrial plant production [15].

Figure 6.
Characteristics in the M_1V_2 generation of Curcuma hybrid 'Laddawan', (A) normal plant, (B) variegated leaves, and (C) light green leaves [11].

Figure 7.
Characteristics in M_1V_3 generation on tissue culture of Tillandsia cyanea after acute gamma irradiation, (A) normal plant, (B) light green leaves, (C) dwarf, (D) giant, and (E) variegated leaves [12].

Figure 8.
Bioreactor, (top) vertical sample vessels, and (below) horizontal sample vessels.

5. Application of molecular biology techniques for plant breeding

According to the article mentioned above, chemical and physical mutagens can cause genetic material alteration and have been widely used in plant mutation breeding. Variety identification or classification is an important step for plant breeding including the investigation of genetic variation and relatedness which are critical aspects of the maintenance of biodiversity. The identification method often uses genetic markers as a specific identifier that have been developed and used as a modern technique for characterization genotypes of organisms. A genetic marker is divided into three types [16–21].

1. Morphological marker

This marker is a physiological indicator that can be observed through general appearance and can be observed in morphological traits, such as plant height, canopy width, flower color, size or shape of the flower, flowering time, harvesting time, etc. It may be called naked-eyed polymorphism. These traits are mainly expressed by the controlling gene, so they can be used as a genetic marker. However, this marker has limited use, due to the morphology of plants often varying according to the changing environment, thus, the observed traits may cause errors in species identification. In addition, some plants are closely related species, the external characteristic comparison cannot be easily distinguished. Therefore, other indicators are required to improve the species classification accuracy.

2. Protein marker

The development of protein markers is to help identify the differences of plant species using different protein molecules which are components of plants to be examined. Protein markers are widely used to verify genetic purity for seed production. However, protein markers also have important limitations. The number of genes or proteins used for testing is small and genes that use for investigation must be expressed. Therefore, the investigation is necessary to select the optimal plant tissue and growth stage from gene expression. Moreover, the expression effect depends on the environment as well. This results in increasing the chance of the variation detecting in protein levels being undervalued. In addition, the use of protein markers in plant species detection is also limited because proteins are products of gene expression. It was found that the environment influences the expression of genes. If the environment is not suitable for plant growth, some genes cannot be expressed and proteins cannot be synthesized. Therefore, the results may vary and the classification between species that are closely related to each other can be impossible.

3. DNA marker

DNA marker is a DNA sequence that is used as a unique marker of an organism and can be inherited by the next generation. Each plant species has its own unique arrangement of nucleotides or polymorphisms of the DNA molecule that makes difference. So, it can be used as a marker to construct DNA profiles and assign the breeding lines to the various heterotic groups as well as variety identification. Using DNA as a marker to identify varieties of organism genetic material pattern which has a unique pattern to each individual can be done by a molecular and genetic technique which are commonly known as DNA profiling or

DNA fingerprint. DNA markers can be classified into three types according to the principles and techniques used for development.

- Hybridization based marker

This marker is developed based on the hybridization principle, which is the process of combining DNA probe and DNA target with complementary base pairs, such as restriction fragment length polymorphism (RFLP).

- Polymerase chain reaction-based marker

DNA markers are developed based on the principle of the rapid amplification of DNA as known as polymerase chain reaction (PCR), such as randomly amplified polymorphic DNA (RAPD), amplified fragment length polymorphism (AFLP), simple sequence repeat (SSR), inter simple sequence repeat (ISSR).

- Sequence-based marker

A developed marker based on the identification of nucleotides in DNA, such as single nucleotide polymorphism (SNP) (**Figure 9**).

Sometimes protein markers and DNA markers can refer to molecular markers. Molecular markers are a very useful tool to assess genetic diversity and relationships. The application of molecular markers in plant mutation breeding can help to accelerate the breeding program, increase accuracy and save cost and labor. They can apply to the study of genetic diversity, phylogeny, polymorphism analysis, and cultivar identification [24–28]. These markers have their own advantages and disadvantages, the appropriate marker selection depends on the purpose of use, the specific of each marker, facility, financial, time and knowledge. The comparison of some widely used DNA markers is shown in **Table 1** [19, 29, 30].

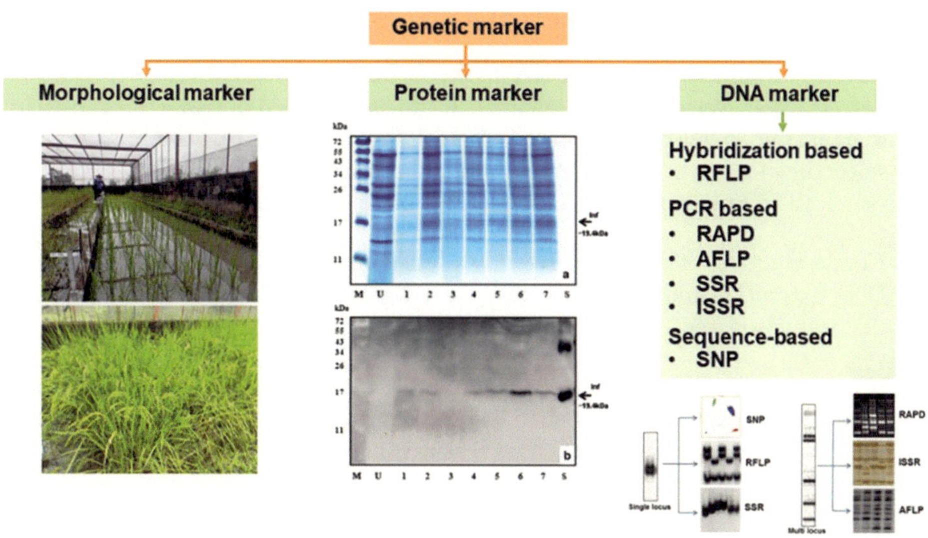

Figure 9.
The overview of genetic markers [22, 23].

Technique	RFLP	RAPD	AFLP	SSR	ISSR	SNP
Amount of required DNA	10 μg	20 ng	50-100 ng	50 ng	50 ng	50 ng
Quality of required DNA	High	High	High	Low	Low	Very high
Reproducibility	High	High	Medium	High	Medium-High	High
Genome abundance	High	Very high	Very high	Medium	Medium	High
Polymorphism level	Medium	Very high	High	High	High	High
Cost	High	Low	High	High	High	Variable

Table 1.
Comparison of some widely used DNA markers.

The development of DNA technology has grown rapidly over the past 10 years. There is a lot of new knowledge that can be applied widely in agriculture especially in plant breeding with an important purpose to create new varieties with good characteristics, such as high yield, disease resistance, pest resistance, and unfavorable conditions resistance. DNA markers are constantly being developed. It is widely used to identify the difference or diversity of organisms whether in terms of quantity or quality trait and is applied as a tool by plant breeders in many fields. Examples of the application of DNA markers in plant breeding are identification and purification of plant varieties, genetic assessment, genetic linkage mapping construction, and gene mappings, such as quantitative trait locus (QTL) and marker-assisted selection (MAS) [31–35]. DNA markers can distinguish more accurately and precisely than using plant morphology which often varies according to the environment. While protein marker is also limited, as it is based on gene expression. The advantages of DNA markers are as follows:

1. High accuracy – DNA marker is a tool that directly selects the desired plant genotype. Therefore, it is more accurate and precise than the selection from phenotypes that are related to the environment.

2. Large number of markers available for use

3. Constant results – The use of DNA marker to help in cultivar selection can be performed at every stage of plant growth and can use seed or tissue from any parts to be examined due to DNA remaining unchanged. Therefore, plant breeders can select the mutants at an early stage, which shortens the selection process. That plant can continue to grow. This is useful in cases when MAS is used because

4. Nondestructive – This is because only a small part of the plant is collected for DNA extraction and examined using the DNA marker. The interested mutant that is investigated can continue growing. There is no need to replant.

5. Inheritance classification – Some DNA markers, RFLP and SSR, are codominant markers that can be distinguished by both heterozygous and homozygous genotypes. While some DNA marker, such as RAPD, is not possible to distinguish between heterozygous genotype and homozygous genotype.

6. Able to select many desired characteristics simultaneously – The use of DNA markers can help select multiple desired traits at the same time and during the early growth stage. It saves time, labor, cost, and planting area.

6. Application of gamma irradiation on crop varieties

Rapid population growth, environmental pollution, and climate change have greatly affected human survival on Earth. The unsustainability of food, medicine, herb, and fuel supply chains is becoming a major problem for people around the world. Agricultural crops are very important to provide food, which is one of the most basic needs of human beings. How to feed the world's consumption poses great challenges to farmers and policymakers. To date, the method of artificial mutagenesis to obtain new biological cultivars, therefore, becomes a major challenge for breeders, and developing strategies to increase the genetic variability has demanded the attention of several research groups. The first paper on the technology of irradiation for mutation breeding employed X-rays in inducing mutations in maize and barley by Stadler in 1928 [36]. The first commercial mutant crop generated by irradiation technology is *Nicotiana tabacum* or "chlorine type" [37]. The leaves of these plants are used in the production of cigars, chewing tobacco, and nicotine replacement products. Plant breeders have been encouraged to use mutation breeding as one of the "peaceful uses of atomic energy." The main strategy in mutation-based breeding has been to upgrade the plant by altering traits to enhance their productivity and quality. In 2022, there are more than 3,391 cultivars developed from mutation breeding and registered in FAO/IAEA mutant database (**Table 2**). The mutant varieties database was split into three categories, crop plants, ornamental plants, and others were 47%,

Category	Records
Crop plants	1,592
Ornamental plants	711
Others	1,088
Total	3,391

Table 2.
The categories of FAO/IAEA mutant varieties database.

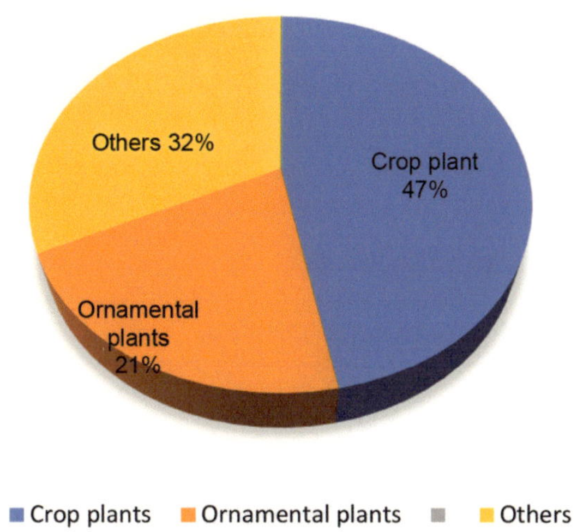

Figure 10.
Pie chart showing the percentage of categories plant data on FAO/IAEA mutant variety database.

21%, and 32%, respectively (**Figure 10**). Among these new varieties, about 1,703 cultivars were induced by gamma-ray irradiation [38].

Gamma irradiation mutagenesis technology has been widely used in crop mutation for a long time, partly due to restrictions on other techniques, such as hybridization, cross-breeding, and transgenic plants. Gamma radiation is able to penetrate the target material thoroughly and cause DNA damage in the chromosome. The DNA damage is difficult to repair effectively and correctly, leading to the generation of mutant plant varieties. To date, the IAEA reported that there are nearly 300 gamma irradiators from around the world [39].

Many crops, such as food, herbal medicine, flower, and ornamental, have been created using gamma irradiation mutagenesis technology for trait improvement. This can support food security, through enhanced characteristics, or increased resistance to environmental stress. Many plant varieties have been developed using gamma irradiation and have been released and widely cultivated worldwide.

Food crops are most important for human survival. Many countries around the world are looking for techniques to improve crop diversity. The mutagenic effects of gamma-rays in plants have been a particularly important issue concerned by the breeders. There are many mutated crops varieties were reported in the FAO/IAEA database (**Table 3**). Wheat, rice, and soybean are the most important crops for the breeders, about 22%, 19%, and 15%, respectively, resented in mutated crop varieties in the database (**Figure 11**). The strategies for genetic improvement of the crops are the increase in production, improvement of nutrition, and higher resistance to unfavorable conditions, such as drought and salinity. However, the high-yield crops are considered by the breeders. Both total doses, dose rate and multiple biological endpoints on crops after exposure to gamma-ray were investigated. The rice seeds of the non-waxy variety 'Toyonishiki' were exposed to 20 kR of gamma-ray. Two out of 20,000 panicles produced on these plants had waxy grain, and one of these brought forth a new commercial variety, "Miyuki-mochi." The yield of the new variety was

Common name of crop plants	Records
Wheat	265
Rice	230
Soybean	182
Barley	106
Maize	89
Groundnut	79
Mungbean	39
Chickpea	27
Pea	30
Lentil	19
Sugarcane	13
Sorghum	18
Others	117

Table 3.
List of mutated crop plant varieties from FAO/IAEA mutant varieties database.

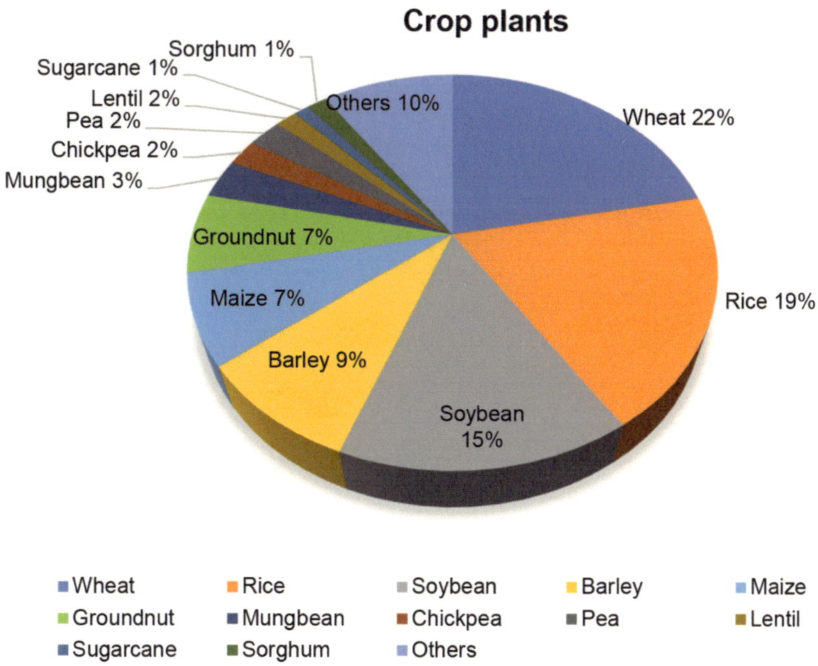

Figure 11.
Pie chart showing the percentage of crop plants on FAO/IAEA mutant variety database.

reduced by 8% as compared with the original variety, but the heading date, culm length, panicle length, and the number of panicles of the new variety were almost the same as those of the original variety. If compared with the old leading waxy variety, "Shinano-mochi," "Miyuki-mochi" is superior in yield by 15% and has high resistance to lodging and rice blast [40]. The seeds of five Japanese rice cultivars were irradiated with 250 Gy of gamma-ray (over 20 hr) and five high-yielding mutants were obtained [41]. High salinity in the soil is one of the major abiotic stresses leading to the reduction of rice yield. The seeds of "Dongan" rice were exposed to gamma-ray and selected for salt tolerance. They introduced the systemic procedures for the selection of salt tolerance rice plant mutants, and two promising mutant lines, ST-87 and ST-301, were finally selected [42].

Medicinal plants are important for disease prevention and in human healthcare for a long time. To date, numerous studies have been carried out globally to verify their efficacy and some of the findings have led to the production of plant-based medicines. The global market value of medicinal plant products exceeds $100 billion per annum [43]. Medicinally important plants are enhancement of secondary metabolite production, both *in vivo* and *in vitro* by gamma irradiation is a current area of interest. In the common herb babchi (*Psoralea corylifolia*), phenolic compounds in whole plants/plant parts irradiated with γ-ray at 20 kGy increase as high as 32-fold. The capsaicinoids, a phenolic compound in paprika (*Capsicum annum L.*) increased by about 10% with 10 kGy. The *in vitro* studies show all three types of secondary metabolites are reported to increase with γ-irradiation. Stevioside, total phenolic, and flavonoids content were slightly increased in 15 Gy-treated callus cultures of stevia (*Stevia rebaudiana* Bert.). In terpenoids, total saponin and ginsenosides content was increased 1.4- and 1.8-fold, respectively, with 100 Gy for wild ginseng (*Panax ginseng* Meyer) hairy root cultures.

In alkaloids, camptothecin yield increased as high as 20-fold with 20 Gy in callus cultures of ghanera (*Nothapodytes foetida*). Shikonins increased up to 4-fold with 16 Gy in suspension cultures of purple gromwell (*Lithospermum erythrorhizon S.*) [44].

Flowering plants and ornamental plants are two types of plants held in a special place in the history of humanity since ancient times. The sight of beautiful fresh flowers has a positive effect on the viewer. The use of such plants for decorative and aesthetic purposes has grown in floriculture. The report on the global market provides

Common name of ornamental plants	Records
Chrysanthemum	285
Rose	67
Dahlia	36
Carnation	28
Alstroemeria	35
Begonia	18
Azalea	15
White lupin	14
Hibiscus	14
Tulip	9
Achimenes	8
Canna lilies	8
Scotch broom	9
Lily	6
Others	82

Table 4.
List of ornamental plant varieties from FAO/IAEA mutant varieties database.

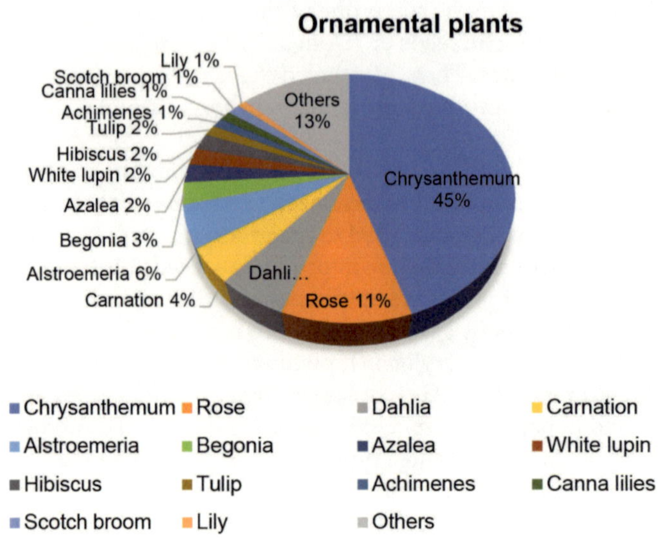

Figure 12.
Pie chart showing the percentage of ornamental plants on FAO/IAEA mutant variety database.

a holistic analysis, market size and forecast, trends, growth drivers, and challenges of flower, and the ornamental plant has been closely monitored and it is poised to grow by 27.18 billion dollars during 2021–2025 [45]. The mutation of flowering and the ornamental plant has become more successful owing to additional changes in phenotypic characteristics, heterozygous nature, and high mutation frequency producing a large number of new plant varieties. Gamma irradiation mutagenesis was used to improve the phenotype of flower and ornamental plants for a long time. In 1954, the first mutant propagated variety was released as a tulip variety with better flower shade and arrangement [46]. To date, there are various ornamental plants were recorded in FAO/IAEA mutant varieties database (**Table 4**). The three most of release mutant plants were chrysanthemum, rose, and dahlia (**Figure 12**). During 1970–1997, FAO/IAEA mutant varieties database shows the list of officially released

Scientific name	Common name	Mutant variety	Country of release	Year of release	Main improved characters
Dhalia sp.	Dhalia	Adagio	France	1970	Flower color
		Bichitra	India	1978	Plant architecture
		Huanghuan	China	1978	Shortness
		Jyoti	India	1978	Plant architecture
		Meiguizi	China	1989	Flower color
		Twilight	India	1978	Plant architecture
Dianthus caryophyllus	Carnation	Bonitas	GDR	1985	Semi dwarfness
		Chaichompon	Thailand	1983	Flower color
		Galatee lonvego	France	1982	Fusarium
		Loncerda	France	1983	Fusarium
		Maiella lonchabi	France	1982	Fusarium
		Scrlet Bell	Japan	1983	Flower color
		Sim Feu Follet	France	1972	Flower color
Gerbera jamesonii	Gerbera	Raisa	Poland	1993	Flower color
Gladiolus sp.	Gladiolus	Shobha	India	1988	Flower color
		Tambani	India	1991	Flower color
Hibiscus sp.	Hibiscus	Anjali	India	1987	Flower color
		Purnima	India	1979	Variegated leaves
Nelumbo nucifera	Lotus	Dandinyuge	China	1997	Flower color
		Dianezhuang	China	1983	Earliness
Rosa sp.	Rose	Beijingzhichun	China	1990	Flower color
		Beiyumudan	China	1986	Flower color
		Bridal Sonya	Japan	1985	Flower color
		Caiyemingxin	China	1986	Leaf morphology
		Paula	USA	1960	Flower color
		September wedding	Canada	1983	Flower color

Table 5.
List of officially released mutant flowering and ornamental plants from FAO/IAEA mutant varieties database [47].

Common name	Scientific name	Applied dosages	Treated materials	Results
Orchid	*Dendrobium sonia*	10 to 200 Gy	Protocorm-like bodies	Untreated orchids' survival rate and weight were higher compared to treated plants
Tuberose	*Polianthes tuberosa*	5, 10, 15, 20, 25, and 30 Gy	Bulbs	Variegated leaves and mutant cultivars were found with 20 Gy
Tuberose	*Polianthes tuberosa*	2 and 4 KR	Bulbs	Early sprouting, early spike emergence, and 50% flowering were noticed at 2 kR than 4 kR
Gladiolus	*Gladiolus grandifforus*	15, 30, 45, and 60 Gy	Corms	Low doses of gamma irradiation encouraged the vase life of a flower and changed floret colors
Gladiolus	*Gladiolus hybrida*	0.5 to 5.0 kR	Corms	Earliest sprouting was observed in 3.0 kR
Gladiolus	*Gladiolus spp.*	25, 40, 55 and 70 Gy	Corms	Spike length, number, and size of florets were decreased at 70 Gy
Gladiolus	*Gladiolus hybrida*	1 to 7 kR	Corms	2 and 3 kR proved better stimulation overall treatment
Gerbera	*Gerbera jamesoni*	1.5, 2, 2.5, 5, 10, 15, 20, 30 Gy	In vitro shoots	In radiated plants with 5 Gy doses showed moderate resistance to powdery mildew
Hibiscus	*Hibiscus Rosa-sinensis*	10, 15, 20, and 25 Gy	Nodal segments	5 Gy was effective for controlling Phytoplasma and the survival rate of plants
Marigold	*Glebionis segetum*	20, 40, 60, 80, and 100 Gy	Seeds	20 Gy changed disk color and 40 Gy changed floret and disk color
Sunflower	*Helianthus annuus*	100 to 900 Gy	Seeds	100 and 200 Gy have a positive effect on plant height and root length, LD 50 have found at 500 Gy
Blushing philodendron	*Philodendron erubescens*	70, 100, 150Gy	Rooted cuttings	The color composition of leaves ranged from 0–10% dark bluish-green, 60–90% strong yellow-green and 10–30% brilliant greenish-yellow
Gerbera Daisy	*Gerbera jamesonii*	10, 20, 30, 40, 50, 60 Gy	Seeds	Gradual formation of a number of shoot and fresh weight was declined by increasing irradiation
Bougainvillea	*Bougainvillea glabra*	500 to 2000 Krad	Stem cuttings	Survival rate was 94% at 500 Krad and at 2000 Krad sprouting was delayed
Chrysanthemum	*Dendranthema grandiflora*	5, 10, 20, 30 and 40 Gy	Shoot culture	Flower color was detected by 10 Gy irradiation
Chrysanthemum	*Chrysanthemum moriflium*	10, 15, and 20 Gy	Flower bud	10 and 15 Gy can be used for inducing genetic variability
Chrysanthemum	*Chrysanthemum moriflium*	0.5, 1, 2, and 5 Gy	Shoot culture	Nuclear DNA content was comparatively less at low dose rates

Table 6.
Effect of exposure of gamma irradiation on different economically important flowers and ornamental plants during 2008–2019 [48].

mutant flower and ornamental plants induced by gamma irradiation, as shown in **Table 5** [47]. The effect of exposure to gamma irradiation on different economically important flowers and ornamental plants is summarized in **Table 6** [48].

7. Conclusions

The technology of radiation, especially gamma irradiation is a faster tool and environmentally friendly. Plant biotechnology, both tissue culture and molecular biology, plays an important role in shortening the time of mutation breeding. Roughly, tissue culture is new mutant lines synthesized and molecular biology is an analyzed tool. Most of the studies on mutagenesis in plants are using gamma-ray with changes in some characteristics that disappeared in the parents. For developing new plant varieties applying gamma radiation or chemical mutagen to *in vitro* explants is a useful and worldwide method. The mutants can be selected and propagated to produce numerous plantlets that will be further accepted. Therefore, useful changes can provide improving new varieties, new species, and sometimes new genera. Several types of plants including food and medicinal plants and ornamental plants have been improved by these methods. Thus, radiation-induced mutation breeding is a remarkable method that can lead to genetic variations, resulting in superior mutant cultivars with new and useful traits.

And all of these are rapidly artificial evolution by humans.

Conflict of interest

The authors declare no conflict of interest

Author details

Vichai Puripunyavanich*, Lamai Maikaeo, Mayuree Limtiyayothin and Piyanuch Orpong
Thailand Institute of Nuclear Technology, Nakhon Nayok, Thailand

*Address all correspondence to: vichaip@tint.or.th

References

[1] Szmigiera M. 2022. Available from: https://www.statista.com/statistics/262618/forecast-about-the-development-of-the-world-population/

[2] Joint FAO/IAEA. Plant Mutation Breeding and Biotechnology. Plant Breeding and Genetics Section, Division of Nuclear Techniques in Food and Agriculture. Vienna, Austria: International Atomic Energy Agency; 2011. p. 595

[3] Puripunyavanich V. Plant Mutation Breeding for Crop Improvement. Gamma Irradiation: Properties Effects and Development of New Materials, Chapter 3. USA: NOVA Publishing; 2021. p. 150. ISBN: 978-1-53619-980-2

[4] Sharma A, Singh SK. Induced mutation- a tool for creation of genetic variability in rice (*Oryza sativa* L.). Department of Genetics and Plant Breeding, Institute of Agricultural Sciences, Banaras Hindu University, Varanasi, Uttar Pradesh, India. Journal of Crop and Weed. 2013;**9**(1):132-138

[5] Suprasanna P, Mirajkar SJ, Patade YV, Jain SM. Induced mutagenesis for improving plant abiotic stress tolerance. In: Tomlekova NB, Kozgar MI, Wani MR, editors. Mutagenesis: Exploring Genetic Diversity of Crops. Wageningen: Wageningen Academic Publishers; 2014. pp. 349-378

[6] IAEA. Manual on mutation breeding. Third edition. Joint FAO/IAEA Programme. Nuclear Techniques in Food and Agriculture. 2018. p. 297

[7] Acquaah G. Conventional plant breeding principles and techniques. In: Al-Khayri J, Jain S, Johnson D, editors. Advances in Plant Breeding Strategies: Breeding, Biotechnology and Molecular Tools. Springer: Cham; 2015. pp. 115-158. DOI: 10.1007/978-3-319-22521-0_5

[8] Jompuk P. Nuclear Technology and Agriculture. Bangkok: Kasetsart University; 2010. p. 288

[9] Frank MH, Chitwood DH. Plant chimeras. The good, the bad, and the 'Bizzaria'. Developmental Biology. 2016;**419**(1):41-53

[10] Sukin N, Limtiyayothina M, Jompuk P. Inducing genetic diversity of Anubias nana using gamma rays. Agricultural Natural Resources. 2020;**54**:85-90

[11] Tosri C, Chusreeaeom K, Limtiyayotin M, Sukin N, Jompuk P. Comparative effect of high energy electron beam and 137Cs gamma ray on survival, growth and chlorophyll content in curcuma hybrid 'Laddawan' and determine proper dose for mutations breeding. Emirates Journal of Food and Agriculture. 2019;**31**(5):321-327. DOI: 10.9755/ejfa.2019.v31.i5.1942

[12] Limtiyayothin M, Taychasinpitak T, Jompuk P. Gamma ray - induced in vitro mutations in bromeliad: Tillandsia cyanea. Khon Kaen AGR Journal. 2018;**46**(5):983-990

[13] Topoonyanont N, Nuamjaroen P, Thongthaksin W, Maitadchan S, Ampawan R, Pookmanee T, et al. Low-cost curcuma micropropagation system improvement by means of adapting temporary immersion bioreactor. 2005. p. 88

[14] Dabbhadatta Y. Plant tissue culture technology by bioreactor system. In: 16th NSTDA Annual Conference [Internet]. 2021. Available from: https:// www.nstda.

or.th/nac/2021/2021/03/05/agro4-tissue-culture-bioreactor/. [Accessed: February 24, 2022]

[15] Temporary immersion bioreactor system, new tissue culture innovation [Internet]. 2022. Available from: https://researchex.mju.ac.th/agikl/index.php/knowledge/43-another/172-vegetpotion. [Accessed: February 24, 2022]

[16] Collard BCY, Jahufer MZZ, Brouwer JB, Pang ECK. An introduction to markers, quantitative trait loci (QTL) mapping and marker-assisted selection for crop improvement: The basic concepts. Euphytica. 2005;**142**:169-196

[17] Jiang GL. Molecular markers and marker-assisted breeding in plants. In: Andersen SB, editor. Plant Breeding from Laboratories to Fields. London: Intech Open; 2013. pp. 45-83

[18] Dar AA, Mahajan R, Sharma S. Molecular markers for characterization and conversation of plant genetic resources. Indian Journal of Agricultural Sciences. 2019;**89**(11):1755-1763

[19] Iqbal MZ, Jamil S, Shahzad R, Bilal K, Qaisar R, Nisar A, et al. DNA Fingerprinting of crops and its application in the field of plant breeding. Journal of Agricultural Research. 2021;**59**(1):13-28

[20] Chesnokov YV, Kosolapov VM, Savchenko IV. Morphological genetic markers in Plants. Russian Journal of Genetics. 2020;**56**:1406-1415

[21] Mirzaei S. Application of molecular markers in plant species; An overview. Central Asian Journal of Plant Science Innovation. 2021;**1**(4):192-200

[22] Luchakivskaya Y, Kishchenko O, Gerasymenko I, Olevinskaya Z, Simonenko Y, Spivak M, et al. High-level expression of human interferon alpha-2b in transgenic carrot (*Daucus carota* L.) plants. Plant Cell Reports. 2011;**30**:407-415

[23] Kardlrvel P, Senthilvel S, Geethanjali S, Sujatha M, Varaprasad KS. Genetic markers, trait mapping and marker-assisted selection in plant breeding. In: Bahadur B et al, editors. Plant Biology and Biotechnology: Volume II: Plant Genomics and Biotechnology. India: Springer; 2015. pp. 65-88

[24] Walsh BM, Hoot SB. Phylogenetic relationships of Capsicum (Solanaceae) using DNA sequences from two noncoding regions: The chloroplast atpB-rbcL spacer region and nuclear waxy introns. International Journal of Plant Sciences. 2001;**162**:1409-1418

[25] Garris AJ, Tai TH, Coburn J, Kresovich S, McCouch S. Genetic structure and diversity in *Oryza sativa* L. Genetics. 2005;**169**:1631-1638

[26] Schneider K, Kulosa D, Soerensen TR, Möhring S, Heine M, Durstewitz G, et al. Analysis of DNA polymorphisms in sugar beet (*Beta vulgaris* L.) and development of an SNP-based map of expressed genes. Theoretical and Applied Genetics. 2007;**115**(5):601-611

[27] Solouki M, Mehdikhani H, Zeinali H. and Emamjomeh. Study of genetic diversity in Chamomile (*Matricaria chamomllia*) based on morphological traits and molecular markers. Scientia Horticulturae. 2008;**117**(3):281-287

[28] Selvakumari E, Jenifer J, Priyadharshini S, Vinodhini R. Application of DNA fingerprinting for plant identification. Journal of Academia and Industrial Research. 2017;**5**:149-156

[29] Agarwal M, Shrivastave N, Padh H. Advances in molecular marker techniques and their applications in

plant sciences. Plant Cell Reports. 2008;**27**:617-631

[30] Nadeem MA, Nawaz MA, Shahid MQ, Doğan Y, Comertpay G, Yıldız M, et al. DNA molecular markers in plant breeding: Current status and recent advancements in genomic selection and genome editing. Biotechnology and Biotechnology Equipment. 2018;**32**(2):261-285

[31] Teng S, Qian Q, Zeng D, Kunihiro Y, Fujimoto K, Huang D, et al. QTL analysis of leaf photosynthesis rate and related physiological traits in rice (*Oryza sativa* L.). Euphytica. 2004;**135**:1-7

[32] Perumal R, Krishnaramanujam R, Men MA, Katilé S, Dahlberg J, Magill CW, et al. Genetic diversity among sorghum races and working groups based on AFLPs and SSRs. Crop Science. 2007;**47**:1375-1383

[33] Miklas PN, Kelly JD, Beebe SE, Blair MW. Common bean breeding for resistance against biotic and abiotic stress: From classical to MAS breeding. Euphytica. 2006;**147**:105-131

[34] Wang J, Lydiate DJ, Parkin IA, Falentin C, Delourme R, King GJ. Integration of linkage maps for the amphidiploid *Brassica napus* and comparative mapping with *Arabidopsis* and *Brassica rapa*. BMC Genomics. 2011;**12**:20

[35] El-Mansy A, Diaa A, El-Moneim DA, Alshamrani SM, Alsafhi FA, Abdein MA, et al. Genetic diversity analysis of tomato (*Solanum lycopersicum* L.) with morphological, cytological, and molecular markers under heat stress. Horticulturae. 2021;7:18

[36] J. SL. Mutations in barley induced by X-rays and radium. Science (80-). 1928;**68**:186-187

[37] Solanki R, Gill R, Verma P, Singh S. Mutation breeding in pulses: An overview. 2011:85-103

[38] FAO/IAEA mutant data base. Available from: https://mvd.iaea.org/#!Search?page=1&size=500&sort by=Name&sort=ASC&Criteria[0][field]=CommonName&Criteria[0][val]=307 [Accessed February 24, 2022]

[39] Peeva A. Where Can You Find Industrial Irradiation Facilities? Visit a New Online Database. Available from: https://www.iaea.org/newscenter/news/where-can-you-find-industrial-irradiation-facilities-visit-a-new-online-database. [Accessed: February 24, 2022]

[40] Toda M. Breeding of new rice varieties by gamma-rays. Gamma F Symposium. 1980;**18**:73-82

[41] Kato H, Li F, Shimizu A. The selection of gamma-ray irradiated higher yield rice mutants by directed evolution method. Plants. 2020;**9**:1-16

[42] Song JY, Kim DS, Lee MC, et al. Physiological characterization of gamma-ray induced salt tolerant rice mutants. Australian Journal of Crop Science. 2012;**6**(3):421-429. DOI: 10.3316/informit.360742565763762

[43] Sofowora A, Ogunbodede E, Onayade A. The role and place of medicinal plants in the strategies for disease prevention. African Journal of Traditional Complementart Alternative Medicine. 2013;**10**(5):210-229. DOI: 10.4314/ajtcam.v10i5.2

[44] Vardhan PV, Shukla LI. Gamma irradiation of medicinally important plants and the enhancement of secondary metabolite production. International Jounal of Radiation Biology. 2017;**93**(9):967-979. DOI: 10.1080/09553002.2017.1344788

[45] Global Flower and Ornamental Plants Market 2021-2025. Available from: https://www.reportlinker.com/p05889315/Global-flower-and-ornamental-plants-

market.html?utm_source=GNW. Published 2021. [Accessed February 24, 2022]

[46] Broertjes AM. Applied Mutation Breeding for Vegetatively Propagated Crops. 1st ed. Elsevier Sci; 1988. p. 12

[47] Maluszynski M, Nichterlein K, Zanten L van, Ahloowalia BS. Officially Released Mutant Varieties - the FAO/ IAEA Database. International Atomic Energy Agency (IAEA): IAEA; 2000. Available from: http://inis.iaea.org/search/search.aspx?orig_q=RN:32006534

[48] Anne S, Lim JH. Mutation breeding using gamma irradiation in the development of ornamental plants: A review. Flower Research Journal. 2020;**28**(3):102-115

Section 4

Green Nanoparticle Synthesis

Chapter 11

Green Preparation of Fe_2O_3 Doped Gum Acacia Derived Porous Carbon/Graphene Ternary Nanocomposite as a Supercapacitor Electrode

Vijayasree Haridas, Zahira Yaakob and Binitha N. Narayanan

Abstract

The extended applications of the supercapacitor are possible with the attainment of a wide potential window since then it can exhibit high energy density too. Thus, organic electrolytes are more feasible in supercapacitors due to the accessibility of wide potential windows and the resultant higher storage/release of energy. A high-performance supercapacitor electrode material is prepared here via an eco-friendly procedure using a combination of Fe_2O_3, gum acacia derived porous carbon, and a ball-mill synthesized graphene for the first time. The synergistic action of the metal oxide and the carbon materials provided excellent specific capacitance values to the ternary nanocomposite. An appreciable specific capacitance of 433 F/g has been displayed by the composite coated glassy carbon electrode at a current density of 6 A/g in tetraethylammonium tetrafluoroborate—acetonitrile electrolyte at a wide potential window of 2.5 V. The material showed outstanding cyclic stability of 109% of the initial specific capacitance after 5000 repeated cycles.

Keywords: ternary nanocomposite, graphene, Fe_2O_3, acacia, green synthesis, supercapacitor

1. Introduction

Supercapacitors are potential energy storage systems that revolutionize conventional energy storage devices by their exceptional energy density, power density, promising cyclic stability as well as rate capability [1–6]. Based on the charge storage mechanism, supercapacitors are broadly classified into electrochemical double-layer capacitors (EDLC), pseudocapacitors, and hybrid capacitors. The faradaic reaction is present in the pseudocapacitors, while electrostatic interaction acts in the working of EDLCs, and hybrid capacitors are a combination of both of the above [1–7]. The power density and shelf life shown by the supercapacitors are higher compared to the batteries; while the energy density is found to be lower [1–6]. To enhance the performance, the development of modified electrodes having high specific capacitance and energy density is required without sacrificing the power density and cyclic stability of EDLC, which is a challenge for the scientific community.

The planar covalently bonded hexagonal 2D materials like graphene act as EDLC due to its surprising electrical conductivity, high surface area, etc. [8, 9]. Graphene is a perfect candidate that provides a good conducting network with a theoretical specific capacitance value of 550 F/g [10]. Similarly, porous carbon with a graphitic structure also functions as a good electrode material due to its outstanding electrical conductivity as well as high surface area [11]. The porous nature of the material additionally provides fast movement of the electrolyte ions. The preparations of such porous networks are found to be difficult as they require costly methods like chemical vapor deposition and electro-spinning methods [12–14]. Without disturbing the quality and supercapacitor performance, the development of a porous carbon network in a cost-effective manner is an interesting aspect.

Among the different methods of preparation of graphene, ball-mill-assisted exfoliation of graphite has the advantage of the easiness of preparation under mild conditions [15]. In addition, edge functionalization of graphene with milling agent provides synergistic properties advantageous in various applications. The use of naturally occurring biopolymers as milling agents is highly recommendable due to their low cost, eco-friendly nature, and easy availability [16, 17]. Here we use gum acacia for the ball-mill exfoliation of graphite and in addition, it takes the role of the precursor for porous carbon. Gum acacia, also called gum Arabic, is found in different species of Acacia, for example, Acacia arabica, Acacia babul, etc. [18]. It is a highly branched biopolymer composed mainly of high molecular weight glycosidal acid (Arabic acid) [19]. It has medicinal applications as well as it is used in food as an additive, thickening agent, emulsifier, etc. It is also used as a binder in paints, photography, printmaking, ceramics, etc. [18].

The operating voltage of a supercapacitor increases the energy density since it is related to the square of the operating voltage as evident from the equation, $E = ½CV^2$, where, E is the energy density, C is the specific capacitance, and V is the operating voltage [20]. Electrolytes are one of the vital parts, deciding the performance of supercapacitors. Organic electrolytes like tetraethylammonium tetrafluoroborate ($TEABF_4$), acetonitrile (AN), propylene carbonate, etc. enhance the operating voltage and provide greater specific capacitance as well as specific energy by avoiding the complications caused by the splitting of water in the aqueous electrolytes [21]. Kesavan and co-workers developed nitrogen-doped graphene for high energy density supercapacitors in 1 M $TEABF_4$/AN, and obtained a specific capacitance value of 103 F g^{-1} at a current density of 0.5 mA cm^{-2} [22]. Kovalska et al., performed supercapacitor studies using a gel electrolyte lithium bis(oxalate)borate in propylene carbonate displaying a capacitance of 78 μF/cm^2 using graphene-based supercapacitor [23]. Mostly for commercial purposes, supercapacitors are developed using organic electrolytes. But compared to the aqueous electrolyte, organic electrolytes reduce the electrolyte conductivity proceeding slower diffusion of the electrolyte to the electrode [21]. Therefore, developing a better electrode material with suitable functionalization reform the demerit caused by the organic electrolytes.

Hou et al., synthesized nitrogen-doped porous carbon nanosheets from natural silk and obtained a specific capacitance value of 242 F/g at a current density of 0.1 A/g [24]. Wang and co-workers prepared nitrogen-doped porous carbon from silkworm excrement and utilized it for high-energy-density symmetrical supercapacitor having an energy density of 138.4 Wh kg^{-1} and lithium-ion hybrid electrochemical capacitors of energy density 242.2 Wh kg^{-1} [12]. Liu et al., synthesized graphene-like porous carbon nanosheets from salvia splendens and displayed good capacity retention from 1 to 100 A/g [25].

The introduction of metal oxides in the carbon matrix improves the supercapacitor performance due to pseudocapacitance behavior in addition to the EDLC nature of carbon species [26]. The transition metal oxides like Fe_2O_3, MnO_2, RuO_2, $NiCo_2O_4$, etc. improve the supercapacitor performance of graphene-like materials by adapting pseudocapacitive behavior [27–30]. The non-toxic nature and low cost together with high theoretical capacitance promote the use of Fe_2O_3 as a supercapacitor electrode material [31]. The conducting carbon network on Fe_2O_3 can provide excellent supercapacitor properties to the combination [31, 32].

Herein, we have developed a Fe_2O_3 decorated biomass-derived porous carbon/graphene ternary nanocomposite in an eco-friendly and cost-effective manner. Environmentally benign gum acacia is used here for the exfoliation purpose. The composite was characterized and supercapacitor performance studies were done via cyclic voltammetric (CV), galvanostatic charge-discharge (CD), and electrochemical impedance spectroscopic (EIS) studies in 1 M tetraethylammonium tetrafluoroborate-acetonitrile ($TEABF_4$/AN). The ternary composite showed excellent supercapacitor performance with a specific capacitance value of 433 F/g at a current density of 6 A/g and outstanding cyclic stability of 109% of the initial specific capacitance after 5000 repeated cycles.

2. Experimental

2.1 Materials

Graphite flake (Sigma Aldrich Chemicals India Pvt. Ltd.), $FeCl_3.6H_2O$ (Loba Chemie), gum acacia (Loba Chemie), ammonia (Nice Chemicals Pvt. Ltd.) ethanol (98.5%), and polyvinyl alcohol (PVA, Loba Chemie) of reagent grade were used as such without purification. Electrolytic water (using electrolytic water purifier-double stage water purification system—type II water, W3T324491, EVOQUA Water Technologies) was used throughout the electrochemical experiments. Deionized water was used for the material preparation.

2.2 Preparation of Fe_2O_3/porous carbon graphene ternary composite

For the exfoliation of graphite, the planetary ball-mill procedure was used with 9 balls of 1 cm diameter and 5 balls of 2 cm diameter. The ball to powder weight ratio is fixed to be 4:1. Forty-nine gram of gum acacia and 1 g of graphite were mixed well and the resulting mixture is then dry milled for 30 hours. To the ball-milled mixture, 75 ml of water was added and then wet-milled for 3 hours. The mixture was then recovered from the ball mill and 425 ml of water was added which was further sonicated (Bath Sonicator, 6.5 L, PCI Analytics Ltd.) for 1 hour. Centrifugation is carried out to remove unexfoliated graphite. 6.77 g of $FeCl_3.6H_2O$ dissolved in 50 ml water was then added to the graphene dispersion under sonication. Ammonia solution was then added until basic pH and the obtained solution was kept overnight. Hydrothermal treatment was given to the dispersion in a tightly closed Teflon container at 120°C for 18 hours. The treated solution was transferred to a dialysis membrane (Himedia Dialysis Membrane-50, 14.3 mm diameter, 1.61 ml/cm approximate capacity) and dialyzed using water by stirring for a day until chloride ions were completely removed. It was dried and calcined for 3 hours in a tubular furnace at 350°C in a crucible closed with aluminum foil. The ternary system is further designated as Fe_2O_3-PC/graphene. Binary systems without Fe_2O_3 and without graphene are represented as PC/graphene and Fe_2O_3-PC, respectively, where PC indicates porous carbon.

2.3 Material characterization

X-ray diffraction (XRD) measurements of the materials were conducted by an advanced X-ray powder diffractometer (Bruker AXS D8 with CuKα radiation 0.15406 nm) in a 2θ range of 10–90°. Fourier transform infrared (FTIR) spectra were measured using a Perkin Elmer Spectrum TwoL1600300 FTIR Spectrometer. Raman analysis was conducted to analyze the defective nature of the prepared composite using JASCO NRS-4100 Spectrometer of 532 nm wavelength laser light. To study morphology, transmission electro microscopic images were taken using a high-resolution transmission electron microscope (TEM/JEM 2100). To investigate the elemental composition and nature of bonding, X-ray photoelectron microscopic analysis was performed by X-ray photoelectron spectroscopy with Auger electron spectroscopy module (PHI 5000 Versa Prob II, FEI Inc.) with C1s as internal standard.

2.4 Electrochemical measurements

The electrochemical studies were recorded using CHI-760E Electrochemical Analyzer (CH Instruments, USA) with the techniques cyclic voltammetry (CV), galvanostatic charge-discharge (GCD), and electrochemical impedance spectroscopy (EIS). All of the measurements were done in 1 M $TEABF_4$/AN electrolyte using a modified glassy carbon electrode (GCE) as the working electrode, Pt wire as the counter electrode, and Ag/AgCl as the reference electrode. The studies were conducted by drop coating the dispersion on GCE as prepared by sonicating 70 wt% composite, 15 wt% carbon black (Phillips Carbon Black Limited), and 15 wt% PVA in 60 vol% ethanol-water mixture. The specific capacitance values were calculated from the following equation based on CD studies [33].

$$\text{Specific capacitance} = I\Delta t / A\Delta V \quad (1)$$

Where I is the current (mA), t is the discharge time (s), ΔV represents the potential window (V), and A is the area of the electrode of the CD measurement. EIS measurements were conducted with an amplitude of 5 mV in a frequency range of 0.01–10,000 Hz.

3. Results and discussion

A ternary composite of Fe_2O_3 with carbon nanostructures is prepared here for its use as a supercapacitor electrode material. The composite preparation was attained via ball-mill-assisted exfoliation of graphite with gum acacia and further hydrothermal treatment of the obtained graphene dispersion with the iron oxide precursor. The as-prepared colloidal mixture is dried and heat-treated for the formation of Fe_2O_3

embedded porous carbon on the highly conducting graphene sheets. The porous nature of the carbon derived from gum acacia can be a result of the release of gaseous materials such as H_2O, CO_2, etc. during the heat treatment at 350°C. The different stages in the preparation are schematically illustrated in **Figure 1** and detailed in the experimental section, from where it is clear that the preparation procedure follows a green strategy throughout.

The formation of iron species can be explained as follows. During preparation, to the graphene-gum acacia dispersion $FeCl_3$ is added followed by ammonia. Initially, $FeCl_3$ interacts with basic ammoniacal medium to form $Fe(OH)_3$ precipitate as observed during preparation. In hydrothermal treatment, Fe^{3+} interacts with polysaccharides in the medium. The alcoholic moieties of polysaccharides exist as

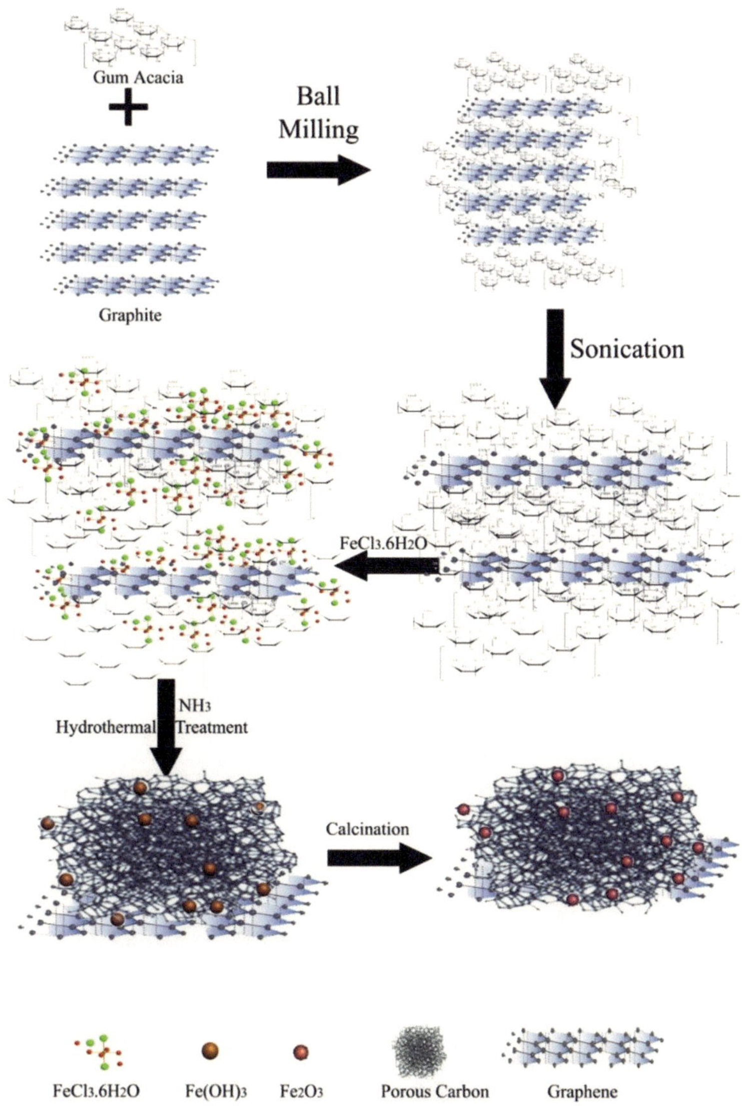

Figure 1.
Schematic representation showing the formation of Fe_2O_3—PC/graphene.

deprotonated in basic medium and these negative species coordinate with Fe^{3+} via electrostatic interaction forming complexes. Since, after hydrothermal treatment, a colloidal solution is obtained, the iron can exist as FeOOH covered by the polysaccharide as in agreement with reports [34]. Dialysis and further heat treatment convert the colloidal iron species to Fe_2O_3 as evident from various characterization techniques. In addition, during calcination, the polysaccharides from gum acacia got transferred to a porous carbon network embedding the Fe_2O_3 nanoparticles. These materials lay on the conductive graphene sheets to work as an efficient ternary composite electrode material.

3.1 Material characterization

X-ray diffraction patterns of the ternary composite and Fe embedded carbon shown in **Figure 2** displayed broad peaks centered around 2θ values of 24° and 43° indicating the presence of amorphous carbon [35]. Graphitic (002) diffraction is observed as a sharp band at 26.5° [36]. This band is found to be absent on the sample prepared without graphene. Weak bands around 33°, 35.7°, 54°, 62.5°, and 72° in the ternary composite as well as on iron embedded amorphous carbon indicate diffraction from the (104), (110), (116), (214), and (119) planes of α-Fe_2O_3 in the composites [37, 38]. The results suggest the coexistence of amorphous carbon, graphene, and α-Fe_2O_3 in the ternary composite.

FTIR spectra of the samples Fe_2O_3-PC and Fe_2O_3-PC/graphene (**Figure 3(a)**) show peaks corresponding to C-OH (~3400 and 1150 cm^{-1}), C-H (2916 cm^{-1}), C=O (1708 cm^{-1}), CH_2 (1380 cm^{-1}), C-O-C (1056 cm^{-1}), and Fe-O (~580 cm^{-1}) suggesting the presence of oxygen-containing functionalities on the carbon that can bind

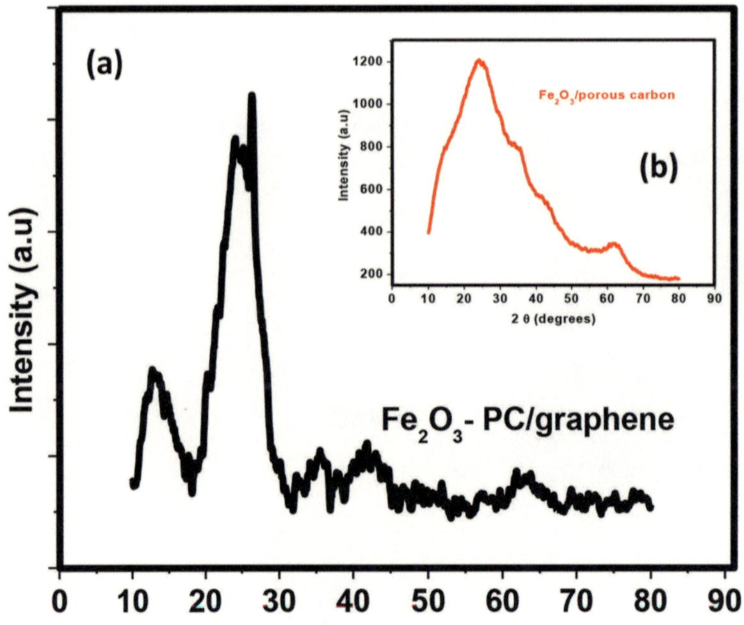

Figure 2.
XRD patterns of (a) Fe_2O_3/-PC graphene nanocomposite. And (b) Fe_2O_3-PC.

DOI: http://dx.doi.org/10.5772/intechopen.103080

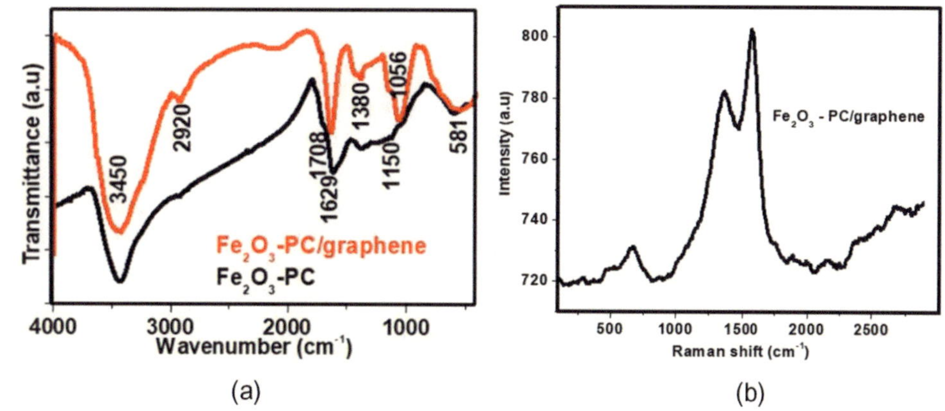

Figure 3.
(a) FTIR spectra of Fe_2O_3/-PC graphene nanocomposite and Fe_2O_3-PC, and (b) Raman spectrum of Fe_2O_3/-PC graphene nanocomposite.

with Fe_2O_3 [39–43]. The presence of aromatic carbon stretching of graphene (C=C, 1629 cm^{-1}) is also indicated in the ternary composite [44].

Raman spectrum of the ternary composite (**Figure 3(b)**) displayed two major peaks corresponding to the D and G bands of carbon matrix at 1350 and 1580 cm^{-1} [45, 46]. The I_D/I_G value of 0.71 indicates the presence of disordered graphitic structure in the composite [47]. The broad bands around 700 and 2850 cm^{-1} are indicative of the hematite and graphitic 2D bands, respectively [45, 48].

The morphology of the ternary composite is investigated from the TEM images (**Figure 4**). The Fe_2O_3 nanoparticle embedded porous carbon structure on graphene sheets is well evident in the images. HRTEM image of the Fe_2O_3 nanoparticle embedded carbon displayed the lattice fringe at 0.26 nm corresponding to the (110) plane of α-Fe_2O_3 [49]. Graphene prepared with the assistance of gum acacia displayed a sheet-like

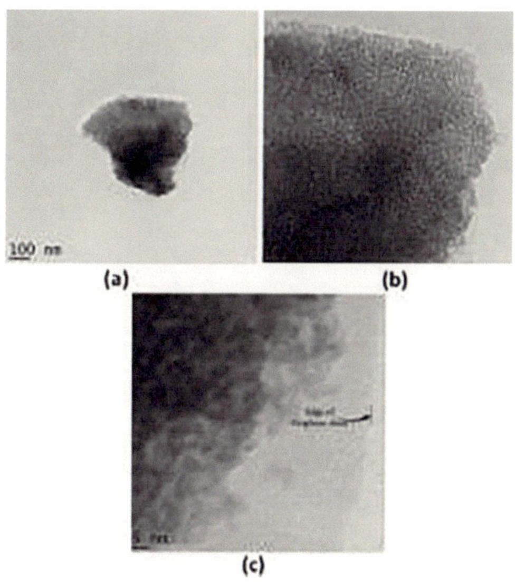

Figure 4.
(a)–(c) TEM images of Fe_2O_3/-PC graphene nanocomposite.

graphene structure; in addition, torn carbon sheets are also visible with a porous texture (**Figure 5**).

XPS wide scan spectra indicate the presence of C (74.72 at %), O (22.68 at %), and Fe (2.59 at %) in the ternary composite (**Figure 6(a)**). Deconvoluted C1s (**Figure 6(b)**) show the interaction between Fe_2O_3 and carbon as evident from the peak at 283.69 eV. The peak at 284.8 eV indicates the sp3 C-C bond, and the aromatic C=C is confirmed by the peak at 283.99 eV [50]. The slight shift in the reported values of sp3 C-C bond may be a result of the interaction of some of the sp3 C with Fe. The presence of oxygen moieties is validated from the peak at 286.89 eV indicating the presence of C-O-C/C-OH functionalities [50]. In the XPS profile of O1s (**Figure 6(c)**), the peaks at 532.57 eV, and 531.95 eV indicate the C-O-C and C-OH groups in the ternary composite [51]. A well-specified band noticed at 530.09 eV corresponds to Fe-O-C indicating binding of Fe with carbon via oxygen [52]. The

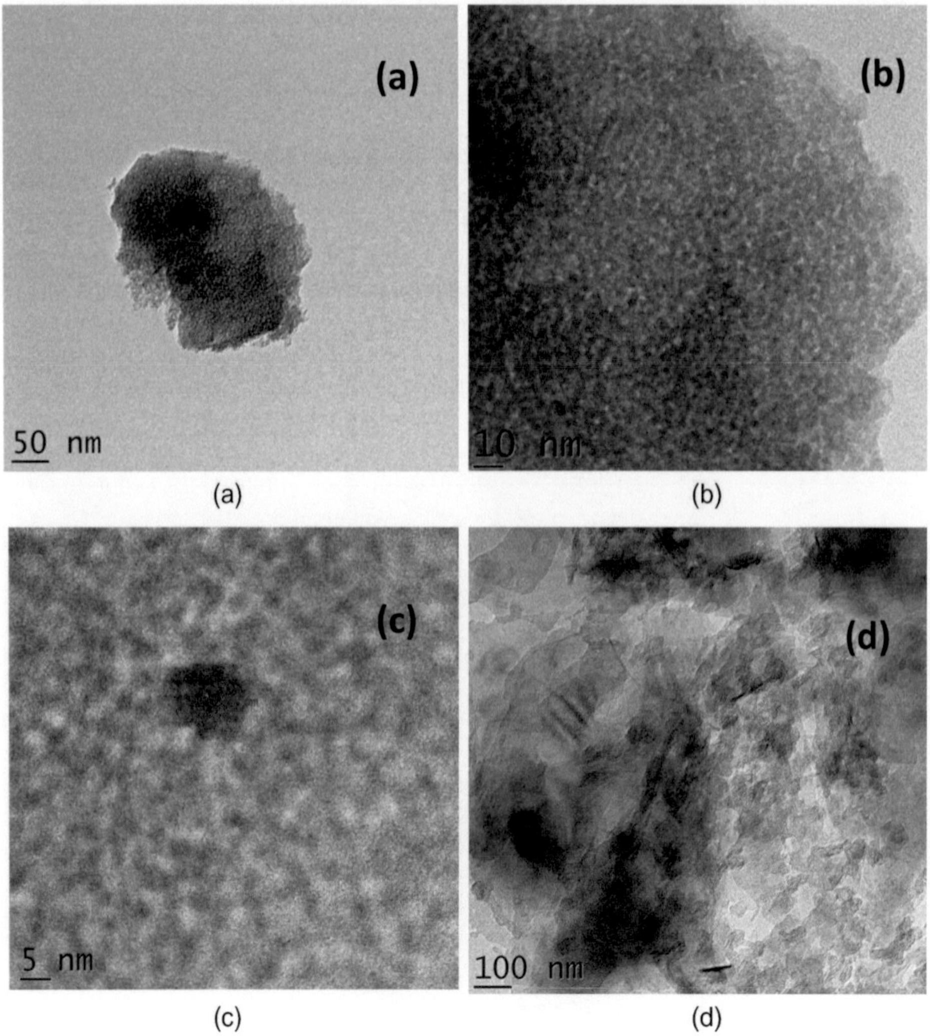

Figure 5.
(a)–(c) HRTEM of Fe_2O_3/-PC and (d) of PC/graphene.

band at 530.73 eV is indicative of Fe-O in Fe_2O_3 [53]. The Fe-O binding in Fe_2O_3 and its Fe-C interaction is additionally evident from the band at 531.88 eV [54]. The peak at 532.66 eV further confirms the C-OH functionalities in the composite [55].

3.2 Supercapacitor performance evaluation

The supercapacitor performance of the materials is evaluated using CV and CD measurements. CV curves are shown in **Figure 7(a)**, which illustrate both electrochemical double-layer capacitance (EDLC) as well as the pseudocapacitive nature of the materials. The ternary composite displayed a very high current in comparison with GCE, Fe_2O_3-PC, as well as PC/graphene systems. This indicates a synergistic effect of the ternary system enhancing the electrochemical performance of each of the components. Both the EDLC and redox behavior are displayed in the CV curves [56, 57]. A wide potential window has been attained by the electrodes as a result of the use of organic electrolyte, which can provide high energy density to the system [58].

The charge-discharge curves of GCE, Fe_2O_3-PC, and PC/graphene are shown in **Figure 7(b)**. The specific capacitance value of Fe_2O_3-PC (77.4 F/g) was found to be lower than that of PC/graphene (83 F/g) that can be due to the lower conductivity of Fe_2O_3, whereas the high diffusion of electrolyte on the porous carbon and high

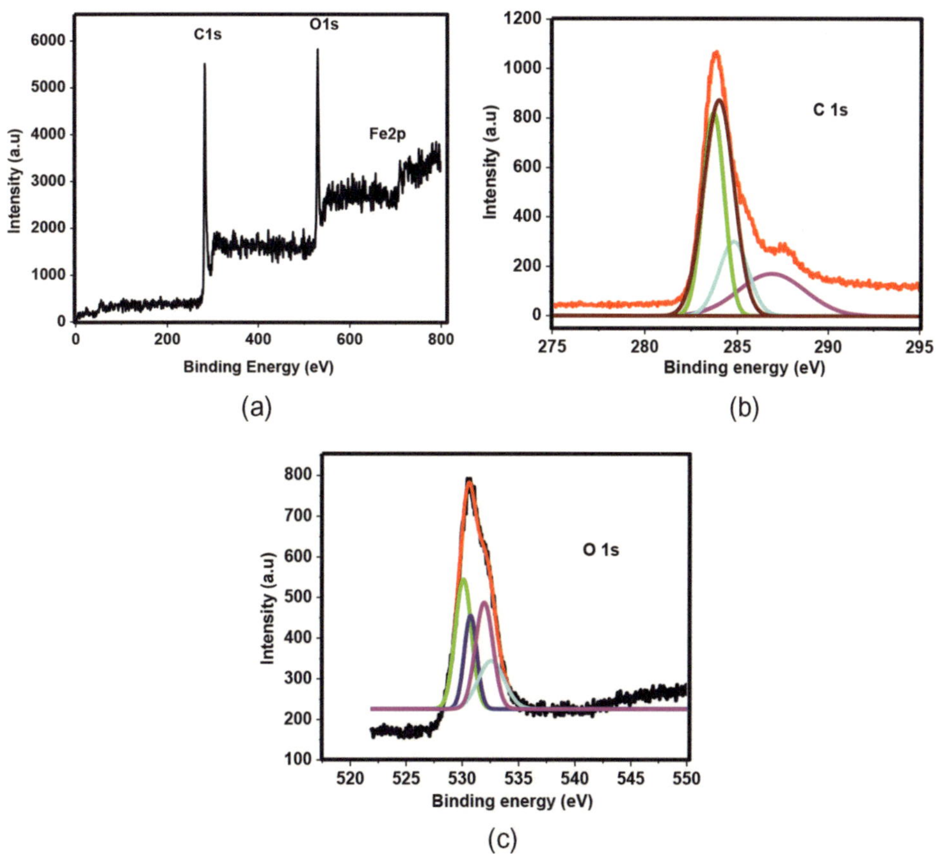

Figure 6.
XPS spectra of Fe_2O_3-PC/graphene nanocomposite (a) wide scan spectra, (b) C 1 s, and (c) O1s.

conductivity of graphene are the reasons of the enhanced specific capacitance of PC/graphene [59]. The specific capacitance of Fe_2O_3-PC/graphene (433 F/g) measured at a current density of 6 A/g is found to be excellent as a result of the synergy between the individual components in the hybrid structure. All the systems showed an initial IR drop, but the discharge time for the ternary nanocomposite is found to be slower thereafter. The ternary composite shows good specific capacitance values of 433, 97, 88, 84, 23, 14, and 12 F/g at current densities of 6, 7, 9, 10, 25, 50, and 100 A/g, respectively. The synergistic behavior attained by the ternary composite is composed of high pseudocapacitance of Fe_2O_3, fast electrolyte diffusion on porous carbon, and the highly conducting high surface area graphene sheets with good EDLC [60]. The galvanostatic CD curves of Fe_2O_3-PC/graphene-modified electrodes at different current densities are shown in **Figure 7(c)**. As expected, the discharge time and thus the specific capacitance values decrease with an increase in the current density [61].

The stability of the electrode in repeated CD cycles was evaluated at a current density of 10 A/g to check the suitability of present ternary systems in electronic devices (**Figure 8(a)** and **(b)**). The performance of the Fe_2O_3-PC/graphene was slightly

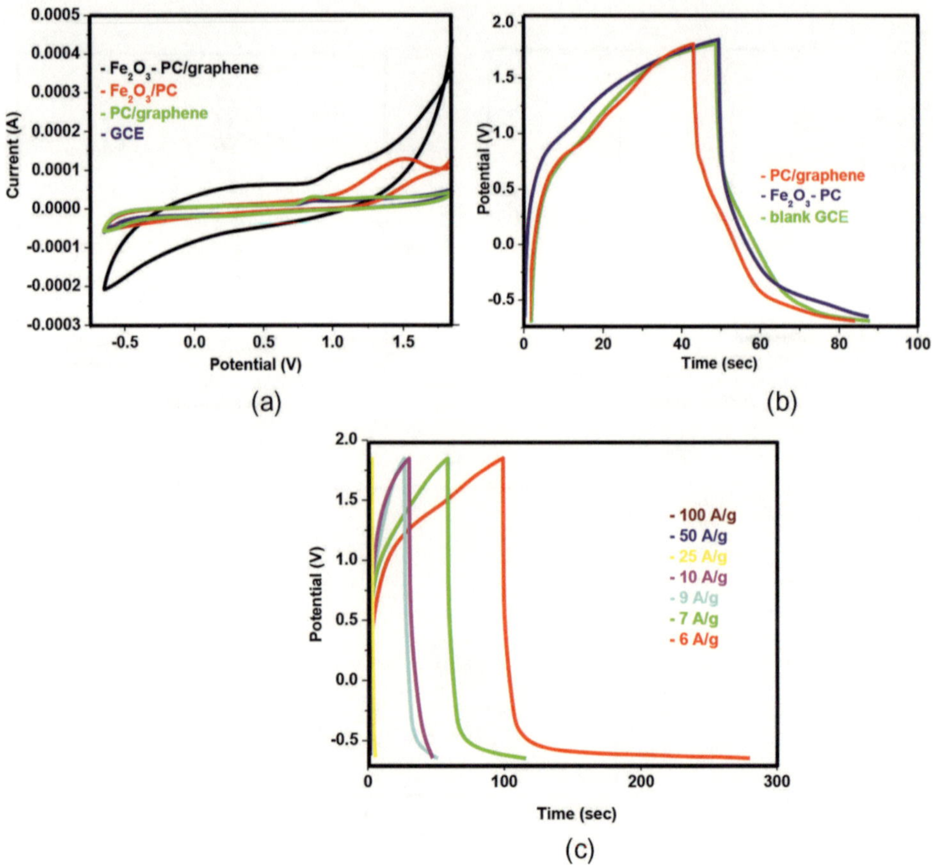

Figure 7.
(a) Cyclic voltammograms of Fe_2O_3-PC/graphene, Fe_2O_3-PC, PC/graphene and GCE in 1 M $TEABF_4$/AN, (b) GCD curves of Fe_2O_3-PC, PC/graphene and GCE, and (c) GCD curves of Fe_2O_3-PC/graphene at different current densities.

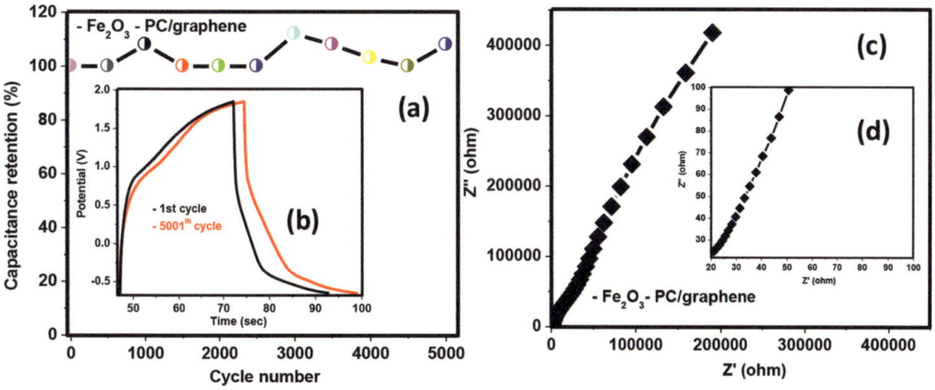

Figure 8.
(a) Capacitance retention versus the number of charge-discharge cycles, (b) comparison charge-discharge cycles of first and 5001th cycle and EIS spectrum (c), and (d) of Fe_2O_3-PC/graphene nanocomposite.

improved to 109% of its initial performance after 5000 repeated runs, which is highly promising and potential quality of a supercapacitor electrode material.

EIS of the ternary composite (**Figure 8(c)**) indicates Warburg impedance and absence of a semicircle at high-frequency region (**Figure 8(d)** is indicative of the absence of charge transfer limitations in the composite electrode as well as its high conductivity [62, 63]. The inclination of the EIS plot towards the Y-axis suggested enhanced diffusion of electrolyte ions, which can be resultant of the porous nature of carbon in the composite [64].

Table 1 indicates a comparative supercapacitor performance evaluation of the present ternary composite with other carbon-based electrodes using organic electrolytes (details of the electrode materials are given in the appendix **Table A1**). The better specific capacitance of the Fe_2O_3-PC/graphene is well evident from the data.

Sl. No.	System	Current density/ Scan rate	Specific capacitance	Electrolyte	% Retention @ no. of cycles	Ref
1	NAC@Gr	1 A/g	228 F/g	1 M $TEABF_4$/AN	93.64 @ 10,000	[65]
2	ResFaGO-A	10 A/g	105 F/g	1.5 M Et_4NBF_4	72 @ 10,000	[66]
3	NPCFs	50 mA/g	306 F/g	1 M $TEABF_4$/AN	—	[67]
4	GMS	1 mV/s	125 F/g	1 M Et_4NBF_4/PC	100 @ 600	[68]
5	RGO/V_2O_5	0.1 A/g	384 F/g	$LiClO_4$/PC	82.2 @ 10,000	[69]
6	Graphene-CNT composites	1 A/g	110 F/g	1 M $TEABF_4$/PC	—	[70]
7	Co_3O_4-graphene	1 A/g	424.2 F/g	1 M $LiPF_6$	—	[71]
8	PCNS	1 mV/s	120–150 F/g	1 M $TEABF_4$/AN	90 @ 10,000	[72]
9	PNSC	2 A/g	210 F/g	1 M $TEABF_4$/PC	95 @ 2000	[73]
10	Fe_2O_3-PC/ graphene	6 A/g	433 F/g	1 M $TEABF_4$/AN	109 @ 5000 cycles	This work

Table 1.
The specific capacitance of carbon-based graphene nanocomposites using organic electrolytes.

4. Conclusions

A perfectly green procedure is reported here for the preparation of a ternary system having Fe_2O_3, porous carbon, and graphene as constituents. Gum acacia takes the dual role of exfoliating agent to graphite and the precursor of porous carbon. The material characterization revealed the porous nature of Fe_2O_3 embedded carbon and the strong interaction between the components. Electrochemical studies revealed the electrochemical double-layer capacitance, pseudocapacitance, and the conducting nature of the composite leading to high specific capacitance values. Excellent cyclic stability of 109% is offered by the nanocomposite even at a high current density of 10 A/g after 5000 continuous charge-discharge measurements.

Acknowledgements

Vijayasree Haridas acknowledges UGC, New Delhi, India for UGC-SRF. The authors thank Sree Neelakanta Govt. Sanskrit College Pattambi and the University of Calicut for providing the facilities to carry out the research work. SAIF KOCHI, India is acknowledged for XRD and TEM analyses. The authors express gratitude to ACMS, IIT, Kanpur, for XPS analysis, and PSG Institute of Advanced Studies, Coimbatore for Raman analysis. Technical support for instrument purchase resulting from FIST-2016 grant of Department of Science & Technology, New Delhi, India is greatly acknowledged.

Conflict of interest

All authors declare, there are no financial/commercial conflicts of interest associated with this manuscript.

Appendix

Sl. No.	System designation	Electrode material
1	NAC@Gr	Nitrogen- and oxygen-enriched porous carbon/graphene
2	ResFaGO-A	Carbon-graphene composite under Ar atmosphere
3	NPCFs	Sandwich-like nitrogen-doped porous carbon/graphene nanoflakes
4	GMS	Carbon graphene mesosponge
5	RGO/V_2O_5	Reduced graphene oxide/V_2O_5
6	Graphene-CNT composites	Graphene-carbon nanotube composites
7	Co_3O_4-graphene	Nanoporous Co_3O_4-graphene composite
8	PCNS	Porous carbon nanosheets
9	PNSC	Porous network structured carbon

Table A1.
*Details of various graphene-porous carbon-based modified electrodes (mentioned in **Table 1** of the manuscript) used in the supercapacitor performance evaluation studies.*

Author details

Vijayasree Haridas[1], Zahira Yaakob[2] and Binitha N. Narayanan[1*]

1 Department of Chemistry, Sree Neelakanta Government Sanskrit College Pattambi (Affiliated to University of Calicut), Palakkad, Kerala, India

2 Department of Chemical and Process Engineering, Universiti Kebangsaan Malaysia, Bangi, Selangor, Malaysia

*Address all correspondence to: binithann@yahoo.co.in

References

[1] Conway BE. Electrochemical Supercapacitors: Scientific Fundamentals and Technological Applications. New York: Kluwer-Plenum; 1999

[2] Arico AS, Pruce P, Scrosati P, Tarascon JM, Schalkwijk WV. Nanostructured materials for advanced energy conversion and storage devices. Nature Materials. 2005;**4**:366-377

[3] Chu A, Braatz P. Comparison of commercial supercapacitors and high power lithium-ion batteries for power-assist applications in hybrid electric vehicles initial characterization. Journal of Power Sources. 2002;**112**:236-246

[4] Burke A. Ultracapacitors: Why, how, and where is the technology. Journal of Power Sources. 2000;**91**:37-50

[5] Chen SM, Ramachandran R, Mani V, Saraswathi R. Recent advancements in electrode materials for the Highperformance electrochemical supercapacitors: A review. International Journal of Electrochemical Science. 2014;**9**:4072-4085

[6] Wang G, Zhang L, Zhang J. A review of electrode materials for electrochemical supercapacitors. Chemical Society Reviews. 2012;**2012**(41):797-828

[7] Conway BE, Birss VJ. Wojtowicz the role and utilization of pseudocapacitance for energy storageby supercapacitors. Journal of Power Sources. 1997;**66**:1-14

[8] Geim AK, Novoselov KS. The rise of graphene. Nature Materials. 2007;**6**:183-191

[9] Shan CS, Yang HF, Song JF, Han DX, Ivaska A, Niu L. Graphene: Synthesis and applications. Analytical Chemistry. 2009;**81**:2378-2382

[10] Ke Q, Wang J. Graphene-based materials for supercapacitor electrodes – A review. Journal of Materiomics. 2016;**2**:37-54

[11] Ma C, Xu J, Fan Q, Shi J, Song Y. Synthesis and electrochemical performance of high surface area hierarchical porous carbon with ultrahigh mesoporosity for high-performance supercapacitors. Journal of Solid State Electrochemistry. 2019;**23**:2153-2163

[12] Wang P, Zhang G, Li MY, Yin YX, Li JY, Li G, et al. Porous carbon for high-energy density symmetrical supercapacitor and lithium-ion hybrid electrochemical capacitors. Chemical Engineering Journal. 2019;**375**:122020

[13] Zhong Y, Xia XH, Deng SJ, Zhan JY, Fang RY, Xia Y, et al. Popcorn inspired porous microcellular carbon: Rapid puffing fabrication from rice and its applications in lithium-sulfur batteries. Advanced Energy Materials. 2018;**8**:1701110

[14] Yang XY, Chen LH, Li Y, Rooke JC, Sanchez C, Su BL. Hierarchically porous materials: Synthesis strategies and structure design. Chemical Society Reviews. 2017;**46**:481-558

[15] Balasubramanyan S, Sasidharan S, Poovathinthodiyil R, Ramakrishnan RM, Narayanan BN. Sucrose-mediated mechanical exfoliation of graphite: A green method for the large scale production of graphene and its application in catalytic reduction of 4-nitrophenol. New Journal of Chemistry. 2017;**41**:11969-11978

[16] Mallick H, Sarkar A. An experimental investigation of electrical

conductivities in biopolymers. Bulletin of Materials Science. 2000;**23**:319-324

[17] Chabot V, Kim B, Sloper B, Tzoganakis C, Yu A. High yield production and purification of few layer graphene by gum Arabic assisted physical sonication. Scientific Reports. 2013;**3**(1378):1-7

[18] https://en.wikipedia.org/wiki/Gum_arabic

[19] Bahadur S, Sahu UK, Sahu D, Sahu G, Roy A. Review on natural gums and mucilage and their application as excipient, journal of applied. Pharmaceutical Research. 2017;**5**:13-21

[20] Choi C, Ashby DS, Butts DM, DeBlock RH, Wei Q, Lau J, et al. Achieving high energy density and high power density with pseudocapacitive materials. Nature Reviews Materials. 2020;**5**:5-19

[21] Li SM, Yang SY, Wang YS, Tsai HP, Tien HW, Hsiao ST, et al. N-doped structures and surface functional groups of reduced graphene oxide and their effect on the electrochemical performance of supercapacitor with organic electrolyte. Journal of Power Sources. 2015;**278**:218-229

[22] Kesavan T, Aswathy R, Arul Raj I, Prem Kumar T, Ragupathy P. Nitrogen-doped graphene as electrode material with enhanced energy density for next-generation supercapacitor application. ECS Journal of Solid State Science and Technology. 2015;**4**:88-92

[23] Kovalska EC. Kocabas organic electrolytes for graphene-based supercapacitor: Liquid, gel or solid. Materials Today Communications. 2016;**7**:155-160

[24] Hou J, Cao C, Idrees F, Ma X. Hierarchical porous nitrogen-doped carbon Nanosheets derived from silk for ultrahigh-capacity battery anodes and supercapacitors. ACS Nano. 2015;**9**:2556-2564

[25] Liu B, Yang M, Chen H, Liu Y, Yang D, Li H. Graphene-like porous carbon nanosheets derived from salvia splendens for high-rate performance supercapacitors. Journal of Power Sources. 2018;**397**:1-10

[26] Davies A, Yu A. Material advancements in supercapacitors: From activated carbon to carbon nanotube and graphene. Canadian Journal of Chemical Engineering. 2011;**89**:1342-1357

[27] Wang H, Xu Z, Yi H, Wei H, Guo Z, Wang X. One-step preparation of single-crystalline Fe_2O_3 particles/graphene composite hydrogels as high performance anode materials for supercapacitors. Nano Energy. 2014;**7**:86-96

[28] Lei Z, Zhang J, Zhao X. Ultrathin MnO_2 nanofibers grown on graphitic carbonspheres as high-performance asymmetric supercapacitor electrodes. Journal of Materials Chemistry. 2012;**22**:153-160

[29] Majumdar D, Maiyalagan T, Jiang Z. Recent Progress in ruthenium oxide-based composites for supercapacitor applications. ChemElectroChem. 2019;**6**:4343-4372

[30] Wu Z, Zhu Y, Ji X. $NiCo_2O_4$-based materials for electrochemical supercapacitors. Journal of Materials Chemistry A. 2014;**2**:14759-14772

[31] Zeng Y, Yu M, Meng Y, Fang P, Lu X, Tong Y. Iron-based supercapacitor electrodes: Advances and challenges. Advanced Energy Materials. 2016; **6**:1601053

[32] Tian Y, Hu X, Wang Y, Li C, Wu X. Fe_2O_3 nanoparticles decorated

on graphene-carbon nanotubes conductive networks for boosting the energy density of all-solid-state asymmetric supercapacitor. ACS Sustainable Chemistry & Engineering. 2019;7:9211-9219

[33] Jeong KH, Lee HJ, Simpson MF, Jeong SM. Electrochemical synthesis of graphene/MnO_2 Nano-composite for application to supercapacitor electrode. Journal of Nanoscience and Nanotechnology. 2016;**16**:4620-4625

[34] Somsook E, Hinsin D, Buakhrong P, Teanchai R, Mophan N, Pohmakotr M, et al. Interactions between iron(III) and sucrose, dextran, or starch in complexes. Carbohydrate Polymers. 2005;**61**:281-287

[35] Shang H, Lu Y, Zhao F, Chao C, Zhang B, Zhang H. Preparing high surface area porous carbon from biomass by carbonization in molten salt medium. RSC Advances. 2015;5:75728-75734

[36] Wang G, Yang J, Park J, Gou X, Wang B, Liu H, et al. Facile synthesis and characterization of graphene nanosheets. Journal of Physical Chemistry C. 2008;**112**:8192-8195

[37] Manikandan A, Vijaya JJ, Kennedy LJ. Structural, optical and magnetic properties of porous α-Fe_2O_3 nanostructures prepared by rapid combustion method. Journal of Nanoscience and Nanotechnology. 2013;**13**:2986-2992

[38] Kumar B, Smita K, Galeas S, Sharma V, Guerrero VH, Debut A, et al. Characterization and application of biosynthesized iron oxide nanoparticles using Citrus paradisi peel: A sustainable approach. Inorganic Chemistry Communications. 2020;**119**:108116

[39] Samargandi D, Zhang X, Liu F, Tian S. Fourier transform infrared (FT-IR) spectroscopy for discrimination of fenugreek seeds from different producing areas. Journal of Chemical and Pharmaceutical Research. 2014;**6**:19-24

[40] Emiru TF, Ayele DW. Controlled synthesis, characterization and reduction of graphene oxide: A convenient method for large scale production. Egyptian Journal of Basic and Applied Sciences. 2017;**4**:74-79

[41] Duygu DY, Udoh AU, Ozer TB, Akbulut A, Erkaya IA, Yildiz K, et al. Fourier transform infrared (FTIR) spectroscopy for identification of Chlorella vulgaris Beijerinck 1890 and Scenedesmus obliquus (Turpin) Kützing 1833. African Journal of Biotechnology. 2012;**16**:3817-3824

[42] Shopska M, Zara P, Zheleva C, Paneva DG, Lliev M, Kadinov GB, et al. Biogenic iron compounds: XRD, Mossbauer and FTIR study. Central European Journal of Chemistry. 2013;**11**:215-227

[43] Kumar B, Smita K, Galeas S, Guerrero VH, Debut A, Cumbal L. One-pot biosynthesis of Maghemite (γ-Fe2O3) nanoparticles in aqueous extract of Ficus carica fruit and their application for antioxidant and 4-Nitrophenol reduction. Waste and Biomass Valorization. 2021;**12**:3575-3587

[44] Yang W, Chen Y, Wang J, Peng T, Xu J, Yang B, et al. Reduced graphene oxide/carbon nanotube composites as electrochemical energy storage electrode applications. Nanoscale Research Letters. 2018;**13**:181

[45] Ferrari AC, Meyer JC, Scardaci V, Casiraghi C, Lazzeri M, Mauri F, et al. Physical Review Letters. 2006;**97**:187401-1-187401-4

[46] Dresselhaus BS, Jorio A, Souzafilho AG, Saito R. Defect

characterization in graphene and carbon nanotubes using Raman spectroscopy. Philosophical Transactions of the Royal Society A. 2010;**368**:5355-5377

[47] Geng X, Guo Y, Li D, Li W, Zhu C, Wei X, et al. Interlayer catalytic exfoliation realizing scalable production of large-size pristine few-layer graphene. Scientific Reports. 2013;**3**:1-6

[48] Song K, Lee Y, Jo MR, Nam KM, Kang YM. Comprehensive design of carbonencapsulated Fe_3O_4 nanocrystals and their lithium storage properties. Nanotechnology. 2012;**23**:505401

[49] He L, Wu H, Zhang W, Bai X, Chen J, Ikram M, et al. High-dispersed Fe_2O_3/Fe nanoparticles residing in 3D honeycomb-like N-doped graphitic carbon as high-performance room-temperature NO_2 sensor. Journal of Hazardous Materials. 2021;**405**:124252

[50] Wang G, Yang J, Park J, Gou X, Wang B, Liu H, et al. Facile synthesis and characterization of graphenenanosheets. Journal of Physical Chemistry C. 2008;**112**:8192-8195

[51] Pedrosa M, Silva ESD, Martínez LMP, Drazic G, Falaras P, Faria JL, et al. Hummers' and Brodie's graphene oxides as photocatalysts for phenol degradation. Journal of Colloid and Interface Science. 2020;**567**:243-255

[52] Zubir NA, Yacou C, Motuzas J, Zhang X, Costa JCD. Scientific Reports. 2015;**4**:4594-4602

[53] Plecenik T, Gregor M, Sobota R, Truchly M, Satrapinskyy L, Kurth F, et al. Physics Letters. 2013;**103**:052601-1-052601-4

[54] Muruganandham M, Amutha R, Ahmmad B, Repo E, Sillanpaa M. Self-assembled fabrication of superparamagnetic highly stable mesoporous amorphous iron oxides. Journal of Physical Chemistry C. 2010;**114**:22493-22501

[55] Hashemi SA, Mousavi SM, Faghihi R, Arjmand M, Rahsepar M, Bahrani S, et al. Superior X-ray radiation shielding effectiveness of biocompatible polyaniline reinforced with hybrid graphene oxide-iron tungsten nitride flakes. Polymers. 2020;**12**:1407

[56] Charoen-amornkitt P, Suzuki T, Tsushima S. Ohmic resistance and constant phase element effects on cyclic voltammograms using a combined model of mass transport and equivalent circuits. Electrochimica Acta. 2017;**258**:433-441

[57] Kim BK, Sy S, Yu A, Zhang J. Electrochemical Supercapacitors for Energy Storage and Conversion, Handbook of Clean Energy Systems. John Wiley & Sons, Ltd; 2015

[58] Khomenko V, Raymundo-Piñero E, Béguin F. High-energy density graphite/AC capacitor in organic electrolyte. Journal of Power Sources. 2008;**177**:643-651

[59] Tian H, Wang T, Zhang F, Zhao S, Wan S, He F, et al. Tunable porous carbon spheres for high-performance rechargeable batteries. Journal of Materials Chemistry A. 2018;**6**:12816-12841

[60] Yu P, Duan W, Jiang Y. Porous Fe_2O_3 Nanorods on hierarchical porous biomass carbon as advanced anode for high-energy-density asymmetric supercapacitors. Frontiers in Chemistry. 2020;**8**:611852

[61] Kumar YA, Sambasivam S, Ahmed S, Hira K, Zeb W, Uddin TNV, et al. Boosting the energy density of highly efficient flexible hybrid supercapacitors

via selective integration of hierarchical nanostructured energy materials. Electrochimica Acta. 2020;**364**:137318

[62] Nagamuthu S, Vijayakumar S, Muralidharan G. Ag incorporated Mn_3O_4/AC nanocomposite based supercapacitor devices with high energy density and power density. Dalton Transactions. 2014;**43**:17528-17538

[63] Liu W, Li X, Zhu M, He X. High-performance all-solid state asymmetric supercapacitor based on Co_3O_4 nanowires and carbon aerogel. Journal of Power Sources. 2015;**282**:179-186

[64] Jeon B, Ha T, Lee DY, Choi MS, Lee SW, Jung KH. Preparation and electrochemical properties of porous carbon nanofiber electrodes derived from new precursor polymer: 6FDA-TFMB. Polymers. 2020;**12**:1851

[65] Zhou S, Xie Q, Wu S, Huang X, Zhao P. Influence of graphene coating on supercapacitive behavior of sandwich-like N- and O-enriched porous carbon/graphene composites in aqueous and organic electrolytes. Ionics. 2017;**23**:1499-1507

[66] Fernandez GM, Urbano JLG, Enterrıa M, Rojo T, Carriazo D. Flat-shaped carbon–graphene microcomposites as electrodes for high energy supercapacitors. Journal of Materials Chemistry A. 2019;**7**:14646-14655

[67] Zhang Y, Taob B, Xing W, Zhang L, Xue Q, Yan Z. Sandwich-like nitrogen-doped porous carbon/graphene Nanoflakes with high-rate capacitive performance. Nanoscale. 2016;**8**:7889-7898

[68] Xiong C, Lia B, Lin X, Liu H, Xu Y, Mao J, et al. The recent progress on three-dimensional porous graphene-based hybrid structure for supercapacitor. Composites Part B Engineering. 2019;**165**:10-46

[69] Liu Z, Zhang H, Yang Q, Chen Y. Graphene/V_2O_5 hybrid electrode for an asymmetric supercapacitor with high energy density in an organic electrolyte. Electrochimica Acta. 2018;**287**:149-157

[70] Jung N, Kwon S, Lee D, Yoon DM, Park YM, Benayad A, et al. Synthesis of chemically bonded graphene/carbon nanotube composites and their application in large volumetric capacitance supercapacitors. Advanced Materials. 2013;**25**:6854-6858

[71] Huang X, Sun B, Chen S, Wang G. Self-assembling synthesis of free-standing Nanoporous graphene–transition-metal oxide flexible electrodes for high-performance lithium-ion batteries and supercapacitors. Chemistry, an Asian Journal. 2014;**9**:206-211

[72] Sevilla M, Fuertes AB. Direct synthesis of highly porous interconnected carbon Nanosheets and their application as high-performance supercapacitors. ACS Nano. 2014;**8**:5069-5078

[73] Liang Y, Liang F, Zhong H, Li Z, Fu R, Wu D. An advanced carbonaceous porous network for high-performance organic electrolyte supercapacitors. Journal of Materials Chemistry A. 2013;**1**:7000-7005

Chapter 12

Recent Advances in the Green Synthesis of Lanthanide-Based Organic Compounds for Broad Application Spectrum in Different Sectors: A Review

Kamna Chaturvedi, Deeksha Malvi, Manish Dhangar, Harsh Bajpai, Ranjan K. Mohapatra, Avanish Kumar Srivastava and Sarika Verma

Abstract

The present review highlights the various green method of synthesis and discrete applications of inner transition compounds. Green chemistry's strategies are developing, producing, and using effective, reliable, and eco-friendly chemical products and processes to manage pollution. In this review, the greener or environmentally sound route for synthesizing lanthanide compounds is discussed briefly. The initial section briefs the fundamental principles of greener chemistry. It further emphasizes in-depth studies of synthesis of the different lanthanide-based complexes and their applications in different dimensions. It includes Green Synthesis of (a) lanthanide-doped nanophosphors, (b) rare-earth zirconates, (c) metal oxide nanoparticles, (d) rare-earth ions-doped nanocrystals-based photoluminescent materials, (e) self-assembled nanospherical dysprosium MOFs, and (f) nucleotide-based lanthanide coordination polymers. The last section dedicatedly reports the scope for the future perspective and recommendation in the novel area of research.

Keywords: green synthesis, rare earths, nanomaterials, Photoluminescent, MOFs

1. Introduction

The growing awareness for a thriving environment and the integrity of the industrial applications have paved the way for the emergence of green chemistry as a distinct subject. Years ago, the industrial sector started on a significant scale. Ciamician, a highly foresighted researcher, recognized that it is now essential to produce chemicals remarkably similar to natural ones [1, 2]. Green synthesis has become increasingly important in modern chemistry. Green or sustainable chemistry's primary goal is to make valuable molecules and provide resources to humans while allowing no environmental damage. Green chemistry contributes to the emergence of novel techniques to prevent ecological degradation by minimizing the number of toxic pollutants and their health effects by commencing with harmless or healthier substances than those now in use. This discipline is presented through a period of rapid advancement and growth. The progression of "green chemistry" has arisen as a surprising move for resolving this challenge and ensuring the environment's security [3]. Green chemistry includes a diverse catalyst, biocatalysis, solvent-free production, and the utilization of microwave and ultrasound, among other things.

Green chemistry has been characterized in a variety of ways by various institutions. According to IUPAC, green chemistry is described as "the discovery, configuration, & application of chemical products and minimizes possible use and generation of toxic chemicals." The Environmental Protection Agency (EPA) defines "chemistry for source reduction" as green chemistry. But, The Organisation for Economic Co-operation and Development (OECD) explained green chemistry as "sustainable chemistry." These concepts of green chemistry appear to be unique in terms of how they are represented, but they all give rise to the same significance and primary objective. Green chemistry's strategies are developing, producing, and using effective, reliable, and eco-friendly chemical products and processes to manage pollution.

Further, the lanthanides belong to the f-block in the modern periodic table, with atomic numbers falling between 57 and 71 (La-Lu). The outermost electron or valence shell electron of lanthanides goes in a 4f subshell. They possess identical chemical and physical properties. The silvery-white soft metals experience various characteristics. With the increase in the atomic numbers, the hardness of the metals also increases due to the rise in their melting and boiling points. They are also paramagnetic. Lanthanides have many scientific and industrial applications. On heating with halogen, lanthanides exhibit fluorescence under UV lights. Its exposure to air gives rise to tarnishing because of oxidation, which is usually very quick in moist air. Lanthanides could burn quickly in the atmosphere and are also highly reactive. They typically react very steadily with O_2 and H_2O but quickly at high temperatures. Other similar properties of lanthanides include their characteristics as potent reducing agents, making ionic compounds rapidly dissolved in acids. Lanthanides and the different ligands are particularly favorable chemical compounds used in developing functional molecular substances. Lanthanides are an attractive essential matter for synthesizing several divalent or trivalent derivatives of lanthanides with organic compounds. Trivalent lanthanide cations act as "hard" acid, in which the bonding of lanthanides with another element is examined predominantly electrostatic and ionic. Therefore, their affinities toward "hard" ligands with oxygen donor atoms will be decisive. As a suitable fluorescent material, lanthanide-doped nanophosphors gained much interest for biological applications due to their inherent lower cytotoxicity.

The present review highlights the various green route of synthesis of lanthanide-based compounds synthesis and their discrete applications in different dimensions. The initial section gives a brief explanation of the fundamental greener synthesis of organic compounds provided by Paul Anastas & John Warner. It further emphasizes in-depth studies of other lanthanide-based complexes obtained by the green method of their synthesis. It includes green synthesis of a) lanthanide-doped nanophosphors, b) rare-earth zirconates, c) metal oxide nanoparticles, d) rare-earth ions-doped nanocrystals-based photoluminescent materials, e) self-assembled nanospherical dysprosium Metal organic frameworks (MOF), and f) nucleotide-based lanthanide coordination polymers. The last section dedicatedly reports the scope for the future perspective and recommendation in the novel area of research.

2. Basic principles of greener chemistry

In 1998, Paul Anastas and John Warner put forward the basic principles of green chemistry for the first time [4]. As reported, green chemistry mainly comprises 12 principles, which are widely accepted for explaining the most outstanding ever-emerging image of green chemistry [5, 6].

i. **Prevention:** It iss a saying, prevention is better than cure. Hence, preventing waste is better than managing or cleaning the waste after being made.

ii. **Atom economy:** The artificial techniques should be delineated to increase the whole material utilized in the procedure into the final product.

iii. **Nontoxic synthesis of chemicals:** The process used for the synthesis should be planned to generate and use substances that show less or no toxic effects to the environment and humankind.

iv. **Design safe substances:** The chemical substances must be prepared to affect their required functions while decreasing toxicity.

v. **Safe auxiliaries and solvents:** The usage of auxiliary substances like separating agents, solvents, etc., have to be proclaimed inessential wherever manageable and less toxic when it is used.

vi. **Designing for energy efficacy:** The demand for energy for synthetic procedures has acknowledged its economic and environmental effects and has to be reduced. The manufactured techniques must be accompanied at medium pressure and temperature if possible.

vii. **Utilization of renewable sources:** Using feedstock and raw materials that are recycled or renewable instead of depleting is a step toward a greener synthesis route. The renewable feedstock is derived from agro-based products, whereas the depleting feedstock is emanated by the fossil fuels like coal, natural gas, and petroleum.

viii. **Reduction of derivatives:** The unessential formation of by-products, i.e. the usage of blocking groups, protection/deprotection, modification of chemical/physical methods for the time being, etc., should be reduced or avoided. These steps require additional reagents and will sow the seeds of waste.

ix. **Minimize waste by avoiding stoichiometric reagents and using catalysts:** Using catalytic reactions should minimize waste production. Catalysts are more advantageous than stoichiometric reagents, used excessively and employed only once. The catalysts are employed in the least quantity and several times carry out a single reaction.

x. **Designing for degradation:** The chemical substances should be prepared to disintegrate the innocuous degradation of products conclusively after their use, and it may not present further in the surroundings.

xi. **Appropriate evaluation for preventing pollution:** The analytical methodologies should be further developed to control, monitor, or evaluate appropriately before generating toxic or dangerous substances.

xii. **Inherent use of safe chemicals for preventing accidents:** The substances mainly used in a chemical procedure should be chosen in such a manner that reduces the potential for accidents (by chemicals), along with the fires, explosions, and releases.

3. Synthesis of various lanthanide-based complexes with their applications in different dimensions

3.1 Green synthesis of lanthanide-doped nanophosphors

Nanocrystal-based semiconductors are seemed to be a useful luminescent material for bio-fluorescence labeling and bio-imaging because of their quantum-sized effects where the absorbance starts, and the emissions shift toward the high energy with the decrease in size. These compounds are prepared under complex, expensive, and dangerous experimental conditions. Recently, studies indicated that they could affect the penetrability of the cell membrane and engender unwanted risky interactions with the human biological system [7]. As a result, the long-term uses of these compounds in the bio-related field have been limited. The higher increase in the recognition of the certainly unfavorable effects of nanomaterials on the well-being of humankind has been directed toward the lanthanide (ln)-doped nanophosphors of lower toxicity in the field of life science. The recent studies introduced a concept of an advanced perspective based on the microplasma approach for the fabrication of crystalline Ln-doped nanophosphors of advanced quality [8]. This approach is perhaps developed for the greener synthesis of different lanthanide-doped/co-doped nanophosphors because it possesses higher elasticity. In 2019, Lin [9] investigated the model of Eu^{3+}-doped yttria for the synthesis of Y_2O_3:Eu^{3+} nanophosphors of different sizes and discrete Eu^{3+} concentrations of doping. The lanthanide (Ln = Tm, Dy, Tb, Eu)-doped nanophosphors are synthesized by the simple microplasma-assisted process. It is demonstrated that the ultra-high pure crystals of Y_2O_3:Eu^{3+} nanophosphors are prepared by aq. sol. of $Ln(NO_3)_3{\cdot}6H_2O$ (Ln = Eu and Y) at meager plasma power consumption (near about 3–5.5 W), except for any involvement of dangerous chemicals. In addition to this, the trivalent Europium ions proved to be efficient, and they are also doped within the yttria matrix homogeneously. The green route's nanophosphors obtained by the electrolytic solution of $Ln(NO_3)_3{\cdot}6H_2O$ shows the regulated photoluminescence properties.

The advanced microplasma-based method reduces the disturbing obstacles in the present situation of techniques for synthesizing lanthanide-doped nanophosphors. In contrast with the addition of surfactants, stabilizers, or solvents to control the in-homogeneous and vigorous hydrolysis reaction induced by the super-saturated alkali precipitants, this technique generally utilizes H_2O as a "soft" source of "OH" for maintaining the milder conditions of hydrolysis [8]. Hence, it is concluded that heating can increase luminescence performance efficiency by regulating the proceeding conditions. Therefore, this method prevents the complex from the time-taking procedures of post-separation/purification and facilitates the complete synthesis system.

3.2 Green synthesis of rare-earth Zirconates

Rare-earth zirconates ($Re_2Zr_2O_7$) have been extensively engaged in different areas of research like disposal of nuclear waste, catalyst, diesel engines, and thermal-barrier-coating compounds [10–14]. As yet, a different synthesis method appeared for the manufacturing of $Re_2Zr_2O_7$. In 2012, Saradhi [15] synthesized the nanocrystalline $Eu_2Zr_2O_7$ using dil. NH_4OH as precipitant by co-precipitation method. In 2017, Zhang and co-workers used glycine as a propellant and ignited to prepare rare-earth zirconates Ni/Ln2Zr2O7 (Ln = Y, Sm, Pr, and La), used as catalysts [16]. Saitzek and co-workers [17] manufactured thin layer of lanthanum zirconate ($Ln_2Zr_2O_7$) using a sol–gel technique. The other methods of synthesis of the zirconates of lanthanides are comprised of evaporation-induced self-assembly process, thermal impedance process, co-ions complexation, cathode plasma electrolysis, and solid-state reaction [18–20]. At present, the grain size, shape, and the rate of purity have been proved to be the influential factors for the specification of its discrete features and effectivity of the nano-compounds [21–23]. Thus, a substantial number of researchers have concentrated on the composition of the nano-compounds with controlling and modifying the required factors. In this study, Zinatloo-Ajabshir [24] synthesized $Ln_2Zr_2O_7$ (where, Ln = Pr, Nd)-based ceramic nanostructures by tea leaf extract (green) as new and effective material by green and straightforward path. Remarkably, the effectiveness of $Ln_2Zr_2O_7$ (Ln = Pr, Nd) was evaluated in the propane-SCR-NOx method. The study introduced the nanostructured $Ln_2Zr_2O_7$ (Ln = Pr, Nd) as a potent novel variety of catalysts that are preferably efficient for the C_3H_8-SCR reaction of nitrogen oxides. The catalytic effectiveness of the produced specimen should be considered concerning its capacity of adsorption, specific surface area, crystallinity, and particle size [25]. The catalytic effectiveness of $Pr_2Zr_2O_7$ for NOx reduction is preferably greater than $Nd_2Zr_2O_7$ to such an extent that the transformation of nitrogen oxide to nitrogen for $Pr_2Zr_2O_7$ is approximately 67% and for $Nd_2Zr_2O_7$ is approximately 56%. The accumulation of carbon monooxide in the exit valve that is generally an unwanted by-product in the operation of SCR-nitrogen oxides-C_3H_8 for $Pr_2Zr_2O_7$ is less than $Nd_2Zr_2O_7$. Therefore, it can be said that $Pr_2Zr_2O_7$ exhibits preferably greater effectiveness for transformation of NOx to N_2.

3.3 Greener synthesis of metal oxide nanoparticles

MO-based lanthanides nanoparticles have recently achieved more attention because of their particular biological application and unique properties in various fields. At present, metal oxide synthesis demands a time-efficient, ecologically sound, and cost-effective process. The biological synthesis is a preferable alternative for the traditional chemical process of MO nanoparticles [26]. Biological synthesis is an environmentally sound green route using yeast, algae, fungi, and bacteria to synthesize MO nanoparticles. The biological manufacturing route brings forth a reliable, affordable, and less toxic method for synthesizing MO nanoparticles with differing composition, shape, size, and physico-chemical effects [27]. Tin oxide is a prominent semiconductor used biomedically among the different metal oxides. SnO_2 has exceptional catalytic properties, a higher charge transfer rate, incredible binding energy, chemical and higher thermal stability, and excellent optical and electrochemical properties. In particular, metal/non-metal-doped SnO_2 played a crucial in biomedical applications. S.A. Khan and co-workers suggested that cobalt-doped tin oxide exhibits more significant antimicrobic activity than un-doped SnO_2 [28].

Likewise, K. K. Nair described silver-doped tin oxide as exhibiting more significant antibiotic activity than pure SnO_2. Recently, biosynthesis is using jujube fruits [29], Ficus Carica [30], and Plectranthusamboinicus [31] for the preparation of SnO_2. Solanum nigrum, known as Makoi or Black Nightshade (h) Kakamachi, is an everlasting shrub mainly appearing inside a timbered or forested region. In medicinal chemistry, Makoi/Nightshade has anti-hyperlipidemic, antioxidant, antiherpetic, and anti-inflammatory properties. This herbaceous plant substitutes anodyne, diaphoretics, diuretics, and expectorants. Here, the present study explains the preparation of Y_2O_3, SnO_2, Sn-doped Y_2O_3, and Y-doped SnO_2 and investigates the purpose of doping to increase antibacterial or antimicrobic motility of Y_2O_3 and SnO_2. The outcomes illustrated that the extracted material drawn out from plants and doping is necessary for synthesizing metal nanoparticles and explaining their biomedical implementations (**Table 1**) [32].

3.4 Greener synthesis of rare-earth ions-doped nanocrystals-based Photoluminescent substances

Photoluminescent substances have been practised globally in the security and anti-counterfeiting field because of their eccentric easy identification, excellent reliability, and difficulty with duplication [33, 34]. This paper introduces a simplistic, greener, and cost-efficient method to manufacture biomass composites that comprise cellulose fibers and Ln-doped nanocrystals for anti-counterfeiting applications [35–37]. The photoluminescent substances were synthesized using in situ Chemical Vapour Deposition (CVD) approach of trivalent lanthanides on the bleach hard-wood semi-fluid cellulose fiber (bhpFibers) plane by adding PVP (poly-vinyl-pyrrolidone) to couple the formation of the bhpFibers-PVP@LaF_3:Eu^{3+} composite. In 2018, Qing Wang [38] synthesized the bhpFibers-PVP@LaF_3:Eu^{3+} composite that possesses effective fluorescence. Its luminescence intensity could be easily managed by changing the incorporation of the moral quantity of trivalent lanthanum and europium ions in a solvent.

Moreover, the complex is utilized as a block to form photoluminescence paper through a vacuum filtration method. This synthesized paper shows well writable and printable properties, high flexibility, and good luminescence. Furthermore, the complete process is modest and free from toxic reagents. The bhpFibers-PVP@LaF_3:Eu^{3+} composite exhibited balanced and robust luminescent property for 10–15 days and pH between 2 and 12. Different Ln-doped/fibers nanocrystal composites and various other trivalent lanthanides can be prepared based on this technique. Hence, this simple and greener approach for manufacturing the photoluminescence-active cellulose fibers retains excellent and promising applications in anti-counterfeiting material on a large scale. The detailed schematic representation is explained in **Figure 1**.

3.5 Greener synthesis of self-assembled Nanospherical dysprosium metal: organic frameworks

Metal–organic frameworks are an advanced category of permeable translucent substances prepared from the multi-dentate organic ligands and metal clusters [39]. An increase in the study of MOFs witnessed in recent times because of its possible applications in magnetism [40], nonlinear optics [41], heterogeneous catalysis [42, 43], gas storage [44], and separation [45], etc. Based on MOF, the sensing materials are predicted to be more favorable than the existing sensor probes because of their exposed active sites, tunable framework compositions, high surface areas, and ease of

S. No.	Compound	Precuror (solanum nigrum plant)	Parameters	Result
1.	Yttrium oxide nanoparticles	Yttrium nitrate	60°C for 60 mins; heat at 350–400°C	The solution changes from green to creamy white.
2.	Stannous oxide	$SnCl_2$	60 C for 60 mins; heat at 350–400°C	The solution changes from dark green to black.
3.	Stannous-doped yttrium oxide nanoparticles	0.1 mM yttrium nitrate (80%), 0.1 mM $SnCl_2$ (20%)	60 C for 60 mins; heat at 350–400°C	The solution changes from dark green to creamy white.
4.	Yttrium-doped stannous oxide nanoparticles	0.1 mM $SnCl_2$ (80%), 0.1 mM yttrium nitrate solution (20%).	60 C for 60 mins; heat at 350–400°C	The solution changes from dark green to brown.

Table 1.
Synthesis of ln-based nanoparticles by ethanolic extraction of solanum nigrum plant.

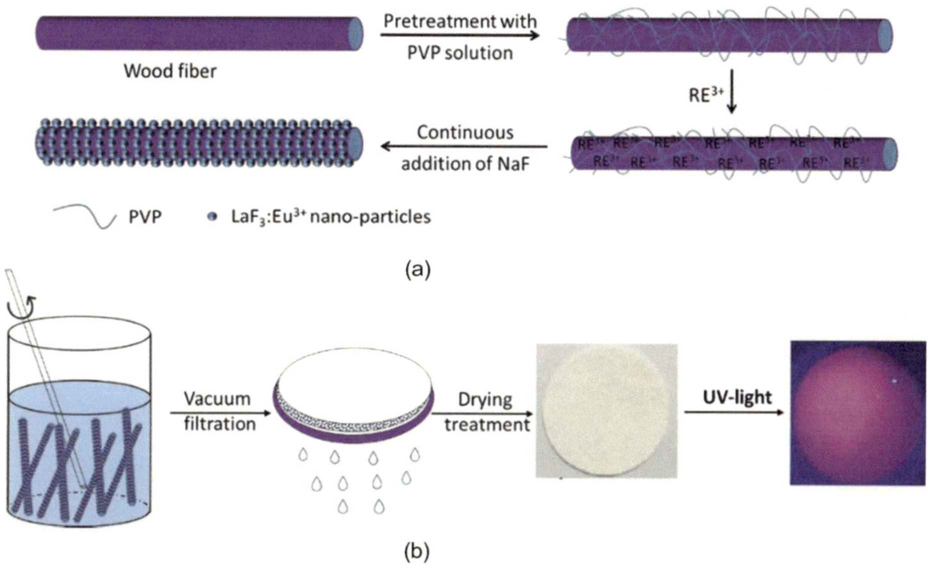

Figure 1.
Diagrammatic illustration of the fabrication of bhpFibers-PVP@LaF$_3$:Eu^{3+} [38].

synthesis [46]. Implementation of MOFs in real-life applications is a significant cause in the present-day research. The main focus was to enhance features like selectivity, sensitivity, latent period, and durability of prepared MOFs to synthesize effective sensory platforms for authentic characteristics. In 2019, Mukherjee and co-workers [47] prepared two new trivalent dysprosium MOFs assimilating πelectron-donating azides performance in ligand vertebrae using the solvothermal method. An ecological and green method has been embraced to scale the prepared substance to nanotechnology. Formerly undetermined Dy (III) carrying nanospherical MOFs thus achieved are used to sense several $-NO_2$-based incendiary devices in the solvent media through fluorescent quenching. On illustrating the potential appropriateness of nanoscale

MOFs for recognizing distinct $-NO_2$-based incendiary devices, the current study also explained the excellent utility of annihilation constants known till date in the permeable substance realm. For attaining the high-performance sensory inquests for reliable and ecological applicative value, ad-libbing the concept of sensors into the vapor state seemed to show excellent outcomes. Hence, researchers considered that the outcomes conferred here obviously expedite a novel approach for the development of green organic–inorganic hybrid nanocomposites to design the new-generation sensory probes. This study illustrates the solvothermal-based preparation of two novel Ln-MOFs, named as $[\{Dy_4(5N_3\text{-}IPA)_6(DMF)_3(H_2O)_4\}(DMF)(H_2O)_2]_\infty$ and $[\{Dy(2N_3\text{-}TPA)_2(H_2O)(CH_3OH)\}]_\infty$ emanated by two $RCOO^-$ ligand named as $2N_3$-TPA (2-azidoterephthalate) and $5N_3$-IPA (5-azidoterephthalate) correspondingly, through confined azide components that are directed toward the pore surfaces having two-dimensional sheet-like arrangement with helical metal carboxylates chain. The two synthesized complexes show good aqueous phase and thermal stabilities. A convenient and straightforward method has been embraced to minimize the synthesized compounds' particle size for the preparation of Nanoscale metal organic frameworks (NMOF). These nanoscale metal–organic frameworks are strongly utilized for fluorescence analysis for the identification of several nitro-analytes like picric acid, nitro-toluene, 4-nitrobenzoic acid, 2, 6-dinitrotoluene, and nitrobenzene, etc. Furthermore, selectivity toward picric acid sensing with another elementally identical nitro-analytes and better recyclability of the operating sensing manifesto revealed excellent potential for these substances considering its authentic applications.

The crystal structure and coordination geometry of both compounds are explained in **Figures 2** and **3**.

3.6 Green synthesis of nucleotide-based inner transition coordination polymers

White-light-emitting substances gained immense recognition due to their applications in various practicable areas, like electrochemical cells, light-emitting diodes,

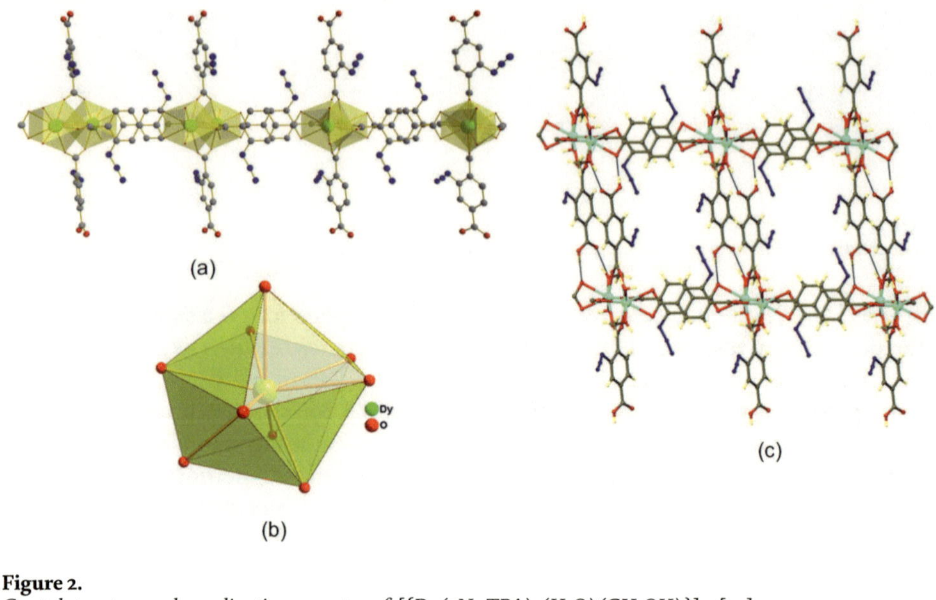

Figure 2.
Crystal structure and coordination geometry of $[\{Dy(2N_3\text{-}TPA)_2(H_2O)(CH_3OH)\}]_\infty$ [47].

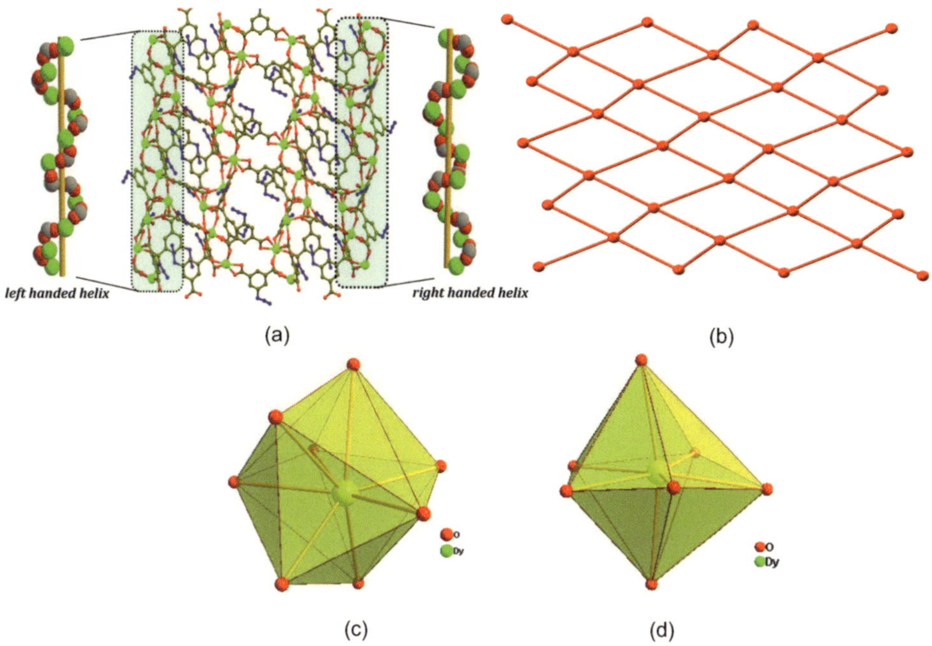

Figure 3.
Crystal structure and coordination geometry of $[\{Dy_4(5N_3\text{-}IPA)_6(DMF)_3(H_2O)_4\}(DMF)(H_2O)_2]_\infty$ *[47].*

and several different varieties of light-emitting devices [48–51]. The advanced white-light-emitting substances noted till today are predominantly concerted on metal–organic hybrid components [52–54], inorganic nanocrystals [55], quantum dots [56], and small-organic molecules [57]. In general, these white-light-emitting materials are achieved by the following three approaches: (1) one-constituent emitting in the entire Vis-region from 400 to 700 nm, (2) two-chromophores carrying constituent emitting orange, yellow or blue, and (3) three-chromophores carrying constituent emitting primary colors [58]. Comparing these three strategies, the outcomes described that the one-constituent white-light-emitting phosphors possess great benefits: lower production cost, higher luminescent efficiency, and higher color displaying index. At the same, the accretion of the one-component substances is efficient of white light fabrication is still difficult. Ln-based phosphors are seemed to be the realm of two-chromophores-assisted white-light-emitting substances during the past decades because of the supremacy of trivalent Ln ions, together with the low toxicity, high photochemical stability, long fluorescence lifetime, sharp emission, and significant Stokes shift.

Presently, the white-light-emitting diodes mainly depend on blue light InGaN chip, and yellow light from Ln-based inorganic materials of $Y_3Al_5O_{12}:Ce^{3+}$ (YAG:Ce) is accessible at a large scale [59]. Though, few defects are persisted because of the absence of red-light-emitting constituent, regulating its multifaceted applications. Currently, metal–organic coordination polymers of lanthanide fascinated much interest for white light administration. It is observed that LMOCP (Ln (III) metal–organic coordination polymers) are highly adaptable with the organic matrices for organic LED characteristics than nitride powders and inorganic metal oxides as organic–inorganic hybrid material [60]. Compared with the crystal MOFs, amorphous LMOCP exhibits wide composition diversity, mild reaction conditions, and high

S.No.	Name of Compounds	Precursors	Method	Applications	References
1	$Y_2O_3:Eu^{3+}$ nanophosphors	aq. sol. of $Ln(NO_3)_3{\cdot}6H_2O$ (Ln = Eu and Y)	Microplasma-assisted approach	Bio fluorescence labeling and bio-imaging	[9]
2	$Ln_2Zr_2O_7$ (Ln = Pr, Nd)	$Pr(NO_3)_3.6H_2O$/ $Nd(NO_3)_3.6H_2O$/ zirconyl nitrate, tea extract	Propane-SCR-NO_x method	Catalyst, thermal-barrier-coating compounds	[24]
3	Stannous-doped yttrium oxide nanoparticles	Solanum nigrum plant extract, 0.1 mM Yttrium nitrate, 0.1 mM stannous chloride solution	Biosynthesis	Antimicrobial, antibacterial	[32]
4	bhpFibers-PVP@LaF_3: Eu^{3+}	Hardwood-bleached pulp, polyvinylpyrrolidone, lanthanum chloride hexahydrate, europium chloride hexahydrate, sodium fluoride	In situ CVD approach	Photoluminescence active, anti-counterfeiting	[38]
5	$[\{Dy(2N_3\text{-}TPA)_2(H_2O)(CH_3OH)\}]_\infty$	5-azidoisophthalic acid, 2-aminoterephthalic acid,	Solvothermal method	Fluorescence-based detection of nitro-aromatic explosives	[47]
6	AMP/Ln-CIP	$TbCl_3{\cdot}6H_2O$, $EuCl_3{\cdot}6H_2O$, $GdCl_3{\cdot}6H_2O$, adenosine-5-monophosphate disodium salt (AMP, 98%), ciprofloxacin and tyrosine.	Green synthesis	White-light-emitting material	[62]

Table 2.
Various lanthanide complexes synthesized by the green method.

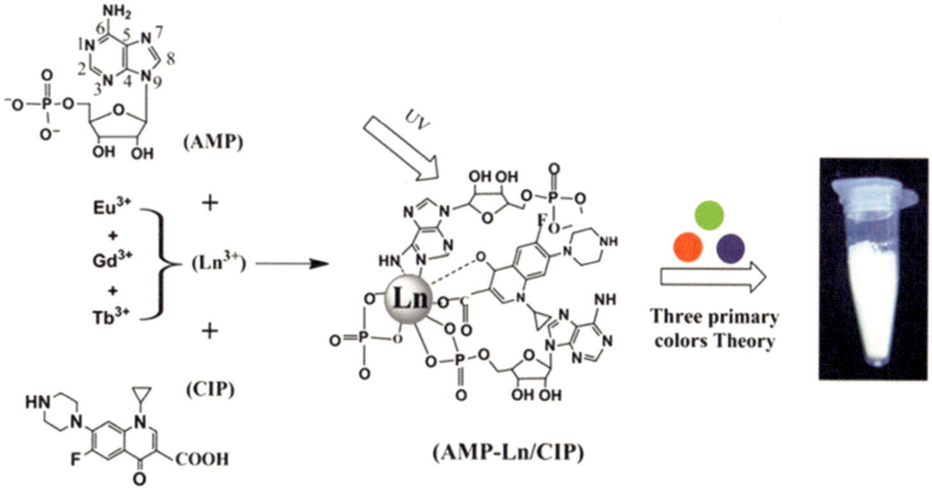

Figure 4.
Diagrammatic representation of the method of synthesis for white-light-emissive AMP/ln-CIP [62].

tailorable properties [61]. Zhang and co-workers synthesized an amorphous Ln (III) metal–organic coordination polymers of adenosine monophosphate (AMP)/Ln-CIP by green route with tunable white light emission by an organic ligand, i.e. adenosine monophosphate (AMP). In the strong structure of AMP/Ln-CIP, Ln = Gd, Eu, and Tb, CIP can coordinate with the trivalent Ln ions with oxygen atoms of the -CO and -COO group for the fabrication of the trivalent Ln complexes. It also fascinates energy in the ultraviolet range for the sensitization of the red emission of trivalent europium ions and the green emission of trivalent terbium ions. On the basis of the three primary colors theory, through a greener approach of preparation, Zhang strongly fabricated a white-light-emissive nanophosphor compound of AMP/$Tb_{0.1}Eu_{0.9}Gd_{99.0}$-CIP. On comparing with the different routes of synthesis of the white-light-emitting constituents, this approach is easy, effortless, and environmentally sound. Hence, it is concluded that the AMP/$Tb_{0.1}Eu_{0.9}Gd_{99.0}$-CIP exhibits several benefits like high quantum yields, long fluorescence lifetime, and high thermostability. Moreover, the homogeneous coordination properties of Tb^{3+}, Eu^{3+}, and Gd^{3+} permit the in situ doping of these metal ions simultaneously into a parent Ln (III) metal–organic coordination polymers concurrently. Hence, it is believed that this synthetic method can be elongated to develop other Ln-based white-light-emitting materials [62]. The diagrammatic illustration of the method of synthesis for white-light-emissive AMP/Ln-CIP is shown in **Figure 4**, **Table 2**.

4. Conclusion and future perspectives

Green or sustainable chemistry's primary goal is to make valuable molecules and provide resources to humans while allowing no environmental damage. Green chemistry contributes to the emergence of novel techniques to prevent ecological degradation by minimizing the number of toxic pollutants and their health effects by commencing with harmless or healthier substances than those now in use. In this chapter, the greener route for the synthesis of Ln-doped nanophosphors, rare-earth zirconates, nucleotide-based lanthanide coordination polymers, self-assembled nano-spherical dysprosium MOFs, rare-earth ions-doped nanocrystals-based photoluminescent materials, and metal oxide nanoparticles is discussed in detail with their applications in different dimensions. The chapter highlights the role of rare-earth elements in the green synthesis of organic compounds. Lanthanides mainly coordinate with other ligands to give an organic compound by environmentally sound route with less toxic effects and no use of dangerous chemical reagents. These methods of synthesizing organic compounds are necessary for the present scenario because they reduce hazardous damages and decrease pollution levels. The products prepared by the green route are also highly adaptable with the organic matrices.

Acknowledgements

Director CSIR-AMPRI Bhopal is also acknowledged for providing necessary institutional facilities and encouragement.

Conflicts of interest

The authors declare no conflict of interest related to this research work.

Author details

Kamna Chaturvedi[1], Deeksha Malvi[2], Manish Dhangar[2], Harsh Bajpai[2], Ranjan K. Mohapatra[3], Avanish Kumar Srivastava[1,2] and Sarika Verma[1,2]*

1 Academy of Council Scientific and Industrial Research, Advanced Materials and Processes Research Institute (AMPRI), Bhopal, MP, India

2 Council of Scientific and Industrial Research, Advanced Materials and Processes Research Institute, Bhopal, MP, India

3 Department of Chemistry, Government College of Engineering, Keonjhar, Odisha, India

*Address all correspondence to: drsarikaverma20@gmail.com; sarika.verma@ampri.res.in

References

[1] Ciamician G. Actions chimiques de la lumière. Bulletin de la Société Chimique de France. 1908;**4**:1-3

[2] Ciamician G. The photochemistry of the future. Science. 1912;**36**:385-394

[3] Albini A, Fagnoni M. 1908: Giacomo Ciamician and the concept of green chemistry. ChemSusChem. 2008;**1**:63-66

[4] Anastas PT, Warner JC. Principles of green chemistry. Green Chemistry: Theory and Practice. 1998;**29**

[5] Anastas PT, Zimmerman JB. Peer reviewed: Design through the 12 principles of green engineering. Environmental Science & Technology. ACS publications; 2003:94A-101A

[6] Anastas PT, Beach ES. Green chemistry: The emergence of a transformative framework. Green Chemistry Letters and Reviews. 2007;**1**:9-24

[7] Mao L, Chen J, Zhang X, Kwak M, Wu Y, Fan R, et al. A promising biodegradable magnesium alloy suitable for clinical vascular stent application. Scientific Reports. 2017;**7**:1-12

[8] Lin L, Starostin SA, Li S, Khan SA, Hessel V. Synthesis of yttrium oxide nanoparticles via a facile microplasma-assisted process. Chemical Engineering Science. 2018;**178**:157-166

[9] Lin L, Starostin SA, Ma X, Li S, Khan SA, Hessel V. Facile synthesis of lanthanide doped yttria nanophosphors by a simple microplasma-assisted process. Reaction Chemistry & Engineering. 2019;**4**:891-898

[10] Du W, Zhao G, Chang H, Shi F, Zhu Z, Qian X. Photocatalytic studies of Ho–Zr–O nano-composite with controllable composition and defects. Materials Characterization. 2013;**83**:178-186

[11] Zinatloo-Ajabshir S, Salavati-Niasari M, Zinatloo-Ajabshir Z. Facile size-controlled preparation of highly photocatalytically active praseodymium zirconate nanostructures for degradation and removal of organic pollutants. Separation and Purification Technology. 2017;**177**:110-120

[12] Naga SM, Awaad M, El-Maghraby HF, Hassan AM, Elhoriny M, Killinger A, et al. Effect of La2Zr2O7 coat on the hot corrosion of multi-layer thermal barrier coatings. Materials and Design. 2016;**102**:1-7

[13] Zinatloo-Ajabshir S, Zinatloo-Ajabshir Z, Salavati-Niasari M, Bagheri S, Abd Hamid SB. Facile preparation of Nd2Zr2O7–ZrO2 nanocomposites as an effective photocatalyst via a new route. Journal of Energy Chemistry. 2017;**26**:315-323

[14] Ren X, Pan W. Mechanical properties of high-temperature-degraded yttria-stabilized zirconia. Acta Materialia. 2014;**69**:397-406

[15] Saradhi MP, Ushakov SV, Navrotsky A. Fluorite-pyrochlore transformation in Eu 2 Zr 2 O 7—Direct calorimetric measurement of phase transition, formation and surface enthalpies. RSC Advances. 2012;**2**:3328-3334

[16] Zhang X, Fang X, Feng X, Li X, Liu W, Xu X, et al. Ni/ln 2 Zr 2 O 7 (ln= La, Pr, Sm and Y) catalysts for

methane steam reforming: The effects of a site replacement. Catalysis Science & Technology. 2017;7:2729-2743

[17] Saitzek S, Shao Z, Bayart A, Ferri A, Huvé M, Roussel P, et al. Ferroelectricity in La 2 Zr 2 O 7 thin films with a frustrated pyrochlore-type structure. Journal of Materials Chemistry C. 2014;**2**:4037-4043

[18] Zinatloo-Ajabshir S, Salavati-Niasari M. Photo-catalytic degradation of erythrosine and eriochrome black T dyes using Nd2Zr2O7 nanostructures prepared by a modified Pechini approach. Separation and Purification Technology. 2017;**179**:77-85

[19] Xu C, Jin H, Zhang Q, Huang C, Zou D, He F, et al. A novel Co-ions complexation method to synthesize pyrochlore La2Zr2O7. Journal of the European Ceramic Society. 2017;**37**:2871-2876

[20] Zinatloo-Ajabshir S, Salavati-Niasari M. Facile synthesis of nanocrystalline neodymium zirconate for highly efficient photodegradation of organic dyes. Journal of Molecular Liquids. 2017;**243**:219-226

[21] Razi F, Zinatloo-Ajabshir S, Salavati-Niasari M. Preparation and characterization of HgI2 nanostructures via a new facile route. Materials Letters. 2017;**193**:9-12

[22] Zinatloo-Ajabshir S, Mortazavi-Derazkola S, Salavati-Niasari M. Sonochemical synthesis, characterization and photodegradation of organic pollutant over Nd2O3 nanostructures prepared via a new simple route. Separation and Purification Technology. 2017;**178**:138-146

[23] Zinatloo-Ajabshir S, Salehi Z, Salavati-Niasari M. Synthesis of dysprosium cerate nanostructures using Phoenix dactylifera extract as novel green fuel and investigation of their electrochemical hydrogen storage and Coulombic efficiency. Journal of Cleaner Production. 2019;**215**:480-487

[24] Zinatloo-Ajabshir S, Ghasemian N, Salavati-Niasari M. Green synthesis of Ln2Zr2O7 (ln= Nd, Pr) ceramic nanostructures using extract of green tea via a facile route and their efficient application on propane-selective catalytic reduction of NOx process. Ceramics International. 2020;**46**:66-73

[25] Zinatloo-Ajabshir S, Morassaei MS, Salavati-Niasari M. Facile synthesis of Nd2Sn2O7-SnO2 nanostructures by novel and environment-friendly approach for the photodegradation and removal of organic pollutants in water. Journal of Environmental Management. 2019;**233**:107-119

[26] Hutchison JE. Greener nanoscience: A proactive approach to advancing applications and reducing implications of nanotechnology. ACS Nano. 2008;**2**:395-402. DOI: 10.1021/nn800131j

[27] Maity S, Roy S, Kumar A. Silver nanoparticles to enhance the efficiency of silicon solar cells. International Journal of Engineering Science. 2013;**2**:101-104

[28] Khan SA, Kanwal S, Rizwan K, Shahid S. Enhanced antimicrobial, antioxidant, in vivo antitumor and in vitro anticancer effects against breast cancer cell line by green synthesized un-doped SnO2 and Co-doped SnO2 nanoparticles from Clerodendrum inerme. Microbial Pathogenesis. 2018;**125**:366-384

[29] Honarmand M, Golmohammadi M, Naeimi A. Biosynthesis of tin oxide (SnO2) nanoparticles using jujube fruit for photocatalytic degradation of organic

dyes. Advanced Powder Technology. 2019;**30**:1551-1557

[30] Hu J, Science JA. Biosynthesis of SnO_2 nanoparticles by fig (Ficus Carica) leaf extract for electrochemically determining Hg(II) in water samples. International Journal of Electrochemical Science. 2015;**10**:10668-10676

[31] Fu L, Zheng Y, Ren Q, Wang A, Deng B. Green biosynthesis of SnO2 nanoparticles by plectranthus amboinicus leaf extract their photocatalytic activity toward rhodamine B degradation. Journal of Ovonic Research. 2015;**11**:21-26

[32] Raj V, Kamaraj P, Sridharan M, Arockiaselvi J. Green synthesis, characterization of yttrium oxide, stannous oxide, yttrium doped tin oxide and tin doped yttrium oxide nanoparticles and their biological activities. Materials Today: Proceedings. 2019;**36**:920-922. DOI: 10.1016/j.matpr.2020.07.032

[33] Huo J, Dong L, Lü W, Shao B, You H. Novel tunable green-red-emitting oxynitride phosphors co-activated with Ce 3+, Tb 3+, and Eu 3+: Photoluminescence and energy transfer. Physical Chemistry Chemical Physics. 2017;**19**:17314-17323

[34] Li W, Zhang H, Zheng Y, Chen S, Liu Y, Zhuang J, et al. Multifunctional carbon dots for highly luminescent orange-emissive cellulose based composite phosphor construction and plant tissue imaging. Nanoscale. 2017;**9**:12976-12983

[35] Chen H, Yan X, Feng Q, Zhao P, Xu X, Ng DHL, et al. Citric acid/cysteine-modified cellulose-based materials: Green preparation and their applications in anticounterfeiting, chemical sensing, and UV shielding. ACS Sustainable Chemistry & Engineering. 2017;**5**:11387-11394

[36] Wang Q, Cai J, Chen K, Liu X, Zhang L. Construction of fluorescent cellulose biobased plastics and their potential application in anti-counterfeiting banknotes. Macromolecular Materials and Engineering. 2016;**301**:377-382

[37] Zhang Z, Chang H, Xue B, Zhang S, Li X, Wong W-K, et al. Near-infrared and visible dual emissive transparent nanopaper based on Yb (III)–carbon quantum dots grafted oxidized nanofibrillated cellulose for anti-counterfeiting applications. Cellulose. 2018;**25**:377-389

[38] Wang Q, Chen G, Yu Z, Ouyang X, Tian J, Yu M. Photoluminescent composites of lanthanide-based nanocrystal-functionalized cellulose Fibers for Anticounterfeiting applications. ACS Sustainable Chemistry & Engineering. 2018;**6**:13960-13967. DOI: 10.1021/acssuschemeng.8b02307

[39] Perry Iv JJ, Perman JA, Zaworotko MJ. Design and synthesis of metal–organic frameworks using metal–organic polyhedra as supermolecular building blocks. Chemical Society Reviews. 2009;**38**:1400-1417

[40] Luzon J, Sessoli R. Lanthanides in molecular magnetism: So fascinating, so challenging. Dalton Transactions. 2012;**41**:13556-13567

[41] Medishetty R, Zaręba JK, Mayer D, Samoć M, Fischer RA. Nonlinear optical properties, upconversion and lasing in metal–organic frameworks. Chemical Society Reviews. 2017;**46**:4976-5004

[42] Gole B, Sanyal U, Mukherjee PS. A smart approach to achieve an

exceptionally high loading of metal nanoparticles supported by functionalized extended frameworks for efficient catalysis. Chemical Communications. 2015;**51**:4872-4875

[43] Gole B, Sanyal U, Banerjee R, Mukherjee PS. High loading of Pd nanoparticles by interior functionalization of MOFs for heterogeneous catalysis. Inorganic Chemistry. 2016;**55**:2345-2354

[44] Ma S, Zhou H-C. Gas storage in porous metal–organic frameworks for clean energy applications. Chemical Communications. 2010;**46**:44-53

[45] Li J-R, Sculley J, Zhou H-C. Metal–organic frameworks for separations. Chemical Reviews. 2012;**112**:869-932

[46] Wang H, Lustig WP, Li J. Sensing and capture of toxic and hazardous gases and vapors by metal–organic frameworks. Chemical Society Reviews. 2018;**47**:4729-4756

[47] Mukherjee S, Ganguly S, Chakraborty A, Mandal A, Das D. Green synthesis of self assembled Nanospherical dysprosium MOFs: Selective and efficient detection of picric acid in aqueous and gas phase. ACS Sustainable Chemistry & Engineering. 2019;**7**:819-830. DOI: 10.1021/acssuschemeng.8b04429

[48] Cho J, Park JH, Kim JK, Schubert EF. White light-emitting diodes: History, progress, and future. Laser & Photonics Reviews. 2017;**11**:1600147

[49] Lee C-L, Cheng C-Y, Su H-C. Enhancing device efficiencies of solid-state near-infrared light-emitting electrochemical cells by employing a tandem device structure. Organic Electronics. 2014;**15**:711-720

[50] Zhang B, Liu L, Xie Z. Recent advances in solution-processed white organic light-emitting materials and devices. Israel Journal of Chemistry. 2014;**54**:897-917

[51] Tang Q, Liu S, Liu Y, He D, Miao J, Wang X, et al. Color tuning and white light emission via in situ doping of luminescent lanthanide metal–organic frameworks. Inorganic Chemistry. 2014;**53**:289-293

[52] Li Y, Li S, Yan P, Wang X, Yao X, An G, et al. Luminescence-colour-changing sensing of Mn 2+ and Ag+ ions based on a white-light-emitting lanthanide coordination polymer. Chemical Communications. 2017;**53**:5067-5070

[53] Wang X, Yan P, Li Y, An G, Yao X, Li G. Highly efficient white-light emission and UV–visible/NIR luminescence sensing of lanthanide metal–organic frameworks. Crystal Growth & Design. 2017;**17**:2178-2185

[54] Cai H, Xu L-L, Lai H-Y, Liu J-Y, Ng SW, Li D. A highly emissive and stable zinc (ii) metal–organic framework as a host–guest chemopalette for approaching white-light-emission. Chemical Communications. 2017;**53**:7917-7920

[55] Fanizza E, Urso C, Pinto V, Cardone A, Ragni R, Depalo N, et al. Single white light emitting hybrid nanoarchitectures based on functionalized quantum dots. Journal of Materials Chemistry C. 2014;**2**:5286-5291

[56] Li L-W. Review of magnetic properties and magnetocaloric effect in the intermetallic compounds of rare earth with low boiling point metals. Chinese Physics B. 2016;**25**:37502

[57] Mazzeo M, Vitale V, Della Sala F, Anni M, Barbarella G, Favaretto L, et al. Bright white organic light-emitting

devices from a single active molecular material. Advanced Materials. 2005;**17**:34-39

[58] Du P, Wang L, Yu JS. Luminescence properties and energy transfer behavior of single-component NaY (WO4) 2: Tm3+/Dy3+/Eu3+ phosphors for ultraviolet-excited white light-emitting diodes. Journal of Alloys and Compounds. 2016;**673**:426-432

[59] Qiu Z, Shu J, Liu J, Tang D. Dual-channel photoelectrochemical ratiometric aptasensor with up-converting nanocrystals using spatial-resolved technique on homemade 3D printed device. Analytical Chemistry. 2018;**91**:1260-1268

[60] Luo Z, Qi Q, Zhang L, Zeng R, Su L, Tang D. Branched polyethylenimine-modified upconversion nanohybrid-mediated photoelectrochemical immunoassay with synergistic effect of dual-purpose copper ions. Analytical Chemistry. 2019;**91**:4149-4156

[61] Shen H, Liu B, Liu D, Zhu X, Wei X, Yu L, et al. Lanthanide coordination polymer-based biosensor for citrate detection in urine. Analytical Methods. 2019;**11**:1405-1409

[62] Zhang Y, Liu B, Shen Q, Wei X, Zhou Y, Zhang Y, et al. Nucleotide-based green synthesis of lanthanide coordination polymers for tunable white-light emission. Green Processing and Synthesis. 2020;**9**:578-585

Chapter 13

Ascorbic Acid-assisted Green Synthesis of Silver Nanoparticles: pH and Stability Study

Katherine Guzmán, Brajesh Kumar, Marcelo Grijalva, Alexis Debut and Luis Cumbal

Abstract

In this chapter, eco-friendly *in situ* synthesis of silver nanoparticles (AgNPs) using a mixture of ascorbic acid and citric acid is introduced. The synthesis conditions of the AgNPs were optimized by adjusting the pH of the reaction mixture. Different spectroscopic and microscopic techniques have been used to characterize the physico-chemical properties of AgNPs. The synthesis of AgNPs was primarily identified by the appearance of yellow colour and confirmed by showing λ_{max} = 409 nm in UV-visible spectroscopy. All characterization techniques reveal that the generated AgNPs were non-aggregated, quasi-spherical shapes with an average size of 22.4 ± 13.2 nm, and face-centred cubic crystalline structures. Infrared spectroscopy confirms the surface of AgNPs covered with -COOH group and shows peaks at 1733, 1759, 3262 and 3633 cm^{-1}. Moreover, synthesized AgNPs at pH 10 were stable for one month with a slight change in size. A straightforward, facile and environmentally-friendly synthesis of highly stable AgNPs may contribute to future engineering applications.

Keywords: green synthesis, silver nanoparticles, ascorbic acid, TEM, UV-Vis spectroscopy

1. Introduction

Nanotechnology's ability to design and manipulate matter at an atomic scale has been encouraging over the last few decades. Being a source of novel products that need to be beneficial for the environment, the metal nanoparticles (MNPs) displayed an unusual characteristics that make them valuable. Metals like silver include remarkable optical and chemical properties in which the synthesis of silver nanoparticles (AgNPs) highly depends on controlling the size, shape, crystallinity, composition and several features ruled by the unique effects of nanobiotechnology [1, 2]. Silver is the most promising metal due to the bactericidal potential used since ancient times and unique properties such as high thermal stability [3]. AgNPs have diverse applications such as environmental protection, catalysis, food industries, medicine and others [4].

However, hazardous and environmentally unfriendly conditions are not demanded to apply for the synthesis of nanomaterial. The protocols followed to synthesize AgNPs should be feasible, cleaner, long-lasting, low-cost and safer. In other words, the green synthesis of AgNPs does not require any physical, electrical or modification processes [5]. In recent years, researchers have been interested to control and improve engineering procedures of the desired nanoparticle features [6]. Some experts have suggested that the AgNPs synthesis key relates to the selection of right parameters that permit the control over the size and shape outcomes [7].

Numerous scientific articles in the field of AgNPs synthesis have been published, and most of the traditional methods are based on the *bottom-up* category such as chemical reduction, light assisted, sol–gel, electrochemical method, etc [8, 9]. Green synthesis of AgNPs using a variety of reducing–stabilizing agent includes gallic acid-chitosan [10], cellulose [11], gelatin [12], starch [13], ascorbic acid [14], quercetin [15], sodium citrate [16], ascorbic acid-gelatin [17], ascorbic acid-sodium dodecyl sulphate [18], tannic acid-sodium citrate [19], plant extracts [20], etc have been reported.

In this study, the *bottom-up* strategy was used to synthesize AgNPs using ascorbic acid as a reductive agent and citric acid as a potent complex stabilizer, effectively. We also looked at the effect of pH on stable AgNPs colloid formation and the particle size. Furthermore, various spectroscopic and microscopic techniques were used to characterize the physicochemical properties of AgNPs.

2. Experimental

2.1 Chemicals

Silver nitrate ($AgNO_3$, 99%) was obtained from Sigma-Aldrich, Germany. Ascorbic acid ($C_6H_8O_6$, 99%, powder/USP/FCC) and sodium hydroxide (NaOH) were purchased from Fisher Chemical, China. Trisodium citrate ($Na_3C_6H_5O_7$, 99%) was acquired from Loba Chemie, India.

2.2 Preparation of AgNPs

The formation of AgNPs from ascorbic acid and sodium citrate was monitored by a colour change in the reaction médium [21]. The final adapted solution for the preparation of the AgNPs was 20 mL of ascorbic acid (0.6 mM) and 20 mL of sodium citrate (3 mM). Before preparation of AgNPs, the pH of these solutions was adjusted to 9, 10 and 11 by the addition of 0.1 M NaOH in a separate flask. One flask was stored as a control (pH = 6.7). Subsequently, 4 mL of 1 mM of $AgNO_3$ was added to each flask under stirring conditions in the water bath at 60–65°C for two hours. Each reaction mixture containing AgNPs was diluted seven times and analyzed in the UV-visible spectrophotometer.

2.3 Characterization of AgNPs

Formation and stability of AgNPs were determined by a UV-visible spectrophotometer, SPECORD® S600 from Analytik Jena, Germany. The mean particle size and stability were analyzed in a dynamic light scattering (DLS) instrument (HORIBA, LB-550 Japan). The shape, size and selected area electron diffraction (SAED) pattern of AgNPs were captured from transmission electron microscopy, TEM (FEI Tecnai, G2 Spirit Twin, Holland). X-ray diffraction (XRD) analyses were performed with a PANalytical brand θ-2θ configuration (generator-detector) x-ray tube, copper λ = 1.54059 Å and EMPYREAN diffractometer. The Fourier transform infrared spectroscopy (FTIR) measurements were performed on a Spectrum 100 IR spectrometer (Perkin Elmer, USA) using the attenuated total reflectance (ATR) technique to determine the possible involvement of functional groups in the synthesis of nanoparticles.

3. Results and discussion

3.1 Visual and UV-vis spectroscopy studies

The initial synthesis of AgNPs was assessed through visual colour and UV-Vis spectrometric analysis. After the addition of $AgNO_3$ to the alkaline reaction mixture, the colour becomes yellowish within 5 mins, showing the reduction of Ag^+ to Ag^0 and the formation of AgNPs [22]. Detection of UV-visible peak between 390 and 410 nm implies the presence of small and spherically shaped AgNPs [23]. Mie [24] said that only one surface plasmonic resonance (SPR) band produces spherical nanoparticles, while two or more SPR bands cause shape variation. The colour arises in nanoparticles due to SPR, a non-linear optical property related to the excitation of electrons and collective oscillation of dipoles under the influence of electromagnetic field of light [2]. Kumar et al., 2014 observed similar SPR result in UV-vis absorption spectroscopy using *Plukenetia volubilis* leaf extracts [25].

The SPR band can provide useful information about the morphology and particle distribution of the synthesized nanoparticles. In order to understand the effect of pH in the synthesis of AgNPs, we performed the reaction at different pH levels (**Figure 1**). The reaction mixture showed a sharp absorption band at pH 9, 10 and 11, while no band was observed at pH 6.7. The SPR band observed for pH 10 was narrower than others (pH = 9 and 11). It clearly showed a narrow size distribution of AgNPs at pH 10 and was more selective [26]. The cause behind this observation was that at pH>9, free H^+ of ascorbic acid and citric acid are consumed, and the amount of free OH^- increases. In **Figure 2**, the maximum absorption band (λ_{max}) for AgNPs at pH 10 was 409 nm and λ_{max} observed at pH 11 was 404 nm, respectively. Synthesis of AgNPs at pH 10 provides a narrow absorption band and seems to be a uniform size distribution. After 4 weeks, its absorption band increases slightly and shifted towards shorter wavelengths (408 nm and 406 nm), which is a typical characteristic of smaller nanoparticles. The redox potential of ascorbic acid depends on the pH, when the pH increases, the redox potential decreases. Increased difference of redox potentials between Ag^+ and the ascorbic acid/ reducing agent makes the reaction faster, and smaller AgNP nuclei are created [14]. The reduction of Ag^+ ions is difficult due to the complex formation with citrate and results in greater stability through the outer surface coverage [27, 28].

To monitor the stability of the AgNPs prepared at pH 10, the SPR absorption spectra of the AgNPs (stored at 23–25°C) were measured for 7, 14, 21 and 28 days (**Figure 3**). At various incubation periods, a slight increase in the intensity of the SPR peaks was observed. In addition, λ_{max} for the SPR peaks of AgNPs have almost remained constant. It clearly explained that the reaction mixture having a pH at 10 is most appropriate to reduce Ag^+ ions into AgNPs and stable over a month. These

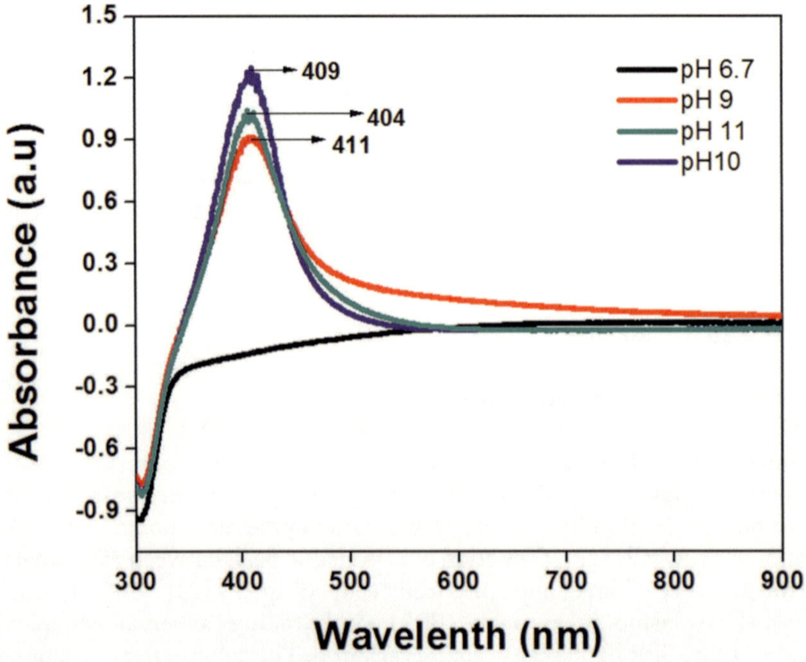

Figure 1.
UV-vis spectrum of AgNPs synthesized at different pH.

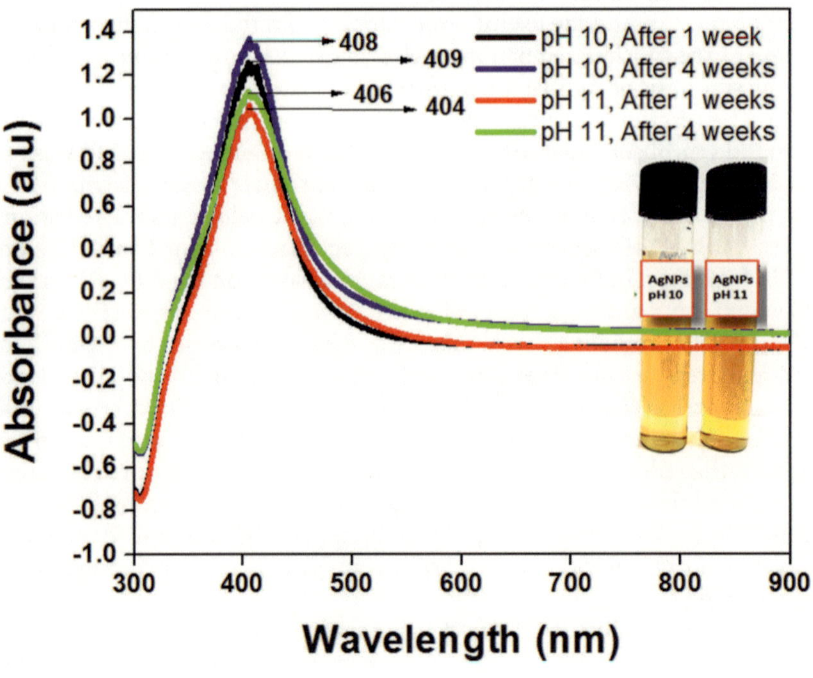

Figure 2.
UV-vis spectrum of AgNPs synthesized at pH 10 and 11 for different incubation times [Inset: visual image of AgNPs synthesized at pH 10 and 11].

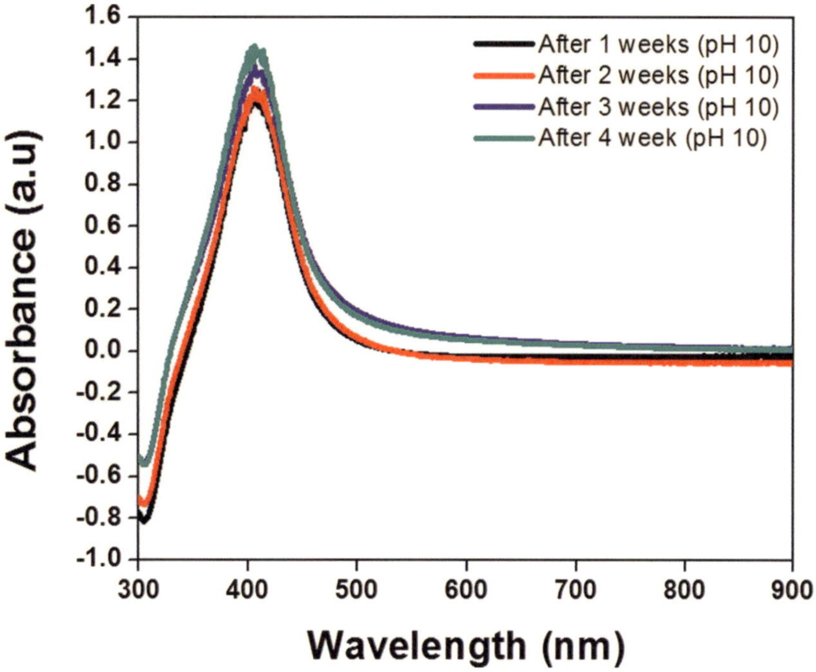

Figure 3.
UV-vis spectrum of AgNPs synthesized at pH 10 and different incubation times.

findings were further confirmed by the DLS analysis. The reduction of Ag^+ ions by ascorbic acid ($C_6H_8O_6$) was suggested as follows [28]:

$$C_6H_8O_6 + 2Ag^+ \rightarrow 2Ag^0 + C_6H_6O_6 + 2H^+$$

3.2 DLS and TEM analysis

DLS tests are performed to investigate the size distribution and stability of the colloidal AgNPs in the aqueous médium. As shown in **Figure 4**, the mean size of the AgNPs synthesized at a pH of 10 over 7 days is 22.4±13.2 nm and the observed polydispersity index (PDI) is 0.3472. PDI > 0.1 confirms the polydispersed nature of nanoparticles in the aqueous phase [2, 10]. These findings were summarized and correlated by the UV-Vis and DLS analysis mentioned in **Table 1**. The DLS data of reaction mixture at pH 10 (λ_{max} = 409 nm) directly indicated that smaller particles were formed with respect to other pH levels. Hence, pH 10 was preferred for the synthesis of stable and smaller diameter particles. This result agrees with UV-vis and TEM results.

In **Figure 5a** and **b**, TEM image clearly showed the presence of non-aggregated and monodisperse AgNPs of an average size of 30 nm. Furthermore, the low magnification TEM images confirm the uniform size and quasi-spherical shape of AgNPs. Slight chage in mean size observed in TEM and DLS analysis may be due to the

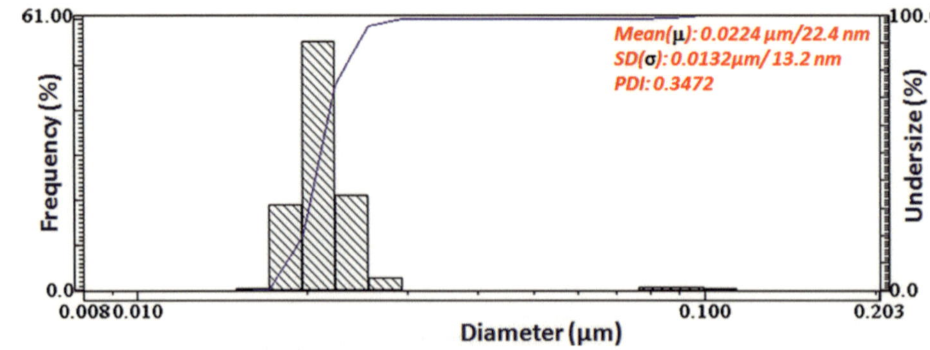

Figure 4.
DLS pattern of AgNPs synthesized at pH 10 for 1st week.

pH of reaction mixture	Particles diameter (μ ± σ) nm Time: 1st week	Particles diameter (μ ± σ) nm Time: 2nd week	Particles diameter (μ ± σ) nm Time: 3rd week	Particles diameter (μ ± σ) nm Time: 4th week	UV-vis λ_{max} (nm)
6.7	117.72 ± 54.3	125.7 ± 203.3	156.5 ± 55.9	158.75 ± 60.3	No sharp peak
9	49.9 ± 59.9	45.34 ± 39.2	59.9 ± 33.4	67.3 ± 33.9	411
10	22.4 ± 13.2	23.1 ± 13.3	25.2 ± 13.9	29.1 ± 11.1	409
11	28.7 ± 17.6	32.2.7 ± 18.1	30.1 ± 16.5	32.3 ± 14.3	404

Table 1.
Particle mean diameter and UV-vis values for different pHs and time intervals.

presence of some impurity in the sample or screening of small particles by bigger ones in the DLS analysis [29]. The SAED pattern of ascorbic acid-induced AgNPs is illustrated in **Figure 5c**. The concentric ring-like diffraction pattern in the SAED pictures clearly indicates that the particles are polycrystalline and spherically shaped [23, 25].

3.3 XRD and FTIR analysis

The crystal structure of the nanoparticles was confirmed with the help of the XRD technique. The XRD profile of the AgNPs synthesized by ascorbic acid and sodium citrate at pH 10 is presented in **Figure 6**. The presence of an intense peak at 38.0696° corresponds to (111) planes and indicates the growth of AgNPs along the (100) directions, while another peak at 44.2506° indexed as (200) reflection plane of metallic silver, respectively (ICSD no: 98-018-0878). The presence of an unindexed peak at 2θ = 31.6858° in the XRD profile indicates the organic/ bioinorganic impurities [8, 10]. These observations confirmed the successful formation of the semicrystalline AgNPs and also justify the outcomes of the SAED.

FTIR spectrum analysis is crucial to describe the processes that take place during AgNPs synthesis. The characteristic bands observed at 1759 and 1733 cm^{-1} correspond to the C=O stretching of citric acid, whereas the presence of C=O stretching vibration at 1653 cm^{-1} confirms the oxidized form of ascorbic acid [19]. The presence of 1559 and 1383 cm^{-1} is characteristic of C=O asymmetric stretching in the -COO^{-} and

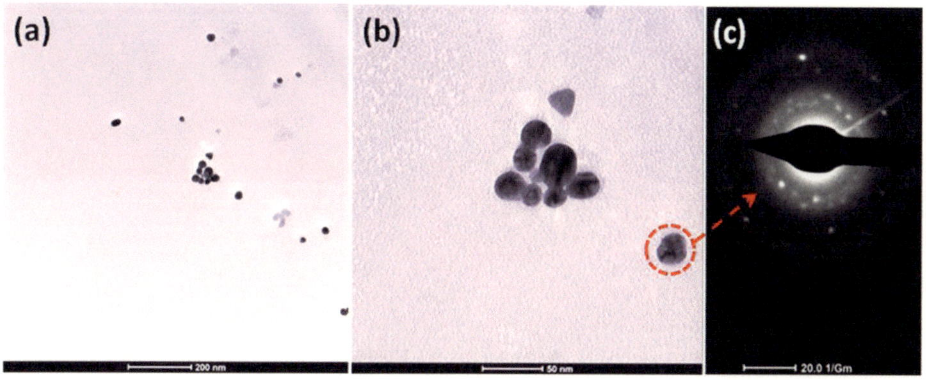

Figure 5.
(a,b) TEM image and corresponding (c) SAED pattern of AgNPs.

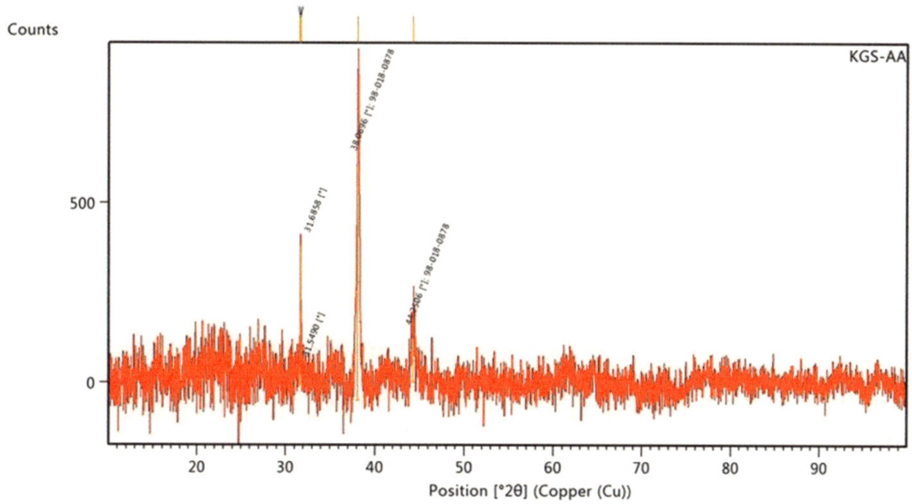

Figure 6.
XRD pattern of AgNPs synthesized by a mixture of ascorbic acid and sodium citrate.

C–OH stretching peaks of citric acid. Both –O–H (multiple) and C-O-C stretch vibration is detected at 3885, 3810, 3633, 3262 cm^{-1} and 1049 cm^{-1} represented for ascorbic acid and organic/citric acid (**Figure 7**) [23]. Furthermore, the bands at 1462 cm^{-1} correspond to C-H bending and CH_2 scissoring of the aliphatic group, while peaks occurred at 915 cm^{-1} for alkenes and 830 cm^{-1} for -OH out of plane deformation/C-C ring stretching. The FTIR results indicate that the citric acid is directly involved in the stabilization and capping of AgNPs, whereas ascorbic acid reduces Ag^+ ions to Ag^0. The UV-vis, TEM and XRD data justify the FTIR analysis statement.

4. Conclusions

In conclusion, the combined use of ascorbic acid and sodium citrate was proposed as a simple and non-toxic method for the green synthesis of AgNPs because it permits

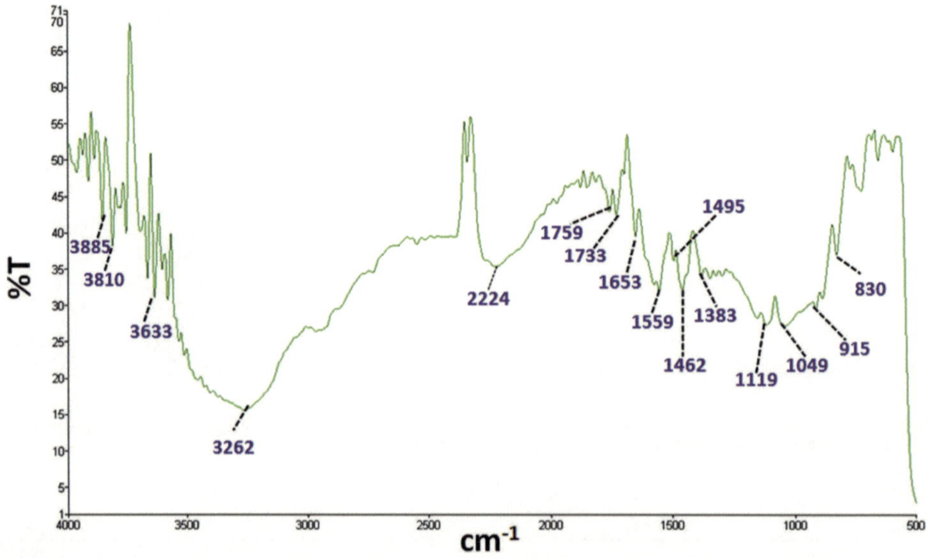

Figure 7.
FTIR pattern of AgNPs synthesized by ascorbic acid and sodium citrate.

the control of the nucleation, growth and stabilization of the synthesis process. It shows λ_{max} = 409 nm at pH 10 and produces monodisperse, quasi-spherical shape AgNPs with an average size of 30 nm under optimized conditions. Additionally, AgNPs were stable for one month, the XRD result confirms that the crystallinity of AgNPs is FCC in nature and FTIR suggested capping of AgNPs by $-COO^-$ group of sodium citrate. Therefore, the suggested environmentally-friendly synthesis of AgNPs will contribute to the future engineering applications.

Acknowledgements

This scientific work has been funded by the Universidad de las Fuerzas Armadas ESPE and Prometeo Project of the National Secretariat of Higher Education, Science, Technology and Innovation (SENESCYT), Ecuador.

Conflict of interest

The authors confirm they have no conflict of interest.

Author details

Katherine Guzmán[1], Brajesh Kumar[1,2]*, Marcelo Grijalva[1], Alexis Debut[1] and Luis Cumbal[1]

1 Centro de Nanociencia y Nanotecnologia, Universidad de las Fuerzas Armadas ESPE, Sangolqui, Ecuador

2 Department of Chemistry, TATA College, Kolhan University, Chaibasa, Jharkhand, India

*Address all correspondence to: krmbraj@gmail.com

References

[1] Singh OV. Bio-nanoparticles: Biotechnological Applications. New Jersey: John Wiley & Sons; 2015

[2] Kumar B, Smita K, Angulo Y, Debut A, Cumbal L. Single-step biogenic synthesis of silver nanoparticles using honeybee-collected pollen. Inorganic and Nano-Metal Chemistry. 2022. DOI: 10.1080/24701556.2022.2081198

[3] Gebauert J, Malissek J, Sonja S, Knauer S, Maskos M, Stauber M, et al. Impact of the nanoparticle–protein corona on colloidal stability and protein structure. Langmuir. 2012;**28**(25):9673-9679

[4] Singh M, Manikandan S, Kumaraguru AK. Nanoparticles: A new technology with wide applications. Research Journal of Nanoscience and Nanotechnology. 2010;**1**(1):1-11

[5] Lok C, Chi-Ming L, Rong C, He Q, Yu Y, Sun H, et al. Silver nanoparticles: Partial oxidation and antibacterial activities. Journal of Biological Inorganic Chemistry. 2007;**12**:527-534

[6] Burda C, Chen X, Narayanan R, Mostafa A, Sayed E. Chemistry and properties of nanocrystals of different shapes. Chemical Review. 2005;**105**(4):1025-1102

[7] Sönnichsen C, Reinhard M, Liphardt J, Alivisatos AP. A molecular ruler based on plasmon coupling of single gold and silver nanoparticle. Nature Biotechnology. 2005;**23**(6):741-745

[8] Kumar B, Smita K, Cumbal L, Debut A. *Ficus carica* (Fig) fruit mediated green synthesis of silver nanoparticles and its antioxidant activity: A comparison of thermal and ultrasonication approach. BioNanoScience. 2016;**6**:15-21

[9] Pacioni NL, Borsarelli CD, Rey R, Veglia AV. Synthetic Routes for the Preparation of Silver Nanoparticles. Switzerland: Springer International Publishing; 2015

[10] Guzmán K, Kumar B, Vallejo MJ, Grijalva M, Debut A, Cumbal L. Ultrasound-assisted synthesis and antibacterial activity of gallic acid-chitosan modified silver nanoparticles. Progress in Organic Coatings. 2019;**129**:229-235

[11] Mochochoko T, Oluwafemi OS, Jumbam DN, Songca SP. Green synthesis of silver nanoparticles using cellulose extracted from an aquatic weed; water hyacinth. Carbohydrate Polymers. 2013;**98**:290-294

[12] Kanmani P, Rhim J-W. Physicochemical properties of gelatin/silver nanoparticle antimicrobial composite films. Food Chemistry. 2014;**148**:162-169

[13] Kumar B, Smita K, Cumbal L, Debut A, Pathak RN. Sonochemical synthesis of silver nanoparticles using starch: A comparison. Bioinorganic Chemistry and Applications. 2014;**2014**:8. DOI: 10.1155/2014/784268

[14] Malassis L, Dreyfus R, Murphy RJ, Hough LA. DonnioB, Murray, CB. One-step green synthesis of gold and silver nanoparticles with ascorbic acid and their versatile surface post-functionalization. RSC Advances. 2016;**6**:33092-33100

[15] Egorova M, Revina AA. Synthesis of metallic nanoparticles in reverse micelles

in the presence of quercetin. Colloids and Surfaces A: Physicochemical and Engineering Aspects. 2000;**168**:87-96

[16] Li H, Xia H, Ding W, Li Y, Shi Q, Wang D, et al. Synthesis of monodisperse, quasi-spherical silver nanoparticles with sizes defined by the nature of silver precursors. Langmuir. 2014;**30**(1):2498-2504

[17] Chekin F, Ghasemi S. Silver nanoparticles prepared in presence of ascorbic acid and gelatin, and their electrocatalytic application. Bulletin Material Science. 2014;**37**(6): 1433-1437

[18] Rastegarzadeh S, Hashemi F. A surface plasmon resonance sensing method for determining captopril based on in situ formation of silver nanoparticles using ascorbic acid. Spectrochimica Acta Part A: Molecular and Biomolecular Spectroscopy. 2014;**122**:536-541

[19] Ranoszek-Soliwoda K, Tomaszewska E, Socha E, Krzyczmonik P, Ignaczak A, Orlowski P, et al. The role of tannic acid and sodium citrate in the synthesis of silver nanoparticles. Journal of Nanoparticle Research. 2017;**19**:273

[20] Kumar B. Green synthesis of gold, silver, and iron nanoparticles for the degradation of organic pollutants in wastewater. Journal of Composite Science. 2021;**5**:219. DOI: 10.3390/jcs5080219

[21] Kim D-Y, Sung JS, Kim M, Ghodake G. Rapid production of silver nanoparticles at large-scale using gallic acid and their antibacterial assessment. Materials Letters. 2015;**155**:62-64

[22] Stamplecoskie K. Silver Nanoparticles: From Bulk Material to Colloidal Nanoparticles. In: Alarcon E, Griffith M, Udekwu K, editors. Silver Nanoparticle Applications. Engineering Materials. Vol. 1. Cham: Springer; 2015. pp. 1-12. DOI: 10.1007/978-3-319-11262-6

[23] Kumar B, Smita K, Debut A, Cumbal L. Extracellular green synthesis of silver nanoparticles using Amazonian fruit Araza (*Eugenia stipitata* McVaugh). Transactions on Nonferrous Metals. 2016;**26**:2363-2371

[24] Mie G. Contribution to the optical properties of turbid media, in particular of colloidal suspensions of metals. Annals of Physics (Leipzig). 1908;**25**:377-452

[25] Kumar B, Smita K, Cumbal L, Debut A. Synthesis of silver nanoparticles using Sacha inchi (*Plukenetia volubilis* L.) leaf extracts. Saudi Journal of Biological Sciences. 2014;**21**(6):605-609

[26] Vigneshwaraṇ N, Nachane RP, Balasubramanya RH, Varadarajan PV. A novel one-pot 'green' synthesis of stable silver nanoparticles using soluble starch. Carbohydrate Research. 2006;**341**:2012-2018

[27] Dadosh T. Synthesis of uniform silver nanoparticles with a controllable size. Materials Letter. 2009;**63**:2236-2238

[28] Singha D, Barman N, Sahu K. Facile synthesis of high optical quality silver nanoparticles by ascorbic acid reduction in reverse micelles at room temperature. Journal of Colloid and Interface Science. 2014;**413**:37-42

[29] Kumar B, Smita K, Awasthi SK, Debut A, Cumbal L. *Capsicum baccatum* (Andean Chilli)-assisted phytosynthesis of silver nanoparticles and their H_2O_2 sensing ability. Particulate Science and Technology. 2022;**40**(6):772-780